T0288992

The book is very well written and several chapters will be, without doubt, extremely useful to the engineering community as they bring together data from diverse sources, and provide a very interesting overview of important topics. It is a real pleasure to discover or rediscover page by page many aspects of this particularly interesting type of concrete whether through background theory or practical guidance.

Dr. Nicolas Roussel,
Reviewer on behalf of the
Educational Activities
Committee of RILEM

Self-Compacting Concrete

Geert De Schutter, Peter J. M. Bartos,
Peter Domone and John Gibbs

Published by
Whittles Publishing,
Dunbeath,
Caithness KW6 6EY,
Scotland, UK
www.whittlespublishing.com

Distributed in North America by
CRC Press LLC,
Taylor and Francis Group,
6000 Broken Sound Parkway NW, Suite 300,
Boca Raton, FL 33487, USA

ISBN 978-1904445-30-2
USA ISBN 978-1-4200-6833-7

Typeset by Compuscript Ltd., Shannon, Ireland.

Printed by Athenaeum Press Ltd.

Contents

Foreword

Many consider self-compacting concrete (SCC) to be the most successful recent innovation in concrete construction – as well as in construction in general. Over a very short period of time SCC has progressed from the initial concepts of desired functions to routine use in competitive construction markets.

The development of a more fluid concrete to facilitate concrete construction has been a goal of the industry for a very long time during which incremental developments occurred. The breakthrough happened in Japan in the mid-1980s when the construction industry, together with academia, started to find new solutions for difficulties encountered with the durability of concrete structures and with quality assurance of complex structures that were difficult to cast. The lack of highly skilled concrete workers added to the rising concern. The solution proposed was to develop concrete which 'compacted itself' through the use of gravity alone.

Research groups, construction companies and the chemical industry took on the challenge and successful solutions were found. The first full-scale pilot project was carried out in 1990. International researchers were inspired by the potential benefits and started adopting the new technology for their own needs. They experimented with local raw materials and collaborated with construction companies that exploited their new SCC knowledge. A RILEM technical committee on SCC was established in 1996 with members from four continents. Japan hosted the first international workshop in Kochi (1998) and the first international RILEM symposium was held in Stockholm in 1999. Biannual SCC symposia have been held in Tokyo, Reykjavik, Chicago and Ghent with ever increasing numbers of participants and of countries represented. Numerous national and regional symposia and seminars on SCC have been held all over the world.

SCC has developed in different directions in the various national construction markets. In some, the main focus is on using SCC as a high-performance concrete for very demanding structures, in others, development targets mainstream concrete for conventional applications. SCC is used for in situ applications as well as for precast production. With increasing knowledge and experience, an improving track record has increased confidence in the market, leading to the development of production plant and equipment specifically adapted to SCC.

It is increasingly understood that SCC is not only an alternative material to traditional vibrated concrete (TVC) but that it also opens the possibility for radical changes to the whole concrete construction process. It is exciting to see how new production facilities, such as precast concrete plants, are now designed to use SCC and fully exploit its potential benefits.

However, the potential gains cannot be achieved without effort. SCC is a more demanding and sensitive technology than TVC and the choice of constituent materials, mix design, mixing process, casting and indeed also material and product design have to be considered. Adequate new skills and education are needed.

Gradually, the increased material cost of SCC is being outweighed by gains through increased productivity, improved performance, better quality as well as an improved working environment. SCC is a more expensive and demanding material but the construction market has now realised that the final hardened concrete products offer a competitive performance-over-price ratio.

SCC is here to stay. It has been a fantastic and rapid journey since the first structured approach was taken to develop a practical high-performance concrete needing only gravity for its compaction. It is now very timely to gather the underlying knowledge and practical experience gained thus far to provide a guide for future work. Thus this book is welcome. The extensive experience of the authors in both research and practice ensures that this will be a valuable publication.

Dr Åke Skarendahl
Managing Director of BIC-Swedish Construction Sector Innovation Centre
Chairman of RILEM TCs on SCC 1996–2006
RILEM President 2003–2006

Preface

Self-compacting concrete (SCC) is an important and significant advance within concrete technology which is having a major impact on concrete practice. Originally developed in Japan with the first significant applications in the early 1990s, it has rapidly been adopted worldwide in construction. SCC is a 'new' material requiring new placing techniques, and is providing novel insights and innovative developments to 'classical' concrete technology. Starting from this basis, we explain how and why the new concept of SCC requires a re-evaluation of traditional material models. This process of rethinking relates SCC technology to classical concrete technology.

This book provides essential information for readers who do not yet have an adequate knowledge of SCC and experience of its applications. However, it is also aimed at experienced readers who can benefit from some fundamental background to rheology, hydration and microstructure, which strongly influence the engineering properties and durability of SCC. Information on the selection of the constituent materials, key characteristics, test methods, mix proportioning, construction processes, engineering properties, production and conformity checking of SCC will all be of use. Practical guidance is also given as to how to select and specify SCC and the key properties required for some typical applications are explained extensively.

Throughout the book it is assumed that the reader has some experience of traditional vibrated concrete (TVC) and has a basic knowledge and understanding of classical concrete technology. Many of the chapters can easily be understood without a very detailed knowledge of cementitious materials, and are therefore accessible to all readers, from practitioners who are involved in daily casting operations on-site to doctoral students desiring an introduction into modern, state-of-the-art concrete technology. For other chapters, for example those dealing with hydration, microstructural aspects and durability, a more advanced knowledge is required, However, the advanced parts of the book are presented such that they may be bypassed without impairing the reader's understanding and the basic principles of SCC.

It is clear from worldwide experience that there is no such thing as an ideal SCC mix. However, this book will be helpful to those who wish to make an SCC appropriate for their specific application with their local materials. The readers who are new to the field will find sufficient information to get acquainted with the subject and to start developing mixes. Those readers who already have experience of using SCC will find additional information on both practical and fundamental issues relevant to SCC.

Unlike many SCC 'guidelines', *Self-Compacting Concrete* not only shows how fresh and hardened SCC behaves, it also explains *why* it does so, whenever such information is available. This combination of practical guidance and a review of underlying principles and explanations makes this book suitable for a wide readership of practitioners, students, architects, engineers, contractors, public authorities and researchers. It is intended to enhance the reader's theoretical and practical knowledge of concrete technology and facilitate the further development and application of SCC worldwide, and the realisation of its many advantages within modern construction technology.

Geert De Schutter
Peter Bartos
Peter Domone
John Gibbs

Acknowledgments

During the preparation of the book several of our colleagues have been of great assistance. We would particularly like to thank:

Dr. Ir. Anne-Mieke Poppe and Dr. Ir. Veerle Boel who made significant contributions to the subject of hydration and the microstructure of concrete discussed in Chapter 8. The additional contributions by Dr. Ir. Katrien Audenaert, Dr. Guang Ye and Ir. Gert Baert are also gratefully acknowledged.

Dr. Ir. Veerle Boel and Dr. Ir. Katrien Audenaert have also provided valuable assistance and contributions during the writing of Chapter 10.

Finally we would all like to acknowledge the contributions of our colleagues on the RILEM technical committees and in the two EU-funded programmes on SCC from whom we have learnt so much, and who have therefore indirectly contributed to this book.

Geert de Schutter
Peter J.M. Bartos
Peter Domone
John Gibbs

About the authors

Geert De Schutter is a senior professor at Ghent University, Belgium. He is conducting research in the field of concrete technology at the Magnel Laboratory for Concrete Research, Department of Structural Engineering. He has received several national and international awards, including the important Vreedenburgh Award in 1998 and the prestigious international RILEM Robert L'Hermite Medal in 2001. He was a partner in the EU-funded 'Testing-SCC' project from 2001 to 2004. From 2004 to 2007 he has been chairing the RILEM Technical Committee TC 205-DSC on Durability of SCC. In September 2007, he organised the 5th International RILEM Symposium on SCC, in Ghent, Belgium.

Peter Bartos is an emeritus professor of civil engineering and former director of the Advanced Concrete and Masonry Centre and the Scottish Centre for Nanotechnology in Construction Materials at the University of Paisley. He holds an honorary professorship at the Queen's University of Belfast. He is a chartered civil and structural engineer who worked on-site and in design before turning to industrial and then university research. Professor Bartos is a past president of the UK Concrete Society where he also established and led the SCC Working Group. He is a fellow of RILEM and chairs the TC197-NCM on Nanotechnology in Construction Materials. He founded and chaired the TC 145-WSM on Workability of Special Concrete Mixes (1992–1999), which first identified the significance of SCC. He was a promoter and then in charge of key tasks of a large European industry-led SCC project (1997–2000), credited with bringing SCC into European construction practice. He was project director of EC 'Testing-SCC' (2001–2004), which supported the development of European standards for fresh SCC. His previous books on the subject include *Fresh Concrete: Properties and tests* (Elsevier, 1992) and *Self-Compacting Concrete in Bridge Construction* (UK CBDG and Concrete Centre, 2005).

Peter Domone is a senior lecturer in concrete technology in the Department of Civil and Environmental Engineering at University College London. He was a member of the RILEM technical committee on SCC from 1995 to 1999, a partner in the EU-funded 'Testing-SCC' project from 2001 to 2004, and a member of the UK Concrete Society Working Group on SCC.

John Gibbs is Technical Advisor to the European Ready-Mixed Concrete Organisation. Previously a technical manager in the ready-mixed industry, then Deputy Director of the Advanced Concrete & Masonry Centre at Paisley University, he was a member of the two European projects on SCC, and of the RILEM technical committee on casting SCC.

Chapter 1
Introduction and glossary of common terms

1.1 Introduction

Concrete is without any doubt a fascinating building material. In one way, it is very simple: anyone can mix water, cement and aggregates, cast it in moulds of almost any shape and finally obtain an artificial stone with some strength. In another way, it is an extremely difficult material: no one completely understands its complex behaviour both when fresh and when hardened. This ambiguity makes concrete both the most used building material in the world and a material which creates many problems when not properly designed or placed.

One of the key issues for traditional concrete is that external energy has to be provided to compact it. This can be obtained on site from vibrating pokers and in concrete factories from vibrating tables or alternative methods. Concrete practice has shown that for on-site casting, vibration is not always carried out as it should be. This is quite understandable – no one really likes to handle a vibrating tool for a whole day – but it is surely harmful to the quality of the final structure. The quality of nonvibrated traditional concrete is far lower than its intrinsic quality when properly compacted. The loss in strength may be acceptable in some cases, but the decrease in durability can often be much more significant, leading to accelerated degradation processes such as reinforcement corrosion, frost damage, sulfate attack etc.

SCC is not new. Highly specialised, expensive and difficult to control mixes for underwater placement were successfully produced a very long time ago. However, it was not until the late 1980s that advances in admixtures allowed us to increase the fluidity of the fresh mix sufficiently to make it self-compacting but still cohesive enough to avoid the hitherto inevitable segregation. In simple terms, this concrete fills the formwork like a viscous liquid and does not need any external compaction energy (see Figure 1.1). There is some debate about whether the 'modern' SCC is a new material or just a new placing technique; it can be shown to be both. The requirement for self-compaction demands a composition which is significantly different from the composition of traditional vibrated concrete (TVC), as will be explained in Chapter 6. Differences in composition inevitably lead to changes in hydration, microstructure, and the behaviour of the cementitious material in service.

Even if SCC is considered to be a new material, it remains a cementitious material. Consequently, the basic concrete technology is still largely valid. This book assumes that most readers will have a basic knowledge of concrete technology. There is therefore a focus on the differences between TVC and SCC, in both the fresh state and when hardened. The knowledge needed for SCC could be seen as providing an extension to the 'classical' concrete technology. Basic principles are still valid, but

Figure 1.1 *SCC flowing through a grid of reinforced bars.*

historical 'laws' or 'formulae' may have to be extended to take new parameters into account. As an example, it seems that the water/cement ratio does not explain the behaviour of SCC to the same extent as it does for traditional concrete. However, as will be shown in Chapters 8 and 10, a parameter that has been well known in classical concrete technology since the 1940s, capillary porosity, is still useful for predicting the behaviour of SCC.

Developments in concrete technology such as high-performance concrete and reactive powder concrete in recent decades are very interesting, and increase our knowledge. However, it seems that most of these developments are only important for special applications, and not for mainstream concrete practice. This is different with SCC. Although in the first instance SCC was often considered a high-performance concrete on account of its unusual properties when fresh, and early applications tended to exploit its high potential strength, it was soon produced for nearly every strength class. It may still take some time for SCC to replace TVC and become the mainstream concrete for all in situ placing but within the precast industry this has already happened and in some countries almost their entire production is SCC.

The building industry has always been very traditional and even conservative. Concrete technology forms an indispensable part of the construction process, whose productivity has only been slowly increasing over the last few decades, while other sectors of industry have achieved great improvements, sometimes massive leaps forward. The introduction of SCC provides an important opportunity to advance and make up some of the lost ground. The development of SCC already exploits innovative research. It was the 'surface chemistry' which enabled the development of SCC by the synthesis of a new generation of superplasticisers, using nanoscale molecular manipulation to obtain the best performance. In this way, SCC represents a successful and significant application of nanotechnology to a common building material.

The development of SCC has led to substantial improvements in the environment and working conditions: there is less energy consumption, less vibration, greater productivity, less noise, decreased absence of workers due to sickness etc. The positive impact of SCC is so great that it is considered to be a major development and a major improvement for our society at large. Within Europe, this was illustrated by the fact that the first European Research Project on Self-compacting Concrete, which effectively introduced SCC to Europe, was the first construction-related project ever to reach the final round of the prestigious EU Descartes Prize competition.

The use of SCC currently amounts to only a few percent of the global, annual concrete production. Indeed, the traditional 'grey image' of concrete: dirty, difficult, dangerous and dull is still somewhat commonly held, albeit that this is changing. When SCC becomes increasingly familiar and more widely used it will contribute to this change significantly, enhancing the traditional knowledge and extending the practical uses of concrete.

In order to do so, there is a fundamental need to learn about the SCC technology. Knowledge and training are an indispensable prerequisite for a successful and significant adoption of SCC anywhere, from the smallest to the largest contractor, consultant or architect. Improving knowledge about SCC is the main aim of this book.

1.2 Glossary of common terms related to self-compacting concrete

Addition
A fine-grained inorganic material. Includes two types (EN206-1): inert or nearly inert additions (type I) and pozzolanic or latent hydraulic additions (type II). Quantities are usually 5–50% per mass of powder.

Admixture
A material added in small quantities at time of mixing, usually 0.2–4% per unit mass of powder, to modify the properties of fresh and/or hardened concrete.

Air content
Volume of air-voids in fresh or hardened concrete usually expressed as percentage of total volume of the mix. TVC is usually considered fully compacted when its air content is below 1.5%, excluding any entrained air.

Air-entrainment
Intentional introduction of air uniformly distributed in very small bubbles/voids into concrete primarily to improve frost resistance. Volume of entrained air usually varies between 3–6% of the total volume of concrete.

Binder
Cement and type II addition combined.

Bingham material
A material which when subjected to shear stress behaves as an elastic solid until the yield stress is reached, after which there is a linear relationship between shear stress and

rate of strain (flow velocity). The slope of the linear relationship is the plastic viscosity. Fresh cement paste and concrete show such behaviour under certain conditions.

Blocking 'step' B_J
The difference (if any) between levels of concrete inside (centre) and outside (outer edge) of the J-ring. Result of the J-ring test. (Also sometimes called the 'step height'.)

Compaction
The process in which the volume of trapped air in fresh concrete is reduced below 1–2% of the total volume of concrete, usually by mechanical means such as vibration. The term is interchangeable with 'consolidation' (USA only).

Consistence
Measure of ease by which fresh concrete can be placed. It is the same as and interchangeable with Consistency and Workability.

Consistency
See Consistence.

Filler
See Addition.

Filling ability
The ability of the fresh mix to flow under its own weight and completely fill all spaces in formwork. It is sometimes referred to as 'flow' or 'fluidity'.

Fines
See Powder.

Finishability
The ability of fresh concrete to satisfactorily respond to a specific type of surface finish treatment.

Flow
See Filling ability.

Flow-rate
The speed at which a sample of a fresh mix spreads horizontally outwards from the base of a slump cone until it reaches a selected radial distance, usually a concentric ring of 500 mm diameter. It is indicated by a 'flow-time', such as t_{50}, interchangeable with t_{500}, measured in seconds.

Flow test
Test for consistence of TVC. Result is based on dimensions of a horizontal spread of the concrete over a baseplate after removal of a mould 200 mm high and a number of controlled jolts. This is not to be confused with the slump-flow test.

Flow-time
Usually the time taken for the concrete flow in an Orimet, V- or O-funnel test; an indication of the level of filling ability and rate of flow.

Flowability
See Filling ability.

Fluidity
See Viscosity and/or Filling ability.

J-ring test
Test method used primarily for assessment of passing and filling ability of fresh SCC. Results are expressed as **blocking 'step' B_J** (mm), **spread SF_j** (mm) *or* **flow-time** t_{J50} (s) (= t_{J500})

L-Box test
Test method primarily for the assessment of the passing ability of fresh SCC. The primary result is the passing ratio.

Mortar
The fraction of concrete containing aggregate of maximum size ≤ 4 mm (5 mm).

Newtonian fluid
A fluid, which shows a linear relationship between shear rate and shear stress, passing through the origin in a simple shear flow. Fresh SCC mixes of high filling ability can display this type of behaviour. The slope of the linear relationship is the Newtonian viscosity (or just viscosity).

Orimet test
Test method primarily for assessment of filling ability and flow-rate of fresh SCC.

Paste
Mixture of powder and water, with or without admixtures.

Penetration (segregation test)
Method for in situ assessment of static segregation resistance.

Plastic settlement
Settlement of a concrete immediately after placing while it is still in a fresh state.

Powder content
The total amount of all materials with maximum particle size usually smaller than 125 µm, expressed in mass of powder per unit volume of concrete. This usually includes cement and one or more additions. Sometimes also called the 'fines content', but this should not be confused with fine aggregate (i.e. sand) content.

Proprietary concrete (mix)
Concrete for which the producer assures the performance, subject to good practice in placing being followed. The producer is not required to disclose the composition of the mix.

Rate of flow
See Flow-rate.

Rheology
The branch of science dealing with the deformation and flow of materials under an applied stress. It includes behaviour of both fresh and hardened concrete.

Robustness
The ability of fresh concrete to maintain its properties within narrow limits when the proportions of constituent materials change significantly.

Segregation resistance
The ability of a fresh mix to maintain its original, adequately uniform, distribution of constituent materials (namely aggregate) during transport, placing and compaction. Same as 'stability'.

Sieve segregation test
Test for assessment of resistance to (static) segregation. The result is a segregation index (SI) indicating the proportion of mortar which separates through a sieve within a given period of time.

Self-compacting concrete (SCC)
Fresh concrete which has an ability to flow under its own weight, fill the required space or formwork completely and produce a dense and adequately homogeneous material without a need for compaction.

Settlement column test
The test for resistance of fresh concrete to static and dynamic segregation.

Slump test
The test for consistence of fresh TVC (e.g. EN12350-2). Result is expressed as the depth of the vertical settlement of concrete after removal of a standard (Abrams) mould 300 mm high.

Slump-flow test
The test for filling ability of fresh SCC using a 300 mm tall conical mould (as for the slump test). The primary result is the total spread (SF).

Speed of flow
See Flow-rate.

Spread
The result of the slump-flow (SF) or J-ring (SF_J) tests.

Superplasticiser (SP)
An admixture producing much-increased consistence of a fresh mix without significant retardation or air-entrainment. Also known as 'high range water reducing admixture' (HWRA) in the USA.

Thixotropy
The ability of a material to reduce its resistance to flow (apparent viscosity) with increased flow (shear) or agitation and to regain its original stiffness when at rest, the process being repeatable and reversible.

Traditional vibrated concrete (TVC)
Concrete, which requires vibration to achieve adequate compaction.

V-funnel test
Test for filling ability and speed of flow of fresh SCC.

Viscosity (plastic)
The slope of the linear part of the relationship between shear stress and shear strain rate in a Bingham material. A measure of the resistance to continued flow of a Bingham material.

Viscosity modifying admixture (VMA)
Admixture which increases internal cohesion and segregation resistance of fresh concrete. Same as thickening agent.

Wet sieve segregation test
See Sieve segregation test.

Workability
See Consistence.

Yield point
The point on the stress–strain or shear stress–rate of shear strain diagram which corresponds to the change from elastic to plastic behaviour of a solid material or static to dynamic behaviour of a Bingham fluid.

Yield stress
Stress required to initiate plastic deformation or flow of a material. The stress corresponding to the yield point.

Chapter 2
Self-compacting concrete

2.1 The need for self-compaction

Concrete construction practice has always preferred fresh mixes, which are easy to handle and place. It was also always possible to substantially raise the level of consistence, to make the fresh mix flowable[1]. However, this could only be achieved by adding more water to the mix or by greatly increasing the amount of cement paste. It was also discovered very early in development of 'modern' concrete that the high level of consistence (workability)[2] based on high water content made the fresh mix very prone to severe segregation and greatly reduced the strength of the hardened mix. It became firmly embedded in 'good concrete practice' from the late 1800s until the advent of plasticisers and superplasticisers in the 1970s that fresh mixes of very high consistence, fluid, 'runny' mixes, were synonymous with very poor quality hardened concrete, which was best avoided. The alternative approach to raising consistence (workability) by a much increased cement paste content led to problems due to excessive heat generation during hardening and greatly increased the cost of such concrete.

Fresh concretes of a moderate-low slump therefore prevailed in practical concrete construction[1]. The development of the potential strength of such concretes when hardened then depended critically on achieving full compaction. The most common method for compaction of traditional concrete has been vibration, using internal (poker) vibrators, external vibrators and vibrating tables.

The vibration-based compaction process aimed to 'liquefy' the fresh mix and temporarily increase its consistence so that the air trapped in the mix was able to rise to the surface and escape from the mix. In practice, not all of the trapped air could be expelled and an adequate (sometimes defined as 'full') compaction was assumed to be reached even when 1–2% of air still remained trapped in the mix. There was an additional requirement for that residual air to be uniformly distributed. It was also important to ensure that the volume of any additional 'entrained' air was not substantially reduced and that such air was not expelled from the mix during compaction.

Despite the existence of numerous 'good practice' guidelines emphasising the need for a thorough and effective compaction of traditional mixes, a significant proportion of all concrete produced was never fully compacted. Inadequate compaction has sometimes been apparent on exposed surfaces (see Figures 2.1 and 2.2) but in many other cases it remained hidden within the mass of concrete and reinforcement and only showed up as the principal cause of poor performance (strength, durability) of both plain and especially reinforced concrete.

Figure 2.1 *Honeycombing and many 'cold' joints. Poorly compacted concrete.*

Compaction of concrete, particularly the most common method in which handheld immersion/poker vibrators are used, is very physically demanding work, often carried out in difficult environmental conditions. Lack of workers willing to carry out the placing and vibration of concrete usually leads to untrained labourers being used to carry out the compaction.

Figure 2.2 *Poor compaction of the lower part of a reinforced concrete beam. Severe honeycombing is clearly visible. Bottom reinforcement is poorly protected against corrosion.*

Figure 2.3 *Compaction of two reinforced concrete beams by typical poker vibrators. Medium consistence (workability) concrete (slump of 30–60 mm).*

The compaction of TVC by typical poker vibrators is shown in Figure 2.3. Two reinforced concrete beams of 240 mm × 400 mm cross-section, 4 m long were cast using a fresh TVC of medium consistence (slump 30–60 mm). Each of the beams required between 10–15 min of concentrated effort by an operator to achieve adequate compaction. For comparison, the same beams were also cast using SCC. Placing of each of the beams using SCC took approximately 2 min. Fresh SCC was poured into the formwork directly from a truck-mixer.

Compaction of TVC during casting of floors and slabs is particularly inefficient as the thickness is usually less than 250 mm and the surface area is relatively large. This leads to large teams, such as shown in Figure 2.4, consisting of one nozzleman, three workers with vibrators, two finishers and three workers spreading the concrete.

The compaction process is inherently difficult to supervise. The most important judgement as to whether all the concrete within the radius of action of the vibrator in each insertion has been adequately compacted has to be left to the discretion of the operator of the vibrator. The judgement is purely visual, based on the observation of air bubbles escaping from the surface of the concrete, which tends to settle down a little around the vibrator. The difficulty of visually judging compaction by poker vibrators is illustrated in Figure 2.5.

The operator also has to judge if there was sufficient overlap between the 'radii of action' in each insertion of the vibrator and all of the volume of concrete in a specific structural element or formwork had been subjected to adequate vibration. Moreover, incorrect use of poker vibrators can cause movement and displacement of the reinforcement and assist in any leakage of paste from the formwork, creating patches of honeycombed concrete and unsightly surface defects.

Figure 2.5 shows that the uncompacted 'flow' of the mix sideways during placing (foreground) and the extent of a 'full' compaction produced from a single insertion of

Figure 2.4 *Placing and finishing of TVC (2004). (Reproduced with permission of The Concrete Society.)*

a poker vibrator (background). A slight depression of the concrete surface around the position of the vibrator indicates good compaction, however, it is difficult to judge when it becomes adequate. Good compaction does not appear to extend to the bottom of the 400 mm deep beam, the 'effective' reach of the vibration decreases around the tip of the poker vibrator.

It does not appear to be generally appreciated how considerable is the difference between compacted and uncompacted traditional low-to-medium consistence (workability) concrete. This difference is shown very clearly in Figure 2.6. One batch

Figure 2.5 *Effectiveness of compaction of TVC using immersion (poker) vibrators.*

Figure 2.6 *A single batch of a traditional low-medium slump concrete was used to cast the whole of this wall element. The bottom part was compacted by poker vibrators, the top part was placed without any compaction.*

of concrete had been used to cast a wall element. The concrete in the bottom part of the formwork was vibrated by poker vibrators. The upper part was filled with the mix without any compaction. It is difficult to believe that both parts were cast from one batch of an ordinary traditional mix.

Compaction of fresh concrete of traditional consistence is often claimed to be the most unpleasant and tiring job on a typical construction site. In some countries it has become difficult to find workers prepared to carry this out. Supervision of the compaction process is also inherently difficult and it is therefore not surprising that a substantial proportion of concrete placed worldwide is, in reality, not adequately compacted.

Adequate compaction of fresh concrete is therefore a fundamental requirement for good concrete construction. Lack of adequate compaction remains one of the main causes of the poor performance of hardened concrete, including its surface finish. Good quality concrete cannot be obtained without the mix being adequately compacted when fresh, this being achieved either by self-compaction or by an external compaction process.

The degree of compaction achieved can only be reliably assessed by measurement of the volume of air-filled voids in hardened concrete. It is generally assumed that a TVC after an adequate compaction by vibration will still contain air in up to approximately 1.5% of its volume. The same approach is used to judge the compaction of SCC. Linked to the requirement for a low level of such residual air is the 'homogeneity' of concrete. A 'homogeneous' mix is often sought but there is still no formal agreement as to what constitutes an acceptable degree of homogeneity/uniformity of the composition of the mix.

A completely uniform distribution of all particles and constituents of concrete, a '100%' homogeneity can only be achieved in theory. It would require a totally uniform distribution of all concrete constituents regardless of the size of the sample used to examine it. It is often automatically assumed that concrete discharged from a mixer is 'homogeneous'. Even when an attempt to assess the homogeneity is carried out, it is usually impracticable to test the entire, full-size batch. Only a part of the volume mixed is assessed through samples. Due attention is not always being paid to the size/volume of the samples taken in relation to the maximum size of the aggregate and the volume of the batch assessed, and to the numbers of samples/frequency of sampling. Concrete mixers deliver fresh concrete of varying degrees of 'homogeneity' or 'uniformity[3], which is then practically impossible to improve on during the construction process.

Compaction of concrete by immersion/poker vibrators is often considered as fundamentally beneficial in improving the perceived 'homogeneity' during placing and thus the properties and the overall 'quality' of the resulting hardened concrete. This view has been taken for granted for a very long time, with practically no direct, scientific evidence available to support it. It was only in connection with the introduction of SCC (in Iceland) that an investigation of the uniformity of the key properties of hardened concrete subjected to 'good' compaction by a poker vibrator was carried out by Wallevik and Nielsson.[4] Compacted test specimens were cut into numerous slices after hardening and the distribution of constituents was established. The results of a systematic study of the homogeneity following a 'good' compaction by poker vibrators indicated that the uniformity of the hardened concrete actually decreased. This was a small-scale project focused on local concrete mixes, and its results do not mean that compaction as such is not required. However, it showed that the well-established general notion of the compaction by poker vibrators improving homogeneity may not always apply.

Compaction-related defects are often hidden but they significantly reduce the quality and performance of concrete structures. Such defects are usually discovered when formwork is stripped (see Figures 2.1 and 2.2). What normally follows is a repair of the affected concrete, usually called 'making good' (see Figure 2.7). 'Making good' is never included in the projected costs of concrete construction; after all, it is not expected to occur. Information about the actual amount of such remedial work is therefore difficult to obtain; however, it has been recently estimated that around 30% of all UK concrete placed in situ required some 'making good'.

It has to be taken into account that 'making good' cannot restore the defective concrete to its expected performance even if carried out as well as possible. Such repairs are sometimes technically difficult and if not closely supervised, their results can be 'cosmetic', masking the underlying defects. In such a case a more serious deterioration of the concrete element is only delayed for a short period of time. Poor compaction, which causes the need for making good, decreases quality, causes delays and increases the actual concrete construction costs: yet another reason for use of a SCC mix, which provides a much higher probability of 'getting it right first time'.

Concretes exist which are also produced without compaction, using a completely different approach compared with that for the 'SCC' described above. A technique for placing concrete, sometimes called a 'pre-packed/pre-placed' concrete was developed a long time ago[1]. It is based on the formwork or mould, or any other contained space

Figure 2.7 *Extensive 'making good'. The beam in the foreground has yet to be repaired. (Reproduced with permission of New Civil Engineer.)*

being first filled (packed) with aggregate. The layer of aggregate is then infiltrated by a cement slurry/paste to produce concrete.

Before the advent of plasticisers and superplasticisers, the technique tended to produce concrete of a lower quality as the required high consistence (fluidity) of the slurry was achieved by using a high water/cement (w/c) ratio. Recent advances in plasticising admixtures have permitted very fluid pastes/slurries to be made, which have very high strength and durability when hardened. The process completely avoids the mixing of concrete in traditional mixers. The pastes/slurries needed are now produced in special 'high shear' mixers, which enhance their fluidity. 'High-performance' pastes with very low plastic viscosity and resistance to segregation (bleeding of water), which have very high compressive strength are now available and have begun to be exploited in novel applications.

One such application is in the production of 'slurry-infiltrated-fibre-concrete' (SIFCON) without vibration. SIFCON is a very high (or ultrahigh) performance material produced when a layer of fibres, usually steel fibres, is infiltrated by cementitious slurry. It was developed in the early 1980s[5, 6]. Hardened SIFCON may contain approximately 8–15% by volume of steel fibres, which increase the toughness of this composite by more than one order of magnitude when compared to hardened paste alone. Subjected to compressive stress, the composite may behave in a strain-hardening manner, closer to that of a metal than that of a brittle concrete. Despite its exceptional performance, SIFCON has found only very limited applications because it has to be vibrated intensely in order to guarantee the penetration of the slurry (grout) into the bulk of the packed fibres. In practice, this restricted the production of SIFCON elements to relatively small precast elements attached to powerful vibrating tables. Utilising the latest developments in admixtures for production of SCC, Marrs and Bartos[7] and later Svermova *et al.*[8] demonstrated that SIFCON can be reliably made without vibration and used in in situ concrete construction (see Figure 2.8).

Figure 2.8 *Pouring of a highly fluid but cohesive slurry to produce a 'SIFCONised' knee joint in a concrete frame without vibration. The rest of the frame was cast using 'ordinary' SCC (after Svermova et al.*[8]*).*

2.2 Definition of self-compacting concrete

The definition of SCC appears to be quite simple and obvious from the title itself. It refers to the behaviour of a concrete mix when fresh and it separates fresh concrete mixes, which are self-compacting from those which are not. It is then useful to indicate the essential parameters, which make a fresh mix self-compacting, such as its ability to flow under its own weight, fill the required space or formwork completely and produce a dense and adequately homogeneous material without the need for compaction[9–11].

It is important to appreciate the difference between a genuine SCC and a traditional fresh mix with a very high workability often labelled as a 'flowing' concrete. SCC not only possesses a very high workability, but, unlike the 'flowing' concrete, it does not require compaction of any kind, it completely resists segregation and it maintains its stable composition throughout transport and placing. An SCC mix must therefore be both adequately fluid and cohesive. As the title suggests, SCC does not require any compaction.

Good SCC must be adequately fluid and cohesive (segregation resistant) when fresh, the level of its basic properties depending on the demands of its specific application (construction process, shape/size and properties of final product).

It is still a common misconception that all SCC mixes are automatically special, high-performance concretes in terms of their properties when hardened. Very high-performance concrete can indeed be produced so that it is self-compacting when fresh. However, as the same fundamental relationships between mix design and performance of hardened concrete apply to SCC, as they do for TVC, both low- and high-performance SCC can be produced. Hardened SCC can have a complete range of properties: from a very low to a very high compressive strength, from a very poor to an extremely high durability. It follows that there is no single, unique 'recipe', or specific composition, which produces concrete that is self-compacting when fresh and produces a high-performance material when hardened.

It is also important to appreciate that a fresh mix, which possesses the minimum level of parameters, which define it as self-compacting, may not suit all practical applications where self-compaction is required. Applications require varying levels of the key characteristics which define the self-compaction. There are cases which require a 'higher' than minimum levels of self-compaction. There can also be applications where SCC could not be used because the demands on its key characteristics when fresh would exceed the maximum levels obtainable in practice.

2.3 A brief history of self-compacting concrete

The concept of 'self-compaction' considerably predates that of 'modern SCC' even though it had not been defined as such[1]. There were important applications for concrete in practical situations where compaction was physically impossible.

Examples can be found in underwater concrete construction, foundation works using large diameter concrete piling, deep concrete diaphragm walling etc. Mixes used for such applications therefore had to be self-compacting. Such 'early' SCCs were invariably based on extremely high cement contents. Later, the development of admixtures (plasticisers) in the 1960s helped in the production of these highly specialised mixes. The underwater/foundation concrete approach was tolerated because the well-known adverse side effects of very high cement content of the early self-compacting mixes were substantially mitigated by such concretes being cured underwater, in generally cool conditions which assisted in the dissipation of the high heat of hydration.

Modern SCC dates from the second half of the 1980s when two independent directions of research were pursued, both paving the way for the introduction of SCC as it is currently known[3, 12]. The better-known stream of development of modern SCC originated in Japan. The postwar reconstruction of Japan in the 1950s and 1960s created a building boom where speedy project delivery obscured the importance of quality. Within a decade or two, many reinforced concrete structures started to deteriorate. The situation became serious enough to prompt the government to initiate an investigation into the causes of the widespread unsatisfactory performance of concrete. It supported the setting up of a major research project aimed at a significant improvement of the durability of new concrete construction. The project team, led by H. Okamura of the University of Tokyo, soon indicated that insufficient compaction was the most common cause of deterioration of concrete structures. The project team, including K. Maekawa, K. Ozawa and M. Ouchi proposed a solution by increasing the workability of the fresh mix so much that compaction was no longer necessary: the mix would become 'self-compacting'[12, 14]. Full-scale trials and demonstrations were carried out and in the early 1990s the 'self-compacting' concrete was first used in significant practical applications in Japan[15]. The new, 'modern' SCC enabled an important step towards the improvement of the quality of concrete in general and of its durability in particular.

The other, independent, development was associated with the advent of superplasticisers. This stimulated research into 'flowing' concrete mixes, namely for special applications in which compaction was not possible. Advances in the development of admixtures included 'viscosity modifying agents', which offered a possibility of production of 'nondispersive' concrete for placing underwater. Significant research into the practical workability of 'special' concretes such as 'washout-resistant' fresh concrete for underwater construction was carried out at Paisley, Scotland by a team led by P.J.M. Bartos and in Sherbrooke, Canada by K.H. Khayat and co-workers. The washout-reducing admixtures also provided a 'stabilising' effect on a very highly workable, superplasticised, self-compacting mix, which allowed the traditional underwater mix design to be adjusted. Very high cement contents, typical of the early underwater and foundation concrete mixes were no longer needed. The mixes became suitable for 'normal' concrete applications. They were fluid enough to be placed without the need for compaction by vibration but cohesive enough to avoid segregation. This led to experiments, site trials and limited practical applications in the late 1980s, effectively introducing 'modern' SCC.

Despite its very early introduction and good hard evidence being available in Japan regarding the benefits offered by SCC, its adoption by the Japanese construction industry was slow. The concept did not gain the expected widespread recognition in everyday concrete construction. Instead, a few very large Japanese construction companies, each of which developed their own specific mix designs and practical test methods, exploited the principle of self-compaction. They trained their own supporting technical and structural design staff and 'trade-marked' the SCC mixes with names such as Biocrete (Taisei Co.), NVconcrete (Kajima Co.), SQC concrete (Maeda Co.), etc.[15]. The expertise with SCC was kept within the corporations, which used it to gain competitive advantage in concrete construction. Some of the major projects, such as the construction of the record-holding Akashi-Kaikyo bridge (Figure 2.9) using different varieties of SCC in the late 1990s, remain the most prominent landmark applications of SCC to date.

A combination of additional important factors, including a specific sectorial division of construction, a fragmented ready-mixed concrete supply industry and a significant general slow-down in construction stifled the further advance of SCC in Japan. Research into SCC has continued but the use of SCC as a proportion of the total concrete production has shown little growth and Japan has been overtaken by Europe in the first decade of the second millennium.

There had been a considerable advance in the development of many 'special' and 'high-performance' concretes by the beginning of the 1990s. Despite very encouraging results from laboratory-based experiments, the construction industry was slow to exploit the benefits offered by the very much higher compressive strength, toughness and durability of concretes, which contained special additions, fibres of different types and novel admixtures. The laboratory experiments invariably focused on the ultimate 'special properties' of hardened concrete without paying adequate attention to the practical production process of the many new high-performance cement-based materials. The absence of practical guidelines for the industry, on how to make (batch and mix) and then how to handle, transport and place (and compact) such mixes efficiently and economically, hindered progress.

Action to overcome the practical obstacles to the exploitation of the new high-performance concretes was led by P.J.M. Bartos, who formed an international

Figure 2.9 *Akashi-Kaikyo suspension bridge (clear span of 2 km). Approximately, 500 000 m^3 of SCC was used, mainly for construction of foundations and massive cable-end anchorages.*

RILEM technical committee, with a wide, 'horizontal' brief to investigate production and construction methods and to characterise key properties of a broad range of high-performance and special concretes when fresh. A new technical committee TC145-WSM on *Workability of Special Concrete Mixes* was set up under his chairmanship in 1992[3, 12]. Since then the RILEM has played a key role in the subsequent development of SCC and in promotion of its acceptance worldwide.

Within the range of special concretes investigated by TC145-WSM were high workability, superplasticised flowing concretes and concretes for underwater placement. The results of relevant independent current research worldwide became available and integrated. These included Japanese research, its first practical applications and the work on self-compacting underwater concretes. Two events, which set the stage for introduction of SCC in Europe and started a train of events leading to a gradual spread of SCC technology to the rest of the world, were held in Scotland. The significance of the concrete production process (namely the mixing) and the behaviour of the mixes in fresh state on properties of special and high-performance concretes, including SCC, was emphasised by two international conferences held in Glasgow in 1993[3] and 1996[12]. The very substantial advantages of SCC were specifically highlighted for the first time and the potential impact on concrete construction practice was considered so high that a very large, industry-led (NCC Sweden, GTM–Vinci France and six other partners) research project obtained funding from the European Commission and commenced in 1997[16]. The working group on SCC set up under RILEM TC145-WSM was converted to a new RILEM (TC174-SCC) chaired by Åke Skarendahl and published its guidelines on SCC in 2000[17]. The European SCC project proved that it was indeed practical to build with this material, using a variety of local materials, and that the expected benefits were obtainable in real construction practice.

The project's significance and contribution, not only to the construction process itself but also to improved environmental and health conditions were reflected in its progress to the final of 2002 European Descartes prize, an annual recognition for the greatest achievement in science and technology through EC-supported projects in any subject area. The result helped to spread the awareness of the environmental and health benefits of SCC technology.

A parallel development of SCC took place in the Netherlands, which was not represented in the EC SCC project. A group of Dutch academic researchers and industry, led by J. Walraven of the University of Delft, made a direct approach to one of the Japanese companies with expertise in SCC and adopted their specific approach to SCC production and use. This initiative was later extended into precast concrete in a major training and educational exercise led by W. Bennenk of Eindhoven University, all with considerable success[18]. By 2005 almost all of the precast concrete production in the Netherlands was based on SCC.

SCC offered tangible benefits for which an increasing amount of direct evidence from construction practice was being accumulated. A wider and more rapid acceptance of SCC was clearly hindered by an absence of standardised, or even generally agreed, methods for measuring the key properties of fresh SCC which guaranteed its self-compaction. This was compounded by a lack of independent, practical guidance for working with SCC, namely its specification and verification of conformance on construction sites and wide national differences in the costing of construction projects.

The First European national guidelines on SCC appeared in France[19] and in the Nordic countries in Europe. The matter of international standardisation had to be resolved by a multinational project and in 2001 the European Commission approved and cofunded the 'Testing-SCC' project, led by the ACM Centre, the University of Paisley, Scotland[20]. The project evaluated existing test methods, the final selection being subjected to a large and rigorous prenormative assessment of the performance of the tests in the short list of those suitable for standardisation.

The demand from the industry became so strong that rather than waiting for European or other international standards for the key properties of fresh SCC to appear, national guidelines (discussed in detail in Chapter 11), even national standards[21, 22] were drafted and published, largely based on fragmented information published and without direct expertise and prenormative assessment behind it. The uncoordinated activity in this area threatened to generate obstacles in exploitation of SCC by proposing specifications based on different test methods. This was recognised by EFNARC, a grouping of European companies involved in the provision of materials for concrete production, which set up an international joint project group, which, in parallel with the work of the SCC project, produced the first European guidelines for SCC[10]. Most of the SCC expertise then available was used, including the test methods proposed by the SCC project.

The important role of RILEM in the research and development of SCC continued. It set up a series of biannual International Symposia on SCC, beginning with the first in Stockholm in 1999[17], Tokyo in 2001[23], Reykjavik in 2003[24], Chicago in 2005[25] and Ghent in 2007[26]. The symposia became leading international forums for dissemination of research and development in SCC technology and for its promotion. SCC continued to be a topic of follow-up; RILEM technical committees such as TC188 Casting of SCC led by Å. Skarendahl focused on the production process and TC205 Durability of SCC led by G. De Schutter, focused on the internal structure of hardened SCCs and their long-term performance[27].

Japanese researchers published a few papers[14, 28] on SCC in the USA in the 1990s but the construction industry there, with its own specific business structure, was less receptive to the changes associated with introduction of SCC than in Europe and elsewhere. The situation there began to change following an initiative led by S.R. Shah, as was demonstrated by the large attendance and numerous contributions to the 2nd North-American Conference and 4th RILEM Symposium on SCC held in Chicago in late 2005[25].

Overall, it is possible to conclude that the development of SCC technology is having a positive impact beyond its own boundaries. The need and desire to find out where and how, and if at all, the concrete which is self-compacting when fresh differs from that of TVC has led to a revival of research in all parts of concrete technology. This includes areas where basic relationships were felt to be well established for the TVC, where new scope for innovation may have developed, but where, had it not been for the introduction of SCC, there was not enough impetus to look into it, and to revise/update it, exploiting advances in science and technology in general.

References

[1] Bartos P. *Fresh Concrete: Workability and Tests*, Elsevier Science, Amsterdam, the Netherlands, 1992.

[2] EN 206-1 Concrete – Part 1: Definitions, specifications and quality control, 2000.

[3] Bartos P.J.M., Cleland D.J. (Eds.) *Special Concrete: Workability and Mixing*, E&FN Spon, London, UK, 1993.

[4] Wallevik O.H. and Nielsson I. Self-compacting concrete – a rheological approach. In: Ozawa K. and Ouchi M. (Eds.) *Proceedings of the International Workshop on Self-compacting Concrete*, Japan Society of Civil Engineers, Tokyo, 1999, pp 136–159.

[5] Lankard D.R. and Newell J.K. Preparation of highly reinforced steel fibre reinforced concrete composites. In: *Fibre Reinforced Concrete*, American Concrete Institute SP-81, Detroit, MI, USA, 1984, pp 286–306.

[6] Balaguru P. and Kendzulak J. Mechanical properties of slurry infiltrated fibre concrete (SIFCON). In: Shah S.P. and Batson G.B. (Eds.) Fibre Reinforced Concrete: Properties and Applications, SP-105, American Concrete Institute, Detroit, MI, USA, 1987, pp 247–268.

[7] Marrs D.L. and Bartos P.J.M. Development and testing of self-compacting grout for the production of SIFCON. In: Reinhardt H.W. and Naaman A.E. (Eds.) High Performance Fibre Reinforced Cement Composites (HPFRCC 3), *Proceedings of the Third International RILEM Workshop*, Mainz, Germany, 1999, pp 171–179.

[8] Svermova L., Sonebi M. and Bartos P.J.M. Development of cement slurries for self-compacting SIFCON with silica fume using factorial design model. In: Wallevik O. and Nielsson I. (Eds.) Self-compacting Concrete, RILEM Publications, 2002, pp 741–752.

[9] Concrete Society. Self-compacting concrete – A review, The Concrete Society and BRE, 2005, Technical Report 62, 2005.
[10] European Guidelines for Self-compacting Concrete, Joint Project Group, EFNARC, Brussels, Belgium 2005. www.efnarc.org
[11] Bartos P.J.M, Self-compacting Concrete in Bridge Construction, UK Concrete Bridge Development Group, Camberley, UK, 2005.
[12] Bartos P.J.M, Cleland D.J and Marrs D.L. (Eds.) *Production Methods and Workability of Concrete*, E&FN Spon, London, UK, 1996.
[13] Ozawa K., Maekawa K., Kunishima M. and Okamura H. Development of high performance concrete based on the durability design of concrete structures. In: *Proceedings 'EASEC-2'*, Chiang-Mai, Thailand, 1989, Vol. 1, pp 445–450.
[14] Kuroiva S., Mastuoka Y., Hayakawa M. and Shindoh T. Application of super-workable concrete to construction of a 20-storey building, ACI SP-140, American Concrete Institute, Detroit, MI, USA, 1993.
[15] Hayakawa M. Development and application of super workable concrete. In: *Special Concretes – Workability and Mixing*, Bartos P.J.M. and Cleland D.J. (Eds.) E&FN Spon, London, UK, 1993, pp 183–190.
[16] Grauers M. *et al.* Rational production and improved working environment through using self-compacting concrete. EC Brite-EuRam Contract No. BRPR-CT96-0366, 1997–2000.
[17] Petersson Ö. and Skarendahl Å. (Eds.) *Self-compacting Concrete, Proceedings of the First International Symposium*, Stockholm 1999, RILEM Publications, Cachan, France, 2000.
[18] Bennenk W. SCC as applied in the Dutch precast concrete industry. In: *Proceedings of the Second International Symposium on Self-compacting Concrete*, Tokyo, Japan, COMS Engineering Corporation, Tokyo, Japan, October 2001, pp 625–632.
[19] Association Française de Génie Civil: Provisional recommendations for self-compacting concrete (in French). Documents scientifiques et techniques, July 2000.
[20] Bartos P.J.M. and Gibbs J.C. *et al.* Testing SCC, Final report, EC Growth contract GRD2-2000-30024, http://www.acmcentre.com/scc, 2001–2004.
[21] Norma Italiana UNI 11140 Self-compacting concrete. Specification, characteristics and checking, 2003.
[22] Norma Italiana UNI 11141 – 11145, Test methods for SCC, 2003.
[23] Ouchi M. and Ozawa K. (Eds.) *Self-compacting Concrete, Proceedings of the Second International Symposium*, Tokyo, 2001, COMS Engineering Corporation, Tokyo, Japan.
[24] Wallevik O.H. and Nielsson I. (Eds.) *Self-compacting Concrete, Proceedings of the Third International Symposium*, Reykjavik, 2003, RILEM Publications, Cachan, France, 2002.
[25] Shah S.R. *et al.* (Eds.) SCC 2005 *Proceedings of the Second North-American Conference on Self-consolidating Concrete and Fourth International RILEM Symposium on Self-compacting Concrete*, Chicago, IL, USA, Oct.–Nov. 2005.
[26] De Schutter G. and Boel V. (Eds.) SCC 2007 *Proceedings of the Fifth International Symposium on Self-compacting Concrete*, Ghent, 2007, RILEM Publications, Cachan, France.
[27] De Schutter G. and Audenaert K. (Eds.): Durability of self-compacting concrete, State-of-the-art report, RILEM TC 205-DSCD, RILEM Publications, Cachan, France, 2007.
[28] Okamura H. and Ozawa K. Self-compactable high performance concrete in Japan. In ACI SP-169, American Concrete Institute, Detroit, MI, USA, 1994, pp 31–44.

Chapter 3
Constituent materials

SCC can be made from any of the constituent materials that are normally used for structural concrete. As a general rule, materials that conform to the standards and specifications for use in concrete are suitable, and all the materials discussed in this chapter are assumed to comply with these unless otherwise indicated. However, SCC mixes are less tolerant to variations in material supply than TVC, and so uniformity and consistency of supply throughout the production of the concrete are essential.

3.1 Coarse aggregates

3.1.1 Type

It might be expected that rounded, uncrushed coarse aggregates from, for example, gravel deposits, would be preferable to angular crushed rock aggregates for achieving the high flow properties of SCC. However, these are not readily available in many locations, and crushed rock aggregates are commonly used. An analysis of 63 case studies published between 1993 and 2003 found that crushed rock was used in over three-quarters of these[1]. A little surprisingly, there was no evidence of higher powder contents than in mixes with uncrushed aggregates.

Lightweight aggregate has also been successfully used to produce SCC with a density of 1900 kg/m^3 for large-scale production of precast building panels[2, 3] and mixes with densities down to 1400 kg/m^3 have been developed with expanded clay aggregates which have been prewetted to avoid loss of consistence after mixing[4]. At least two studies have demonstrated the feasibility of using selected recycled aggregate in SCC with careful attention to mix design[5, 6].

3.1.2 Maximum size

Most applications of SCC have used a maximum size aggregate of 16 mm or 20 mm, depending on local availability and practice[1]. 10 mm, 12 mm or 14 mm aggregates have also been used, for example in the production of thin precast structural elements[7]. 40 mm aggregate was used in the large volumes of concrete for the anchor blocks of the Akashi-Kaikyo bridge, a spectacular early use of SCC[8]. There is no evidence of a pattern of variation of powder content with aggregate size – indicating that the aggregate grading is probably a more important factor in mix design[1].

3.1.3 Fine aggregate

As with coarse aggregate, fine aggregates conforming to local standards and practice have generally been found to be suitable for SCC. Of particular importance is the amount of very fine material. Whilst this is not necessarily detrimental, that below a limiting particle size should be considered as being part of the powder fraction for the purposes of mix design. Opinion differs as to the value of the limiting particle size, e.g. 75 μm is recommended in Japan[9] and 125 μm in Europe[10].

3.1.4 Overall aggregate grading

A well-distributed overall grading is desirable, but within this restriction successful SCC has been produced with aggregate of significantly different gradings. For example, Figure 3.1 shows the 11 aggregate gradings considered suitable for SCC by the partners from eight European countries who participated in the Testing-SCC project[11].

Brouwers and Radix[12] have analysed the combined grading curves for all particles (coarse aggregate, fine aggregate, fillers and cement) in some successful SCC mixes and compared these to the particle size distribution given by Funk and Dinger's equation (Equation 3.1)[13]:

$$P(D) = \frac{(D^q - D^q_{\min})}{(D^q_{\max} - D^q_{\min})} \tag{3.1}$$

where $P(D)$ is the fraction passing a sieve size D, $D_{\min}$ is the minimum particle size, $D_{\max}$ is the maximum particle size and q is a constant.

They found that a distribution given by a value of q of 0.25 is appropriate for aggregate for SCC, whereas for TVC a value of 0.5 (which gives the Fuller curve) is more appropriate. The two curves for the overall grading of 20 mm aggregate (with $D_{\min}$ = 0.1 mm) are shown in Figure 3.2. The curve for aggregate for SCC lies approximately in the middle of the set of gradings in Figure 3.1, confirming Brouwers and Radix's conclusions. The need for higher fine aggregate content in SCC is clear.

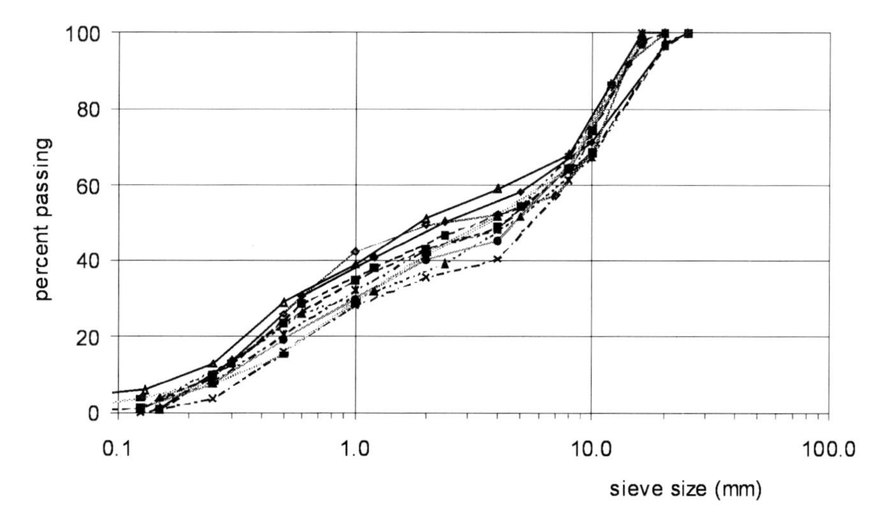

Figure 3.1 *Overall aggregate gradings for typical SCC mixes from Testing-SCC project partners*[11].

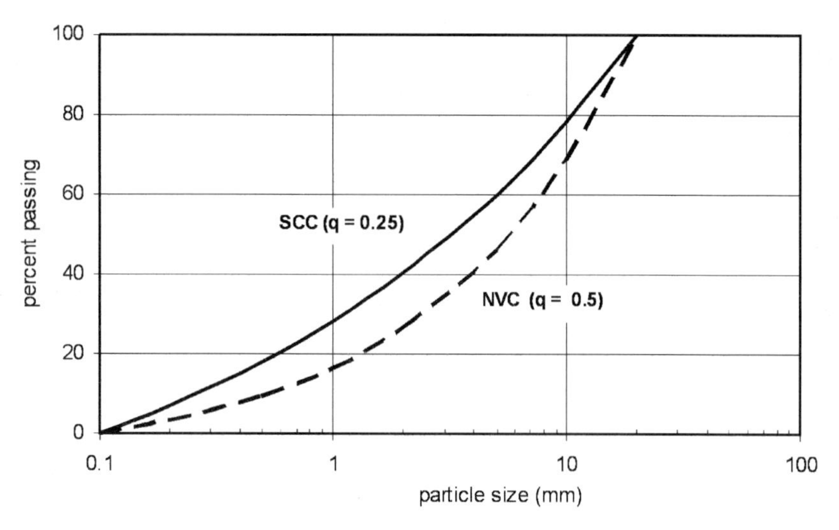

Figure 3.2 *Preferred aggregate grading curves for SCC and TVC (NVC from analysis in*[12]*).*

3.2 Cements and additions

SCC requires a high powder content and a low water/powder ratio and early studies were based on the use of a low heat, high belite content Portland cement[14]. In practice, however, typical Portland cements (e.g. BS EN 197 CEM I[15] or equivalent) have been used in combination with at least one or more additions. The analysis of case studies referred to above[1] found that in over 25% of cases a ternary blend of Portland cement with two additions was used and in about 5% a quaternary blend.

Additions of all common types have been used: nearly inert, such as limestone powder, pozzolanic, such as fly ash or microsilica, and latent hydraulic, such as ground granulated blast furnace slag (the European standard EN-206[16] describes additions as either Type I: nearly inert or Type II: pozzolanic or latent hydraulic, and gives requirements for each). In general the advantages can include:

- control of the strength, particularly where the high strength which would occur with the use of Portland cement alone is not required (we will discuss this further in Chapter 6)
- reduced heat of hydration and therefore reduced risk of cracking from thermal strains
- reduced risk of damage from the alkali–silica reaction associated with the alkali content of the cement
- improved stability and rheological behaviour
- extended consistence retention

Although each type of addition will have different effects on both the fresh and hardened properties, preference for different applications seems to have been mainly based on familiarity with their previous use in TVC and local practice, experience, availability and cost.

The SCC used in many of the early applications of SCC in Japan contained powders which were mixtures of a Portland blast furnace cement and fly ash, i.e. in effect a ternary blend. Subsequent developments, particularly in Europe, made use of limestone powder, which because of the lower water/powder ratios could be added in significant proportions without an unacceptable reduction in strength. Indeed this was an effective way of controlling strength and producing moderate strength SCC, e.g. for housing applications[17]. Some studies have found that varying the overall particle size distribution of the powder by including finer particles, e.g. by incorporating finer-ground limestone powder[18] or ultrapulverised fly ash[19] can lead to increased viscosity and hence stability of the concrete and the production of SCC with satisfactory properties at a lower total powder content. Practice in some regions, particularly in North America, has been to include modest quantities (up to 5%) of microsilica[20].

Since all but the highest strength SCC can benefit from containing significant quantities of inert filler, it is attractive to consider the extent to which waste materials with suitable particle sizes can be incorporated. Clearly there are no standards or specifications available for such materials, and so their use must be approached with caution. Particularly interesting are some promising studies on the use of quarry dust[21–24].

3.3 Admixtures

3.3.1 Superplasticisers

The increasingly widespread availability of superplasticisers in the 1970s led initially to the introduction and use of flowing concrete, which had significant advantages in many production situations[25]. The development of SCC can be seen as a logical step in the exploitation of superplasticiser technology and existing lignosulfonate-, where melamine formaldehyde- and naphthalene formaldehyde-based materials were initially used. Although successful, these materials often experienced compatibility problems with cements such as workability retention[26].

The introduction of the so-called new generation polycarboxylic ether-based admixtures was an important step towards improved and consistent performance. A distinctive advantage is the flexibility of their chemical structure which can be synthesised or modified so that their properties can be tailored to overcome compatibility issues[27] and to meet the varying needs of SCC for different applications.

They achieve dispersion in two ways. First, as with other types of plasticisers and superplasticisers, they attach themselves to the cement particles and impart a negative surface charge thus causing electrostatic repulsion and deflocculation, thus releasing water for increased mobility. In addition, the backbones of their molecules have side chains of a varying length which physically keep the cement particles apart thus allowing water to surround more of their surface area, a mechanism known as steric hindrance[28]. By engineering the length of the molecular backbone and the length and density of the side chains the concrete performance can be significantly altered to advantage, for example by controlling setting and/or consistence retention. Admixtures of this type are now generally the preferred option, and many products are now marketed as being specifically suitable for SCC.

However, the interaction and performance of any combination of powders and admixtures remains critical. Although there are some factors which have been identified as important in this respect, such as the fineness, alkali, silicate, aluminate and sulfate content of the cement and the fineness and carbon content of the additions it is difficult to predict the behaviour of specific products, and therefore establishing this is an essential stage of mix design and development. We will discuss this further in Chapter 6.

3.3.2 Viscosity modifying admixtures

The incorporation of a viscosity modifying admixture (VMA) in addition to a superplasticiser can improve the stability of the SCC and/or make it more robust by reducing its sensitivity to variations in the constituent materials, particularly the moisture content of the aggregate.

There are two broad types of VMA: adsorptive and nonabsorptive[29, 30]. Absorptive VMAs are absorbed onto the surface of the cement particles forming a bridging structure between them. Increasing dosages therefore result in an increase in plastic viscosity and reduced segregation resistance. However, as the superplasticisers and the VMA compete for space on the cement particle surface the effectiveness of the superplasticiser can be reduced. Examples of this type include cellulose-based and acrylic-based water soluble polymers. Nonabsorptive VMAs act on the water alone and increase the plastic viscosity of the concrete by linking their own molecules. Therefore, they do not compete with the superplasticiser for cement surface and hence these have generally been preferred for SCC. Examples commonly used in SCC include water-soluble microbial polysaccharides such as welan gum and diutan gum[31, 32] both of which are compatible with hydration products and are effective at small concentrations of 0.025–0.075% by weight of powder. In addition minerals such as precipitated silica, which is amorphous and has a very fine particle size and large surface area, can also be used.

VMAs can increase both the yield stress and plastic viscosity of the concrete and so an increase in the dosage of the superplasticiser to offset the increase in yield stress may be required. Mixes with VMAs can also exhibit shear thinning, which can enhance flow during placement and thixotropy which enhances the stability at rest after placing[32].

The increased robustness of mixes that contain VMAs is illustrated in Figure 3.3 which has been obtained from the analysis of results from six test programmes on mixes containing a variety of component materials[33]. The change in slump-flow is plotted against the variation in water content of the mix and the slope of the line indicates the sensitivity to variation in water content. The VMAs were, in all cases but one, polysaccharides and mixes with these all have shallower slopes and can therefore be considered to be more robust. The average slope reduced from about 8 mm of slump-flow per kg/m^3 of mix water for mixes without the VMA to about half this value for mixes with a VMA. These values are equivalent to changes in slump-flow of about 150 mm and 75 mm, respectively, for a 1% change in the moisture content of the aggregate in a typical mix.

The effectiveness of a VMA is strongly dependent on the type of superplasticiser, and there are also compatibility problems between some VMAs and some

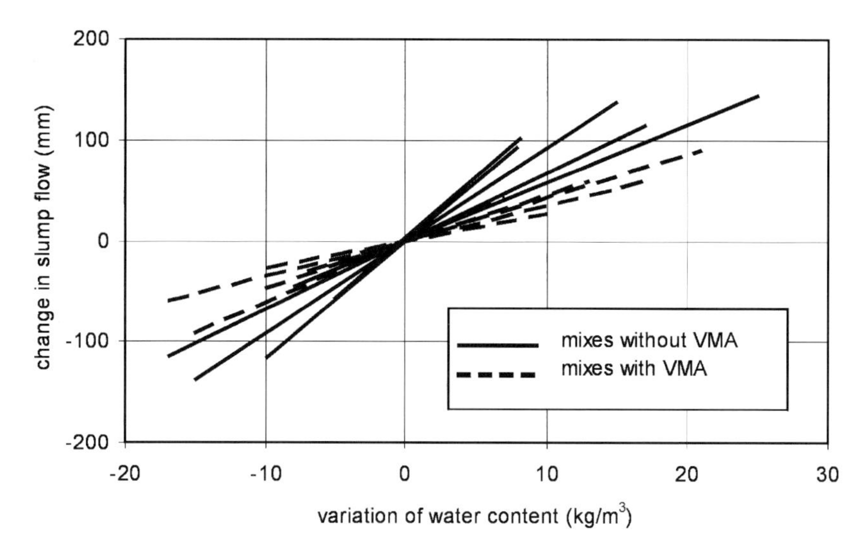

Figure 3.3 *Effect of variation of water content of mixes with and without a VMA on changes in slump-flow*[33].

superplasticisers. Combinations to be avoided include methyl cellulose-based VMAs with naphthalene-based superplasticisers[34].

There are no standards for VMAs for use in concrete, but the UK Cement Admixtures Association have produced guidelines which include a test method to determine their effect on segregation[35]. First, a segregating self-compacting mix without the VMA is produced such that the width of its paste rim at the edge of the concrete after carrying out a slump flow test is at least 50 mm. The mix with the VMA added is then tested and the width of the rim should be reduced to less than 20 mm.

3.3.3 Other admixtures

As with TVC, other admixtures are often used to enhance a variety of other properties, but in some cases these can also affect the properties of the fresh SCC. For example, air-entraining agents will increase the frost resistance of the hardened concrete, but will also contribute to the paste volume and modify its rheology, both of which can be of benefit. Clearly if more than one admixture is used their compatibility must be considered.

Blends of one or more superplasticisers and/or other admixtures such as retarders or air-entraining agents have been marketed. A number of products containing both a superplasticiser and a VMA have also been developed but although useful suffer from the obvious disadvantage of not being able to vary the relative proportions of the two components.

3.4 Fibres

A collaborative European project[36] established the feasibility of producing steel fibre reinforced SCC with up to 60 kg/m^3 of 60 mm fibres in a mix containing 16 mm aggregate with adequate consistence and with a toughness that would be obtained in the equivalent normally vibrated steel fibre reinforced concrete. However, some reduction in passing ability was identified. Broadly similar behaviour was obtained by Khayat and Roussel[37] with 38 mm fibres and 10 mm aggregate.

Grünewald and Walraven[38, 39] have shown that including fibres reduces the filling ability of SCC. However, mixes with satisfactory flow properties were produced with up to 120 kg/m^3 of 30 mm steel fibres or 80 kg/m^3 of 60 mm fibres in mixes with 16 mm aggregate, but with some reduction in the passing ability. Mixes with the longer (60 mm) fibres were also more prone to clustering or nonuniform distribution of the fibres, which therefore adversely affected the hardened concrete performance.

Barragan *et al.*[40] have reported the production of a high-strength SCC (75 MPa) containing 40 kg/m^3 of 30 mm × 0.5 mm hook-ended steel fibres and 12 mm gravel aggregate which was successfully used for casting 24 m × 1 m × 0.08 m wall elements filled from the top.

Busterud *et al.*[41] have produced SCC with 16 mm crushed aggregate and 0.8% by volume (approximately 62 kg/m^3) of 60 mm steel fibres for 150 mm thick house slabs. The mix required increased paste content compared to nonfibre reinforced SCC, which was obtained by the use of an air-entraining agent.

The use of carbon fibre and glass fibre textile reinforcement for SCC has been shown to be feasible[42, 43] and if pursued could lead to wider applications.

References

[1] Domone P.L. Self-compacting concrete: an analysis of 11 years of case studies. *Cement and Concrete Composites* 2006, 28(2) 197–208.

[2] Umehara H., Uehara T., Enomoto Y. and Oka S. Development and usage of lightweight high performance concrete. In: *Proceedings of the International Conference on High Performance Concrete (supplementary papers),* Singapore, American Concrete Institute, Detroit, MI, USA, 1994, pp 339–353.

[3] Umehara H., Hamada D., Yamamuro H. and Oka S. Development and usage of self-compacting concrete in precast field. In: *Proceedings of the First RILEM International Symposium on Self-compacting Concrete,* Stockholm, Sweden, RILEM, Paris, France, 1999, pp 705–717.

[4] Muller H.S., Metcherine V. and Haist M. Development of self-compacting lightweight aggregate concrete. In: *Proceedings of the Second International Symposium on Self-compacting Concrete,* Tokyo, Japan, October 2001, COMS Engineering Corporation, Japan, pp 737–742.

[5] Corinaldesi V. and Moriconi G. The use of recycled aggregates from building demolition in self compacting concrete. In: Wallevik O. and Nielsson I. (Eds.) *Proceedings of the Third RILEM International Symposium on Self-compacting Concrete,* Reykjavik, Iceland, RILEM Publications, Bagneux, France, 2003 pp 251–260.

[6] Tu T-Y., Jann Y-Y. and Hwang C-L. The application of recycled aggregates in SCC. In: *Proceedings of the First International Symposium on Design, Performance and Use of Self-consolidating Concrete SCC,* China, May 2005, Hunin, China, Proceedings PRO 42, RILEM, Paris, pp 145–152.

[7] Fornasier G., Giovambattista P. and Zitzer L. Self-consolidating concrete in Argentina: development program and applications. In: *Proceedings of the First North American*

Conference on the design and Use of Self-consolidating Concrete, Chicago, IL, USA, 2002, pp 439–444.

[8] Furuya N., Itohiya T. and Arima I. Development and application of highly flowing concrete for mass concrete anchorages of Akashi-Kaikyo Bridge. In: *Proceedings of the International Conference on High Performance Concrete (supplementary papers),* Singapore, American Concrete Institute, Detroit, MI, USA, 1994, pp 371–396.

[9] Okamura H. and Ozawa K. Mix design method for self-compacting concrete. Concrete Library of Japan Society of Civil Engineers No. 25, June 1995 pp 107–120.

[10] EFNARC European Project Group The European Guidelines for Self-compacting Concrete: Specification, Production and Use, EFNARC, May 2005 http://www.efnarc.org/pdf/ SCCGuidelinesMay2005.pdf

[11] Aarre T. and Domone P.L. Reference concretes for evaluation of test methods for SCC. In: Wallevik O. and Nielsson I. (Eds.) *Proceedings of the Third RILEM International Symposium on Self-compacting Concrete,* Reykjavik, Iceland, RILEM Publications, Bagneux, France, 2003, pp 495–505.

[12] Brouwers H.J.H. and Radix H.J. Self-compacting concrete: the role of particle size distribution. In: *Proceedings of the First International Symposium on Design, Performance and Use of Self-consolidating Concrete,* May 2005, Hunin, China, Proceedings PRO 42, RILEM, Paris, pp 109–118.

[13] Funk J.E. and Dinger D.R. *Predictive Process Control of Crowded Particulate Suspension Applied to Ceramic Manufacturing* Kluwer Academic, Dordrecht, The Netherlands, 1994.

[14] Okamura H. and Ozawa K. Mix design method for self-compacting concrete. Concrete Library of Japan Society of Civil Engineers, No. 25, June 1995 pp 107–120.

[15] BS EN 197-1:2000 Cement – Part 1: Composition, specifications and conformity criteria for common cements.

[16] BS EN 206-1:2000 Concrete Part 1: Specification, performance, production and conformity.

[17] Petersson Ö. Preliminary Mix Design: Final Report of Task 1, Brite EuRam Contract No. BRPR-CT96-0366, 1998, http://www.civeng.ucl.ac.uk/research/concrete/sccBE.asp

[18] Fujiwara H., Nagataki S., Otsuki N. and Endo H. Study on reducing unit powder content of high-fluidity concrete by controlling powder partice size distribution. Concrete Library of Japan, Society of Civil Engineers, No. 28, December 1996, pp 117–128.

[19] Xie Y., Liu B., Yin J. and Zhou S. Optimum mix parameters of high strength self-compacting concrete with ultrapulverized fuel ash. *Cement and Concrete Research* 2002, 32(3), 477–480.

[20] Khayat H.K. and Aitcin P.-C. Use of self-consolidating concrete in Canada – Present situation and perspectives. In: *Proceedings of International Workshop on Self-compacting Concrete,* Kochi University of Technology, Japan, August 1998, pp 11–22.

[21] Nishio A., Tamura H. and Ohashi M. Self-compacting concrete with high-volume crushed rock fines. In: *Fourth CANMET/ACI/JI Conference on Advances in Concrete Technology*, ACI SP-179, 1998, pp 617–630.

[22] Ho D.W.S., Sheinn A.M.M. and Tam C.T. The use of quarry dust for SCC applications. *Cement and Concrete Research* 2002, 32(4), 505–511.

[23] Westerholm M. and Lagerblad B. Influence of the fines from crushed aggregate on micro-mortar rheology. In: Wallevik O. and Nielsson I. (Eds.) *Proceedings of the Third RILEM International Symposium on Self-compacting Concrete*, Reykjavik, Iceland, RILEM Publications, Bagneux, France, 2003, pp 165–173.

[24] Khrapko M. Development of SCC containing quarry rock dust. In: *Proceedings of the Second North American Conference on the Design and Use of Self-consolidating*

Concrete (SCC) and the Fourth International RILEM Symposium on Self-compacting Concrete, Chicago, IL, USA, October 2005, Hanley Wood, Minneapolis, MN, USA, pp 203–209.

[25] Cement and Concrete Association Superplasticizing admixtures in concrete. Report of Cement Admixtures Association and Cement and Concrete Association Working Party, C&CA, Wexham, UK, 1976.

[26] Aitcin P.C., Joliceur C. and MacGregor J.G. Superplasticizers: how they work and why they occasionally don't. *Concrete International* 1994, 5, 45–52.

[27] Yamada K., Ogawa S. and Takahashi T. Improvement of the compatibility between cement and superplasticizer by optimising the chemical structure of the polycarboxylate type superplasticizer. In: Ozawa K. and Ouchi M. (Eds.) *Proceedings of the Second RILEM International Symposium on Self-compacting Concrete*, 2001, COMS Engineering Corporation, Tokyo, Japan, pp 159–168.

[28] Bury M. and Christensen B.J. The role of chemical admixtures in producing self consolidating concrete. In: *Proceedings of the First North American Conference on the Design and Use of Self-consolidating Concrete,* Centre for Advanced Cement Based Materials, North Western University, Chicago, IL, USA, November 2002, pp 141–146.

[29] Nawa T., Izumi T. and Edamatsu Y. State-of-the-art report on materials and design of self-compacting concrete. In: *Proceedings of International Workshop on Self-compacting Concrete,* Kochi, Japan, August 1998, pp 160–190.

[30] Yammamuro H., Izumi T. and Mizunuma T. Study of non-absorptive viscosity agents applied to self-compacting concrete. In: *Proceedings of Fifth CANMET/ACI International Conference on Superplasticizers and Other Chemical Admixtures in Concrete*, Rome, 1997, ACI SP 173, American Concrete Institute, Detroit, MI, USA, pp 427–444.

[31] Phyfferoen A., Monty H., Skaggs B., Sakata Nm Yania S. and Yoshizaki M. Evaluation of the biopolymer, diutan gum, for use in self compacting concrete. In: *Proceedings of the First North American Conference on the Design and Use of Self-consolidating Concrete,* Chicago, IL, USA, November 2002, pp 147–152.

[32] Khayat K.H. and Ghezal A. Effect of viscosity modifying admixture-superplasticizer combination on flow properties of SCC equivalent mortar. In: Wallevik O. and Nielsson I. (Eds.) *Proceedings of the Third RILEM International Symposium on Self-compacting Concrete,* Reykjavik, Iceland, RILEM Publications, Bagneux, France, 2003, pp 369–385.

[33] Day R. and Holton I. (Eds.) Concrete Society Technical Report No 62 Self-Compacting Concrete: A review. The Concrete Society, Camberley, UK, August 2005.

[34] Khayat K.H. Viscosity-enhancing admixtures for cement-based materials – An overview. *Cement and Concrete Composites* 1998, 20(2-3) 171–188.

[35] Cement Admixtures Association Guidelines for establishing the suitability of viscosity modifying admixtures self-compacting concrete, European Federation for Concrete Admixtures, Knowle, UK, 2004 http://www.efca.info/pdf/CAA_Guidelines_for_SCC-VMA.pdf

[36] Brite EuRam project: Rational production and improved working environment through using self-compacting concrete. Final Technical Report, 2000, http://www.civeng.ucl.ac.uk/research/concrete/sccBE.asp

[37] Khayat K.H. and Roussel Y. Testing and performance of fibre-reinforced, self-consolidating concrete. In: *Proceedings of the First RILEM International Symposium on Self-compacting Concrete*, Stockholm, Sweden, September 1999, RILEM Publications, Bagneux, France, pp 509–521.

[38] Grünewald S. and Walraven J.C. Maximum content of steel fibres in self compacting concrete. In: Ozawa K. and Ouchi M. (Eds.) *Proceedings of the Second RILEM*

International Symposium on Self-compacting Concrete Kochi, Japan, 2001, COMS Engineering Corporation, Tokyo, Japan, pp 137–146.

[39] Grünewald S. and Walraven J.C. Optimisation of the mixture composition of self-compacting fibre reinforced concrete. In: *Proceedings of the Second North American conference on the Design and Use of Self-consolidating Concrete and the Fourth International RILEM Symposium on Self-compacting Concrete,* Chicago, IL, USA, October 2005, Hanley Wood, Minneapolis, MN, USA, pp 393–399.

[40] Barragan B., de la Cruz C., Gettu R., Bravo M. and Zerbino R. Development and application of fibre reinforced self-compacting concrete. In: *Young Researchers' Forum, Proceedings of the International Conference*, Dundee, UK, July 2005, Thomas Telford, London, UK, pp 165–172.

[41] Busterud L., Johansen K. and Dossland A.L. Production of fibre reinforced SCC. In: *Proceedings of the Second North American Conference on the Design and Use of Self-consolidating Concrete and the Fourth International RILEM Symposium on Self-compacting Concrete,* Chicago, IL, USA, October 2005, Hanley Wood, Minneapolis, MN, USA, pp 381–386.

[42] Uebachs S. and Brameshuber W. Self-compacting concrete with carbon fibre reinforcement for industrial floor slabs. In: *Proceedings of the Second North American Conference on the Design and Use of Self-consolidating Concrete and the Fourth International RILEM Symposium on Self-compacting Concrete*, Chicago, IL, USA, October 2005, Hanley Wood, Minneapolis, MN, USA.

[43] Xu S-L. and Li H. Self-compacting concrete for textile reinforced elements. In: *Proceedings of the Second North American Conference on the Design and Use of Self-consolidating Concrete and the Fourth International RILEM Symposium on Self-compacting Concrete*, Chicago, IL, USA, October 2005, Hanley Wood, Minneapolis, MN, USA, pp 409–415.

Chapter 4
Properties of fresh self-compacting concrete mixes

4.1 Introduction to rheology

Rheology is the branch of science dealing with deformation and flow of matter. The word is of Greek origin, referring to *panta rei*, everything flows. The scope of rheology is therefore very wide. In engineering practice, however, it is mainly used to describe the behaviour of materials, which do not follow the laws of deformation and flow, which apply to 'ideal' (simple, elastic, Newtonian) gases, liquids and solids. Engineering materials vary in rheology from those having 'ideal' model behaviour, to those with much more complex rheological performance.

Rheology can be applied to concrete in both its fresh and hardened state, and to all its intermediate stages of ageing and development. Rheology is useful in the assessment of 'plastic' deformations of hardened concrete, namely those occurring near to failure and those due to long-term loading, such as creep. On the other hand, the rheology of SCC is primarily concerned with its behaviour in the fresh state. All concrete produced has to pass through the fresh concrete stage in the period of time between the instant at which all of the constituents of a mix come together in a concrete mixer and the time when it solidifies sufficiently for the laws governing the deformation of solids to begin to apply.

On a macroscopic scale, plain concrete is a composite material consisting of aggregate, cement paste and a small amount of voids filled by air or water. Aggregate alone takes up to approximately 80% of the volume. The cement paste, which itself is a composite, coats and separates the particles of aggregate, forming a 'lubricating' layer which reduces the friction between them and facilitates their movement and re-arrangement. A complex interaction between the paste and the aggregate then controls the flow of a fresh concrete mix and provides the mix with a certain level of workability (consistence).

It is usual to consider fresh concrete as a multiphase material, mainly a dispersion of solid particles (aggregate) in a viscous liquid, represented by fresh cement paste. The paste itself is a multiphase material made up of particles of cement and other particles in the same size range, such as additions (particularly in SCC mixes) suspended in water. Concrete stands out amongst other particulate composites in the extreme range of sizes of the particles present. Additions may comprise particles smaller than 0.001 mm (1 μm) down to 0.000001 mm (1 nm) while the coarse aggregate may include particles greater than 150 mm, a range of 10^9.

Another, less common, approach is to consider fresh concrete first as an assembly of solid particles, which behaves like other granular materials (an analogy with soils is sometimes used). The paste is then added in quantities, which are adequate first to fill the voids between the aggregate particles and second a further amount to 'lubricate' the system and make possible its deformation or movement.

The application of rheology to fresh concrete presents a substantial challenge both in its technical aspects and in the nature of the fundamental assumptions. Because of the progressive physico-chemical changes occurring, the rheological characteristics of fresh concrete also continually change, and in a nonlinear manner. The rate and magnitude of the change depends not only on the composition of the mix, but also on the environmental conditions.

Rheological equations which link the basic rheological characteristics and parameters of a fresh concrete mix rely on several basic assumptions, namely that fresh concrete is:

- a homogeneous material of a uniform composition
- an isotropic material, its properties being the same in all directions
- a continuum, with no discontinuities between any two points/locations within the material

Such assumptions are reasonable in practice for most fluids, the behaviour of which are measured by established rheometry and expressed as basic rheological characteristics. In the case of fresh concrete the degree to which the assumptions listed above are valid varies greatly, depending on the composition of the mix and the scale at which it is observed and assessed. Once the resolution reaches the microscale the material is less likely to represent a continuum. It can hardly be considered as a homogeneous continuum when the scale of resolution drops down to micro- or even nanoscale. The recent increase in the application of nanotechnology to construction materials will necessitate a thorough review of the existing rheological approaches and 'tools' used to describe, measure and predict the behaviour of fresh cement-based materials such as concrete.

The rheological response, the deformation and flow of fresh concrete, depends on the magnitude of the applied shear stress. In general, for a multiphase/composite material such as concrete, the results of measurements at very small strains (deformations) are likely to produce different rheological parameters than those encountered with large strains. The basic rheological characteristics such as yield stress and viscosity enable predictions to be made of the deformations, or flow, due to a given shear stress and, conversely, measurements of deformations and flow may be used to calculate the stresses applied.

Rheometers used to be designed to reach a certain steady rate of shear at which the measurements were made. More recently, it has become possible to preset and program different testing regimes, in which the rate of shear varies in a predetermined pattern. This has brought the test conditions a little closer to reality, but they are still some way from simulating the shear rates obtained in the practical handling and placing of fresh concrete. The problem is now not so much what the rheometers can do, but in finding out what the real shear rates encountered in practice are, and how closely they are simulated in both basic rheometrical work and in the applied, empirical assessment such as in the traditional slump test.

Methods for the measurement of basic rheological characteristics of fresh concretes continue to improve, but it is essential to bear in mind the limitations of each type of test when applied to such a complex and variable material as fresh concrete.

The relationships between basic rheological characteristics, which are often used when practical rheology is applied to fresh concrete, reflect two types of behaviour: Newtonian and non-Newtonian.

4.1.1 Newtonian flow

A 'Newtonian' fluid behaves is a way similar to that of an ideal solid subjected to a shear stress. The shear stress τ produces a shear strain (deformation) γ proportional to the magnitude of the shear stress. The coefficient of the proportionality is the shear modulus G; the relationship can be expressed as:

$$\tau = G.\gamma \tag{4.1}$$

When the shear stress is applied to a Newtonian fluid, it causes the fluid to deform proportionally to the magnitude of stress, but unlike an ideal solid, the deformation of the fluid will continue for as long as the stress is applied.

The rate at which the deformation will continue becomes of significance. It is usually expressed as strain γ per unit of time (s). More generally and using the coefficient of viscosity η, instead of the shear modulus G, a differential equation showing this relationship can be written as:

$$\tau = \eta \, . \, d\gamma/dt \; = \eta.\dot{\gamma} \tag{4.2}$$

Shear in fluids is usually expressed as a relative movement of two parallel 'plates', and a laminar flow is assumed for low shear rates. A continuous shear strain then generates a movement, which has a velocity v.

A Newtonian laminar viscous flow is generated, in which the shear stress is proportional to the velocity gradient D, which is the rate of change of velocity v of the moving plate over a unit length y of distance between the moving plates (see Figure 4.1). The relationship can be expressed as:

$$\tau = \eta \, . \, dv/dy \; = \eta \, .\dot{D} \tag{4.3}$$

The velocity gradient D is equivalent to the rate of change of the shear strain with time $(d\gamma / dt$ or $\gamma^{\cdot})$ and the equation can be re-written as:

$$\tau = \eta.\dot{\gamma} \text{ for shear stress (as above)} \tag{4.4}$$

or

$$\eta = \tau/\dot{\gamma} = \text{ shear stress/shear rate} \tag{4.5}$$

In basic S.I. units the viscosity is therefore measured in $N/m^2/s$ = Pas (pascal second, where 1 pascal = 1 N/m^2).

The viscosity of Newtonian fluids is then determined from measurements of the shear stress produced by the shearing of the fluid at a given rate of shear $\dot{\gamma}$.

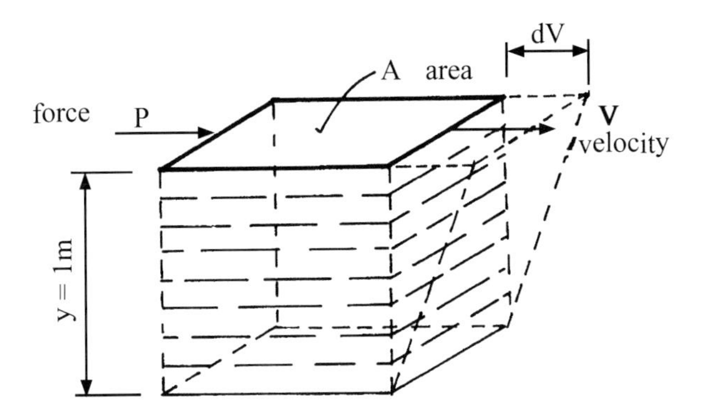

Figure 4.1 *Laminar viscous flow of Newtonian fluid.*

The relationship between the shear stress and rate of shear of a Newtonian fluid is therefore linear, as shown by a 'flow' diagram in Figure 4.2. The flow curves of all Newtonian fluids are straight lines starting at the origin. It follows that the viscosity of a Newtonian fluid (i.e. the slope of the curve) can be determined from a single measurement of a corresponding pair of values of the shear rate and shear stress (see Figure 4.2). The viscosity is independent of the shear stress and shear rate.

There are many ways of determining viscosity, using a range of viscometers/rheometers of different geometries and test conditions. These include rotating coaxial cylinders, capillary tubes, parallel plates, orifices, falling spheres etc.[1, 2].

4.1.2 Non-Newtonian flow

There are many fluids, which do not show the 'ideal' Newtonian behaviour. The viscosity of such fluids is not independent of the shear stress or strain rate applied.

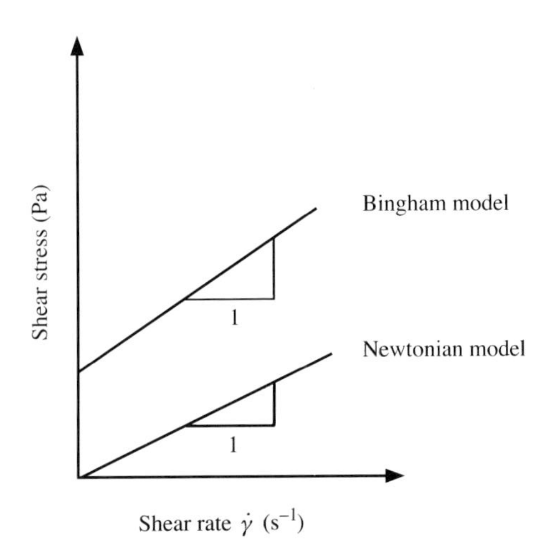

Figure 4.2 *Flow diagram of Newtonian and Bingham viscous fluids.*

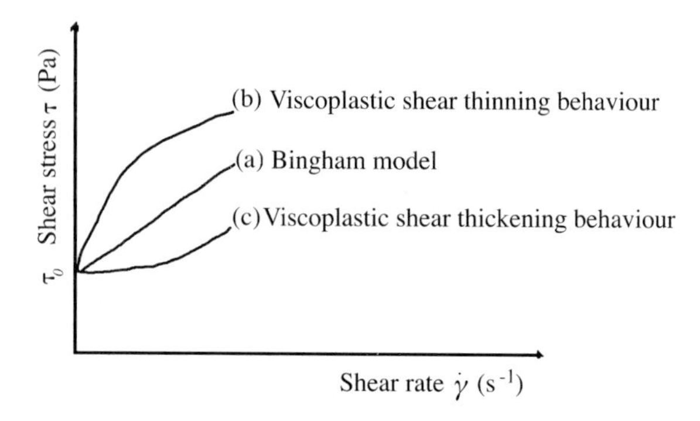

Figure 4.3 *Flow diagram of different types of non-Newtonian fluids.*

Moreover, their behaviour often changes with the time for which the shear is applied and maintained and with its rate of change (decrease and/or increase).

Some non-Newtonian fluids may exhibit a linear relationship between shear stress and shear rate but which does not start until a certain minimum level of shear stress is reached. The value of this minimum stress required to initiate flow is called the yield stress τ_0.

Different types of flow diagrams related to non-Newtonian behaviour are shown in Figure 4.3.

The non-Newtonian fluids, which exhibit a zero yield stress are generally called pseudoplastic. Flow curve (b) in Figure 4.3 shows a decrease of viscosity with shear strain, which indicates a shear-thinning behaviour, as opposed to the shear-thickening behaviour indicated by curve (c). The behaviour of most of the pseudoplastic fluids follows an approximately exponential relationship expressed as:

$$\tau = A \,.\, \gamma^n \tag{4.6}$$

where A is a constant related to the consistence of the fluid and $n < 1$ (shear thinning) or $n > 1$ (shear thickening).

Many non-Newtonian fluids show a change of viscosity when a constant rate of shear is reached and maintained. The rate of decrease of viscosity can reduce with the duration of the shearing and level off. The fluid shows a 'recovery' when the shear rate reduces and stops; the same or a different yield value can be restored, and the whole process may be repeatable. Such fluids are called thixotropic (see Figure 4.4), while anti-thixotropic fluids show an opposite behaviour.

The behaviour of fresh concrete is similar to that of several of the non-Newtonian fluids. In the case of fresh TVCs, which have lower consistence, the model of the Bingham fluid, represented by curve (a) in Figure 4.3, is of particular significance, with the value of yield stress decreasing as the consistence (e.g. value of slump) increases. The relationship:

$$\tau = \tau_0 + \mu \, \dot{\gamma} \tag{4.7}$$

where τ_0 is the yield stress, and μ is the plastic viscosity, which indicates a typical Bingham fluid.

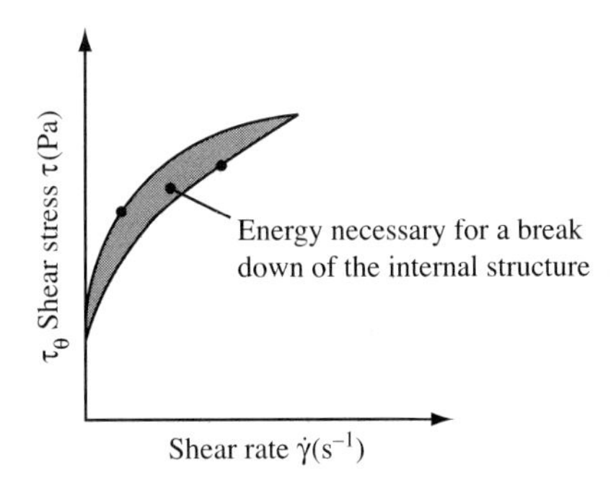

Figure 4.4 *Non-Newtonian fluids showing thixotropic behaviour.*

The introduction of flowing superplasticised concretes, which had high consistence (slump greater than 120 mm), and especially that of SCCs, greatly increased the range of rheological behaviour observed when fresh. While TVCs of low workability behave generally as Bingham fluids, with relatively high yield stresses, typical SCCs exhibit very low or zero yield stress, and often behave as viscous Newtonian fluids.

4.1.3 Rheology and consistence (workability) of fresh concrete

Fresh concrete is a complex composite material, to which different rheological approaches can be applied. The behaviour of a fresh mix can be considered to be that of a solid phase in a granular form with a very wide range of sizes of interacting particles, each lubricated by different amounts of a fluid phase (paste or water), perhaps including a small amount of a gaseous phase (air). The deformation of such a material then depends on the particle packing and particle interaction, with a lesser but still important role for the rheological properties of the fluid phase (paste). The packing of the particles is strongly dependent on shape, size and surface characteristics of the particles and on how they are assembled together during mixing. This approach allows traditional, low-consistence mixes to be considered as granular materials close to engineering soils.

Alternatively, fresh concrete can be treated as a suspension of solid particles of different sizes in a fluid medium (paste) of certain rheological characteristics. Several 'intermediate' situations can be also considered. It should be noted that the 'paste' itself is a suspension/composite though it contains within it a much higher proportion of water (a Newtonian fluid) than when concrete is considered. As the maximum size of the suspended particles is small, the paste lends itself to rheological characterisation more easily than real, 'full-scale' concrete. This is one of the reasons why in recent decades, the focus of basic rheological research in cement and concrete has been overwhelmingly on the paste rather than on concrete, despite the difficulties with the extrapolation of the results of such work into concrete and into concrete construction practice.

Extrapolation of the results obtained from testing pastes would be easier if the test regimes simulated the flow of paste within concrete better. In reality, the paste is always confined within a 'skeleton' of solid particles of aggregate. The geometry of this confinement of the paste and its boundary conditions is variable. The shear rates it is subject to when concrete flows are difficult to estimate because the relative positions/distances (packing) of the aggregate particles also vary during the deformation or flow of the concrete.

It is outside the scope of this book to review the different rheological approaches in detail and compare their merits and limitations, particularly when the paste alone is considered and to review the numerous methods for their measurement. Instead, a brief outline is provided of methods for assessment of basic rheological characteristics of fresh concrete, namely that of the yield stress and plastic viscosity.

4.1.4 Concrete rheometers

The significance of the 'yield value' of fresh concrete became more appreciated in the late 1970s, and attempts were made to develop test methods[3], which would go beyond measurement of useful but largely empirical parameters (e.g. traditional slump) and apply the Bingham model to obtain more basic rheological characteristics from the expression:

$$\tau = \tau_0 + \mu \dot{\gamma} \quad (4.7)$$

where μ is plastic viscosity of fresh concrete (not being vibrated).

To define this relationship it was necessary to carry out tests at two different combinations of shear stress and shear rate, which were simulated by torque, T, and rate of rotation, N, of a coaxial rheometer, hence the need for the 'two-point' testing.

A two-point test apparatus based on measuring the torque required to rotate an impeller in a bowl of concrete at varying speeds was successfully developed, but despite its sound and logical basis, the various versions of the original 'two-point' test apparatus[3, 4] which appeared over more than two decades have achieved only partial success. As the geometries and modes of operation of the apparatus evolved, reliable comparability of the results decreased. The devices proved to be impracticable as means for an on-site assessment or control of the workability of fresh mixes, being too large, expensive and sophisticated for use on average construction projects. This, in turn, curtailed the amount of practical experience obtained, such experience being essential to give this entirely new type of test result (a pair of readings) a meaning, which would be of practical use to engineers and concrete technologists. In its initial form, the two-point test produced two parameters: g which is related to yield stress and h which is related to plastic viscosity.

A change in the mix design of fresh concrete then produced changes in not one but two parameters, often in different directions, as shown in Figure 4.5[4]. It also proved difficult to convert the initial readings of parameters g and h into the basic rheological characteristics of yield stress and plastic viscosity. Such a conversion relied on the selection of an appropriate theoretical rheological model, not necessarily just a simple Bingham relationship, which was difficult to predict before testing. However, relationships relating g to yield stress and h to plastic viscosity were successfully obtained experimentally using suitable pseudoplastic and high viscosity liquids. These calibration factors therefore enabled the two-point test results to be expressed in fundamental rheological units.

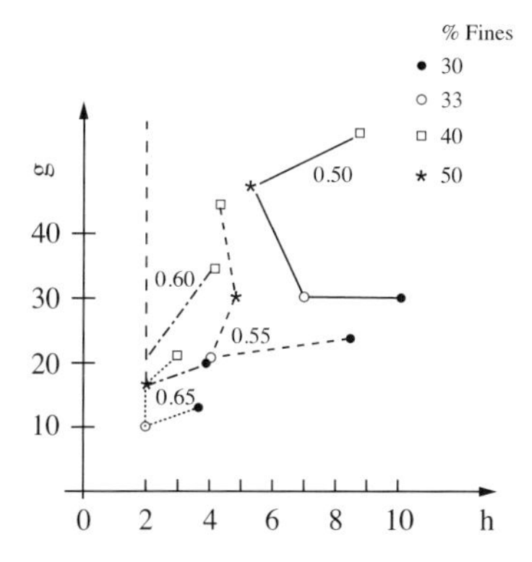

Figure 4.5 *Assessment of rheological characteristics of fresh concrete by a 'BML viscometer'. Note the different directions of change with mix proportions.*

Further technical development, namely that by Wallevik and Gjorv[5] in the late 1990s produced the 'BML viscometer' (see Figure 4.6), currently used in a number of concrete research centres. The apparatus is an automated, computer-operated rheometer, using the principle of a coaxial cylinder rheometer. It is available in several variations, depending on the maximum size of particles in the mix and on its consistence/workability[6]. The BML viscometer is primarily applicable to mixes of medium to very high workability (slump > 80 mm). It has a capacity to apply torque (related to *g*, yield value) over a range of 0.27–27 Nm with rotational velocities varying in the range of 0.1–0.6 revolution per second. The test container for concrete has a volume of 25 l.

The original version of the BML viscometer was developed further when the 'Reiner–Rivlin' and 'plug-flow' theoretical models were introduced for interpretation of the test results, depending on the mode of operation and material tested. As with

Figure 4.6 *'BML viscometer', based on a coaxial cylinder rheometer.*

other similar designs, it is not entirely free of problems associated with the re-arrangement of the packing of particles of coarse aggregate in the vertical shearing zone and settlement/segregation of the particles during a test. Such a re-arrangement widens the spacing of the particles in order to permit the shear-zone to develop, inevitably leading to an increase in the volume of the concrete being tested, a phenomenon usually called 'dilatancy'.

The application of coaxial cylinder rheometry to materials with a high content of very coarse particles is hindered by inherent problems of settlement/segregation and the creation of 'shear/slip' zones during testing. An alternative approach, exploiting the principle of parallel plate rheometry was adopted in the design of the BTRheom apparatus[7] for concrete (see Figure 4.7) at the Laboratoire Central des Ponts et Chaussées (LCPC) in France. Here, the zone of shearing is horizontal, the bottom 'plate' is driven and the rotation/torque transmitted to the top plate is recorded. The top plate can be either free to move upwards (unrestrained shearing), or it is restrained, to avoid the effects of dilatancy caused by re-arrangement of particles of coarse aggregate interlocking across the largely horizontal zone of shearing.

In addition, the measurement could be carried out even when the sample of concrete was vibrated. The apparatus is programmable and computer controlled. It is much less bulky than the BML viscometer.

There are only few established rheometers[8], which are applicable to fluids like fresh concrete, which contain solid particles of up to 25 mm in size. The presence of such coarse aggregate requires large volumes of samples during testing. The sample size depends on the type of the rheometer but it is usually more than 20 l. This tends to make the apparatus large and difficult to move. The very high cost of the equipment, its requirement for specially trained operators and the difficulty of

Figure 4.7 *BTrheom for testing of fresh concrete – based on the principle of a parallel plate rheometer.*

interpreting the test results for practical concrete construction use has restricted the numbers of concrete rheometers in use.

Concrete rheometers differ in the fundamental rheometrical principles they use, and this has cast doubts on the reliability of the measurements of the behaviour of fresh concrete expressed in basic rheological characteristics (yield value, plastic viscosity). However, it was not until recently that a systematic comparative study of the performance of the different types of concrete rheometers was carried out. Two experiments aimed at comparison of rheometers for measurement of basic rheological characteristics of fresh concrete were carried out at Nantes, France in 2000 and at Cleveland, Ohio in 2003[9–11]. The experiments involved several devices, representing three types of a rheometer, all capable of dealing with 'full-scale' concrete:

- coaxial cylinder: Cemagref-IMG and BML
- parallel plate: BTRheom
- impeller/mixing action: the two-point and IBB

The BML rheometer had deeply ribbed coaxial cylinders. The two-point test apparatus was a development of the original version based on the same original principles[12] as the Canadian IBB rheometer[8]. A very rarely used, very large-volume viscometer based on coaxial cylinders, developed at Cemagref-IMG[9] for coarse aggregate suspensions was used at Nantes in an attempt to provide reliable values for comparison with the results from the concrete rheometers. A wide range of fresh concrete mixes (12 in Nantes and 17 in Cleveland), based on different types of binders (cements), aggregate and admixtures were used, including some mortar mixes.

Readings from all the rheometers tested except the IBB were converted into the fundamental rheological characteristics of yield stress and plastic viscosity. The IBB results remained as parameters g (N.m) and h (Nm.s). The behaviour of the concretes tested conformed to the Bingham model.

The values of yield stress and plastic viscosity obtained from each instrument for each of the concretes in both programmes, presented in chronological order of their production are shown in Figures 4.8 and 4.9.

Results shown in Figures 4.8 and 4.9 confirm that the concrete rheometers tested all ranked the wide range of concretes in the same order of yield value and plastic viscosity. The curves following the test results on different concretes rise and fall in a parallel fashion. The yield value was ranked with a higher significance (confidence)[10, 11] that the plastic viscosity, where the results varied more.

Comparisons between pairs of rheometers using different principles were also carried out. It was possible to correlate with high confidence results from tests on all pairings except between the BML–two-point and BTRheom–two-point tests[9–11].

4.2 Key characteristics of fresh self-compacting concrete

It is well understood that the behaviour of fresh SCC differs significantly from that of TVC of moderate consistence, which has to be compacted. What is less well appreciated is the important difference in behaviour of traditional superplasticised,

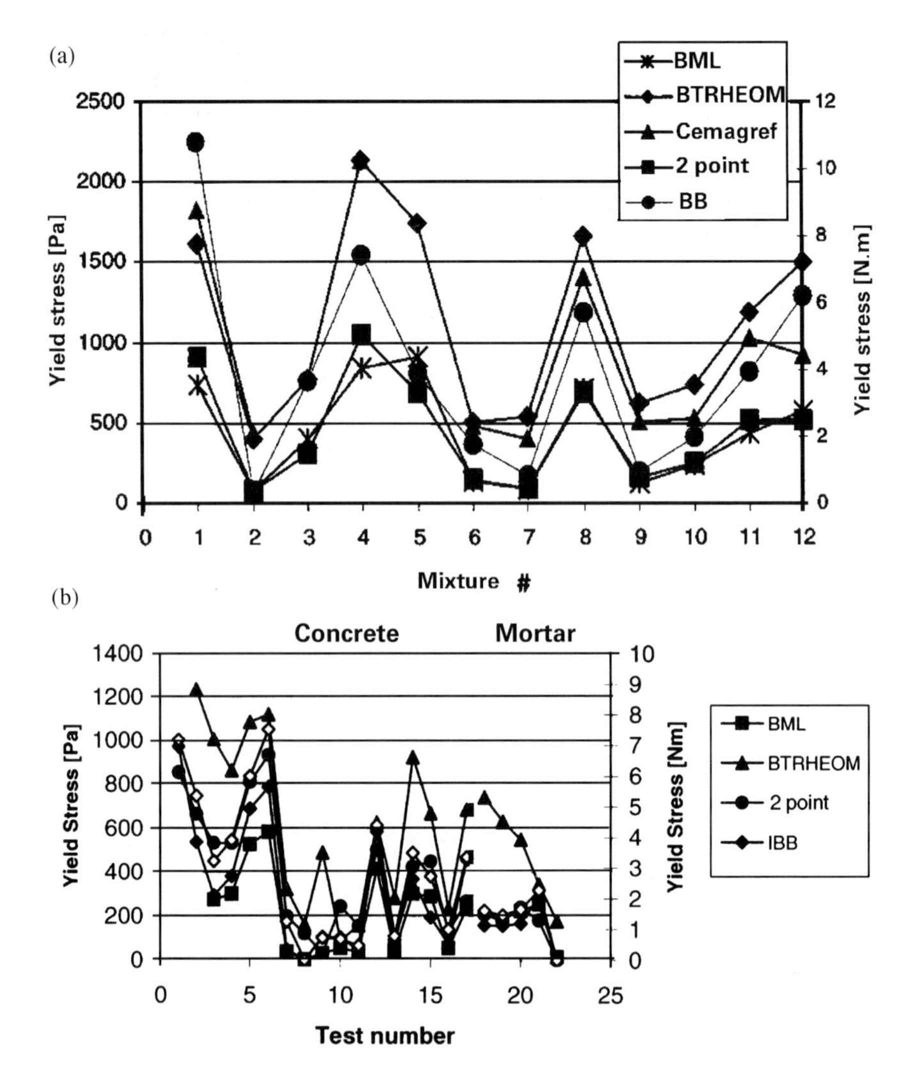

Figure 4.8 *Variation of yield stress measured by different concrete rheometers: (a) Nantes, (b) Cleveland* [9–11].

flowing concretes of a very high consistence (slump of 150 mm up to collapsed slump) and that of a genuine fresh self-compacting mix[13, 14].

These differences mean that none of the well-established and internationally standardised tests for properties of fresh TVC, such as those included in the European EN206/EN12350 series or ASTM in the USA, apply to fresh SCC. The traditional slump test becomes very unreliable when values over approximately 200 mm are recorded. In the case of SCC, it produces a collapsed slump, which cannot be expressed numerically. The traditional slump test cannot distinguish between mixes of very high consistence[8, 15] and therefore cannot be used to define 'self-compaction'.

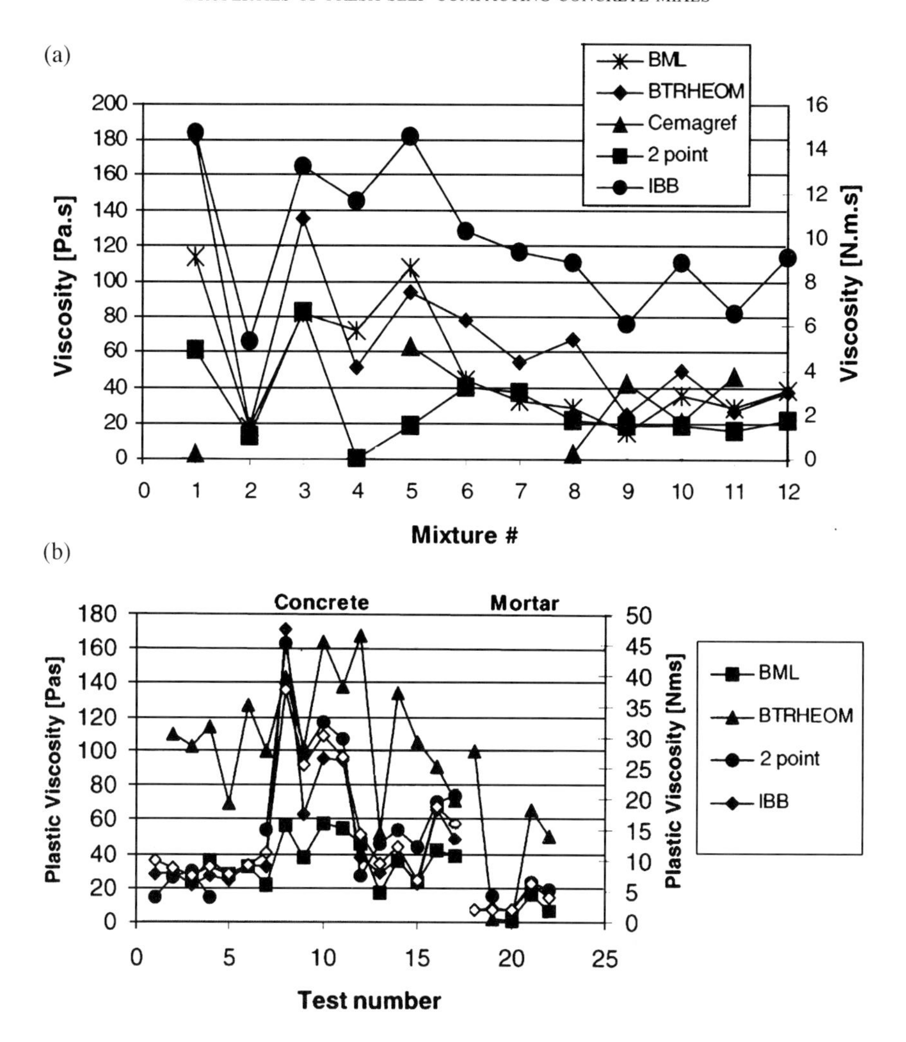

Figure 4.9 *Variation of values of plastic viscosity measured at: (a) Nantes, (b) Cleveland* [9–11].

There are three key properties of fresh SCC:

- filling ability
- passing ability
- segregation resistance

Adequate levels of all three key properties must be reached for fresh concrete to be self-compacting and to remain so during transport and placing by selected means and in given conditions. Work is in progress on the standardisation of test methods to identify and then specify/verify the three key properties of SCC in the fresh state, which are described and discussed in Chapters 5 and 11.

4.2.1 Filling ability

Filling ability is the ability of the fresh mix to flow under its own weight and completely fill all the spaces in the formwork. It is the characteristic often referred to as 'flow' or 'fluidity'. It indicates how far a fresh SCC mix might flow, and how well it would fill formwork and spaces of varying degrees of complexity. The filling ability also governs the 'self-compaction'. The filling ability must be high enough, the mix must be fluid enough, to permit any air introduced in the mixing process, or trapped during placing, to escape and leave behind an adequately compact concrete.

4.2.2 Passing ability

The passing ability determines how well the fresh mix will flow through confined and constricted spaces, narrow openings and between reinforcement. The determination of the passing ability helps to evaluate the level of risk that the flow of the fresh mix will be impaired or even fully blocked by coarse aggregate, which will become wedged or form arches between bars or within narrow passages or openings.

Passing ability is linked to filling ability. In order for concrete to pass freely through reinforcement, it is necessary for the coarse aggregate particles to rearrange their positions within the mix, maintain a degree of separation and not converge, interlock and block the gaps. It has been observed that in a mix with low filling ability the coarse aggregate particles have difficulty in doing so. Such a mix has poor passing ability, even without an excessive content of coarse aggregate. Both an adequate filling ability and passing ability are required for a fresh mix to be adequately self-compacting, and suitable for a given application.

4.2.3 Segregation resistance

Segregation within fresh concrete is a phenomenon related to the plastic viscosity and density of the cement paste. It is strongly linked to two simple assumptions[15, 16]: a solid denser than a liquid tends to sink and a viscous liquid flows with difficulty around a solid. Both assumptions are of significance in the case of fresh SCC. If the density of the aggregate is greater that that of the cement paste and the viscosity of the paste is low, segregation is likely to occur.

Segregation resistance is the ability of a fresh mix to maintain its original (adequately uniform) distribution of constituent materials (namely aggregate) during transport, placing and compaction. It has the same meaning as the 'stability' of a fresh mix. An underlying, fundamental difficulty in the assessment of segregation is the absence of a benchmark, an agreed minimum degree of uniformity of distribution of the constituents of a concrete mix. A completely uniform distribution cannot be achieved in practical construction. It can be only introduced through a theoretical model based on a selected system of the packing of the particles. In reality, the benchmark is set by the uniformity of the mix coming out of a mixer, which does not necessarily mean that it is always adequate. It is difficult to reliably assess when the concrete is still fresh. Instead, indirect assessments by measuring the variability of other characteristics of fresh concrete are usually used and care is taken to separate the variability inherent in the test procedure itself from that associated with the uniformity of the distribution of the constituents of the mix being evaluated. The sampling procedure and the volume/size of samples, compared with the maximum size of the particles (aggregate) in the mix, have profound effects on the results.

The importance of avoiding segregation has long been recognised in the practical technology of fresh TVC. Some attempts have been made to identify the susceptibility of such concrete to segregation when fresh[15] but none have been entirely successful. Concrete construction practice has therefore relied on adherence to general rules of good practice in placing to minimise the risk of segregation. Overcomplicated transport and placing methods of traditional fresh mixes were also to be avoided.

Segregation in TVC is often demonstrated as honeycombing in the hardened concrete, revealed after the formwork has been removed. Honeycombed concrete displays large, usually irregular voids between particles of coarse aggregate, which separated on impact from the mortar/paste. The most vulnerable places are the bases of columns and walls, where it can be the result of an excessive free-fall, which good practice usually limits to a height not greater than about 1.5 m. Honeycombing can be also caused by poor compaction, with the bases of columns and walls being difficult to compact by immersion poker vibrators.

Other forms of segregation include ‘bleeding’ and ‘pressure segregation’. Bleeding will occur when water separates from the mix during placing and rises to the upper, horizontal surface of the cast element, where a thin layer of paste with very high water/cement ratio (laitance) is formed. Pressure segregation can occur during the pumping of concrete, when the cohesion within a mix is insufficient to maintain its original, adequately uniform distribution of aggregate. The more fluid paste or mortar separates and moves ahead, leaving behind a mix with a lack of paste. This can lead to the formation of ‘plugs’ of concrete, which are difficult to pump, or in a severe case, blockage.

Segregation can still occur in a SCC which possesses adequate filling and passing abilities. However, it is unlikely to show in the usual form of honeycombed or uncompacted hardened concrete. Significant voids are unlikely to be present and the segregation takes the form of a nonuniform distribution of aggregate, particularly of coarse aggregate. This may settle at the bottom of the formwork and only be sparsely present at the top, or it can concentrate in locations where passage during casting is more difficult. In either case, the homogeneity of the concrete is reduced.

Evidence that a fresh mix can segregate while remaining basically self-compacting[17, 18] suggests that the effects of such segregation on the fundamental properties of hardened concrete may be less serious than is generally expected. Except in cases of very severe segregation, the high degree of compaction in good SCC is likely to mitigate the effects of the nonuniform distribution of aggregate on the strength and durability of the concrete, particularly when the strength of the paste/mortar is relatively high. More research in this area is required, to exploit the recent development of tests for the assessment of segregation[19, 20]. The results of such work would then provide the hard evidence, which is still needed, to support the development of a rational guidance on the different levels of segregation, backed by evidence from practice. Only then can reasonable limits for static (settlement) and dynamic segregation be established, linked to specific applications.

Different construction methods combined with different demands on the SCC when hardened (namely porosity) clearly require different degrees of segregation resistance in addition to appropriate levels of filling ability and passing ability. Segregation resistance then varies between:

- Minimum resistance: required in cases of casting simple, lightly reinforced concrete elements.
- Maximum resistance: required when casting complex thin wall elements with congested reinforcement, elements containing box-outs and where the mix is required to make frequent changes in its direction of flow. The placing may require 'free-fall' drops of concrete from substantial heights through reinforcement or multiple 'falls'. Higher resistance will be also required when a mix of extremely high filling ability with high rate of flow (low viscosity) is required for an application with long free-flow distances. Pumping to extreme heights will require SCCs, which will not segregate when moved under pressure through tall vertical pipelines.

If the segregation can not only be measured, but also controlled, then opportunities arise for the practical exploitation of this phenomenon. Bellamy and Mackechnie in New Zealand[21] have achieved a breakthrough in this area by developing a system in which a predetermined degree of (static) segregation can be induced in freshly placed concrete. This novel approach is also being exploited in practical construction; traditional 'sandwich' construction elements consisting of insulating material + structural concrete have been replaced by 'integral' concrete elements with 'stratified' internal structures. The properties of such concrete elements vary across their cross-section and provide both the insulating and the structural load bearing functions.

4.2.4 Interactions and additional properties

It is convenient to make theoretical distinctions between filling ability, passing ability, and resistance to segregation. However, the properties are interrelated and in certain conditions the relationships become very strong.

Such complex and variable interrelationships make it difficult to evaluate and rationally compare the performance of the different tests for the key properties which define self-compaction. The response to a test method for one of the key properties of the fresh mix may be affected, to varying degrees, by the other properties of the mix.

Such interaction is demonstrated when a mix has a tendency to segregate (low segregation resistance) which will reduce the ability of the test method to measure the passing ability of the mix. Of course, poor passing ability, blocking, can be caused independently and/or in combination by segregation, aggregate size/concentration and simply by too low filling ability.

Additional properties, which may characterise a SCC mix further and may be required for a given application include:

- flow-rate (speed of flow, 'viscosity')
- finishability

- self-levelling
- washout resistance

4.2.5 Flow-rate

Test methods and apparatus for assessment of filling ability or passing ability of fresh SCC often include an option for a simultaneous determination of the flow-rate (speed of flow)[8, 14, 18, 22] of the concrete tested. This characteristic is sometimes, erroneously, called 'viscosity'. Results of site-tests for filling and passing ability are related to plastic viscosity which can only be properly measured with an appropriate concrete rheometer (see Section 4.1.4) designed to assess 'basic' rheological parameters (shear stress/shear rate; plastic viscosity/yield stress) or 'applied' rheological parameters such as torque and rate of rotation.

The rate of flow is important for planning of very large or complicated pours, where it may significantly influence the overall time for completion of the project. The rate of flow is also important in precasting, where it has a direct influence on the overall productivity of the plant.

4.2.6 Other characteristics

Pumpability

Practically all fresh concrete classed as SCC is pumpable. Difficulties may arise only in extreme cases of very viscous mixes with a small maximum size of aggregate and extremely high contents of fines or fibres. There are no widely used practical tests to assess the pumpability of a fresh mix except by a practical pumping trial when the type of mix or transport/placing conditions are unusual. A 'pressure-bleed' test had been developed[23] but its use is now limited to mix design of special concretes such as sprayed mixes with fibres or mixes with very lightweight, porous aggregate[24, 25].

Finishability

Finishability refers to a successful production of the required types of surface finish, reflecting the relationship between the properties of the fresh mix and the quality of the surface of the concrete placed. Each type of surface finish has its own requirements regarding the characteristics of a fresh mix, namely its consistence/workability and segregation resistance (particularly bleeding). The establishment of universal criteria for properties of fresh SCC mixes which guarantee good finishability remains an unresolved challenge. It is not possible to indicate criteria other than those of an adequate filling ability and segregation resistance. Some guidance on finishability is given in Chapter 7.

There is no established direct test for this property of a fresh mix. In practice, finishability can be reliably assessed only by site trials using the finishing methods, formwork and casting methods proposed for the works. The key characteristics of the mixes that perform well are then recorded and used to maintain the behaviour required for good finishability in a given application.

Self-levelling

SCC*s* with high filling ability and flow-rate are very likely to be self-levelling or nearly so. It is important to check that such mixes have adequate resistance to segregation and hence low bleeding to avoid the formation of a level but weak layer at the top surface. This is important for the construction of slabs, floors and pavement using SCC. There is no established direct test for this property.

Washout resistance

Underwater concretes should be intrinsically self-compacting, with a greater than normal requirement for cohesion, which provides the required degree of resistance to washout. Such mixes usually contain additional VMAs/stabilisers[15], which improve the washout resistance but which reduce the filling ability and the rate of flow of the mix. Slow moving, viscous mixes are used. Specialist tests for washout resistance, such as the MC1 'spray-test'[26] or different 'plunge' tests[27] are available for the assessment of the washout resistance. It is important to ensure that an increased washout resistance does not decrease the filling ability of the fresh mix below the level at which the concrete remains self-compacting.

References

[1] Collyer A.A. and Clegg D.W. *Rheological Measurement*, Kluwer Academic, Dordrecht, the Netherlands, 1995.

[2] Macosko C.W. *Rheology: Principles, Measurements and Applications*, Wiley VCH, Poughkeepsie, NY, USA, 1994.

[3] Tattersall G.H. On the rationale of a two-point test, *Magazine of Concrete Research*, 1973, 84, 169–172.

[4] Tattersall G.H. and Banfill P.F.G. *Rheology of Fresh Concrete*, Pitman, London, UK, 1983.

[5] Wallevik O.H. and Gjorv O.E. Development of a coaxial cylinder viscometer for concrete. In: *Fresh Concrete, Proceedings of RILEM Colloquium*, Chapman and Hall, London, 1990, pp 213–224.

[6] Wallevik O.H. ConTec viscometers for cement paste, mortar and concrete, Multimedia CD ROM, 2005.

[7] De Larrard F., Szitcar J.C., Hu C., Joly M. Design of a rheometer for fluid concretes. In: Bartos P.J.M. and Cleland D.J. (Eds.) *Special Concretes: Workability and Mixing*, E&FN Spon, London, UK, 1993, pp 210–208.

[8] Bartos P.J.M., Sonebi M. and Tamimi A.K. *Workability and Rheology of Fresh Concrete: Compendium of Tests*, RILEM Publications, Cachan, France, 2002.

[9] Ferraris C.F. and Brower L.E. (Eds.) Comparison of concrete rheometers: international tests at LCPC (Nantes, France) in October 2000, NISTIR 6819, NIST, Gaithersburg, MD, USA, 2001.

[10] Sonebi M. Private communication, The Queen's University of Belfast, 2007.

[11] Ferraris C.F. and Brower L.E. (Eds.) Comparison of concrete rheometers: international tests at MBT (Cleveland, OH, USA) in May 2003, NISTIR 7154, NIST, Gaithersburg, MD, USA, 2004.

[12] Domone P., Xu Y. and Banfill P.F.G. Development of the two point workability test for high performance concrete, *Magazine of Concrete Research*, 1999, 51(3), 203–213.

[13] Bartos P.J.M. *Self-compacting Concrete in Bridge Construction – Guide for Design and Construction*, Concrete Bridge Development Group, Technical Guide 7, Camberley, UK, 2005.

[14] Day R.T.U. and Holton I.X. (Eds.) Self-compacting Concrete: a review, Concrete Society Technical Report 62, Camberley, UK, 2005.

[15] Bartos P. *Fresh Concrete: Properties and Tests*, Elsevier Science, Amsterdam, the Netherlands, 1992.

[16] Betancourt G.H. Admixtures, workability, vibration and segregation. *Materials and Structures*, 1988, 21, 286–288.

[17] Grauers M. *et al.* Rational production and improved working environment through using self-compacting concrete, EC Brite-EuRam Contract No. BRPR-CT96-0366, 1997–2000.

[18] Bartos P.J.M., Gibbs J.C. *et al.* Testing SCC, EC Growth contract GRD2-2000-30024, 2001–2004, http://www.civeng.ucl.ac.uk/research/concrete/Testing-SCC

[19] Rooney M.J. Assessment of properties of self-compacting concrete with reference to segregation, Ph.D. thesis, University of Paisley, Scotland, 2002.

[20] Poppe A.M. and de Schutter G. Development of a segregation test at the Magnel Laboratory for Concrete Research, University of Ghent, Ghent, Belgium, 2000.

[21] Bellamy L. and Mackechnie J., University of Canterbury, Christchurch, New Zealand, Private communication, 2006.

[22] Skarendahl Å. and Petersson Ö. Self-compacting concrete: state of the art report. RILEM TC 174, RILEM Publications, Cachan, France, 2000.

[23] Browne R. and Bamforth P. Tests to establish pumpability of concrete. *ACI Journal*, 1977, 19(74) 193.

[24] Austin S.A. and Robins P.J. Wet sprayed concrete technology for repair. Final Report, EPSRC Grant GR/K52829, Department of Civil and Building Engineering, Loughborough University, UK, 1999.

[25] Nolan J. *et al.* Methods for testing fresh light-weight aggregate concrete. The European Commission, Brite EuRam III project BE96-3942/R4, EuroLightCon Contract BRPR-CT97-0381, 1999.

[26] Ceza M. and Bartos P.J.M. Assessment of washout resistance of fresh concrete by the MC1 test. In: Bartos P.J.M., Cleland D.J. and Marrs D.L. (Eds.) *Production Methods and Workability of Concrete*, E&FN Spon, London, 1996, pp 399–413.

[27] Sonebi M., Khayat K.H. and Bartos P.J.M. Assessment of washout resistance of underwater concrete: comparison of CRD C61 and new MC-1 tests, *Materials and Structures*, 1999, 32(218) 273–281.

Chapter 5
Tests for key properties of fresh self-compacting concrete

5.1 Introduction

The fragmented historical development of modern SCC outlined in Chapter 2 and the major differences between the behaviour of fresh SCC and TVC described in Chapter 4 explain the difficulties encountered in the characterisation of fresh SCC. Very early it became obvious that none of the standard or established test methods for workability of fresh TVC[1] were applicable to SCC. Attempts were made to modify these tests[2], which had the advantage of using existing equipment. Numerous new test methods were introduced[2, 3], sometimes on an ad-hoc project basis. There was (and to a significant degree still is) a major problem: the capacity of a test to detect levels of the key characteristics of fresh SCC, which distinguish it from concrete requiring compaction, can be only evaluated on mixes, which have shown to be (or not to be) self-compacting in full-scale casting trials. Such an evaluation requires large resources and carries high costs, and is only possible through special or large projects. If there were no validation of test results against data from full-scale practical casting, the assessment of a test would be unreliable – the performance of a test for a given property would be examined on concrete mixes characterised by results from the test being evaluated.

Considerable progress has been achieved[4] but the absence of standardised, universally established test methods, which are capable of providing reliable, meaningful results using inexpensive, robust test equipment has been a recognised fundamental obstacle to the wider use of SCC in Europe and beyond. This has hindered the increased use of SCC since it is difficult to validate mix designs except by full-scale trials and confidence in the material cannot be increased. The current activities of many national and transnational organisations, such as European CEN and American ASTM are intended to overcome this obstacle and publish first standards as soon as possible (see Chapter 11).

It is important that the standard test methods adopted in different countries should be, if not identical, then at least complementary, and their results easily comparable. It should be noted that none of the tests is without drawbacks, including those most used to date, such as the ubiquitous slump-flow test.

The demand from industry has risen so much, that it has become more important to publish the standards for the most common tests for fresh SCC as quickly as possible, rather than introduce new tests. New tests may perform better (still to be demonstrated) and be more practical, but the cost of obtaining the equipment, training of staff and time required to gain enough practical experience appeared to be commercially unacceptable. It should therefore be noted, that the mere fact of a test having been standardised does not guarantee that it is the best performing test for measurement of a given property. However, standardisation should always mean that the test has been rigorously evaluated and its performance characteristics are known. It should be appreciated that significant modifications to existing tests, including the first standardised ones, may occur and new tests may be introduced in future.

In this chapter, all tests considered for European standardisation and a selection of supplementary tests of practical significance are first described and discussed. In each case, the principle of the test, the equipment and test procedure are described in sufficient detail to enable it to be carried out. Basic dimensions and arrangements are provided, while their exact tolerances can be found in the respective national or international standard specifications referred to. A review of the performance of the tests is provided, with the last part of the chapter focusing on applicability of the tests.

5.2 Sampling

An appropriate sampling procedure must be carefully considered in all cases of testing for the key properties of SCC. This requires an even greater attention when SCC rather than TVC is examined, and is of particular importance when segregation resistance is to be assessed. Standard tests should only be carried out on samples taken in accordance with relevant standard sampling procedures, e.g. with EN 12350-1[5], and the sampling procedure must be carefully considered, agreed and recorded.

Most of the tests for fresh properties (e.g. the slump-flow or J-ring), deal with samples taken from the bulk of the mix being placed. Few tests (e.g. the penetration test for segregation resistance) are applied directly to the bulk of the concrete to test it in situ. The purpose of the test (e.g. for within or between batch variations) must always be taken into account. A decision must be made as to whether enough volume of fresh SCC can be extracted from the bulk in a single scoop or an intermediate composite sample is prepared from several small quantities extracted from different places and usually remixed before testing. It is a common requirement that two or more tests are carried out on the same batch and a composite sample of sufficient volume must be prepared. It is essential to note whether successive test results were obtained from different subsamples from a larger (sometimes composite) sample or from the same sample, which has been retested. This applies to all tests, but especially to the Orimet, V-funnel and L-box tests, where such retesting is easy and may be part of the prescribed procedure.

Two factors need to be considered: the time (from water being added) and the size (volume of concrete tested).

- Samples should be obtained and ready for testing with a minimum of delay. The time which elapses between the concrete leaving the mixer and the commencement of a test must be always recorded. If this exceeds 5 min, the sample should be protected against loss of moisture and a change in temperature. The time factor becomes very important when a composite sample, from which subsamples are extracted and tested directly, has to be prepared first.

- The quantity of concrete required for a sample depends on the type of test to be carried out. It is recommended that the sample has a volume about 10% greater than the net volume required for a particular test. It should be noted that a standard scoop can handle a volume of approximately 2 l (mass of approximately 5 kg) for sampling TVC but it can handle significantly less when used on a fresh SCC.

Collection of a sample prior to testing by simply filling a bucket does not necessarily provide a representative sample of the mix placed in bulk. The mix can change its properties while in the bucket, particularly if it is prone to static segregation/settlement. This may affect both the absolute values and the scatter of results from tests on subsamples, which are simply scooped up or decanted from the bucket.

It is particularly important when SCC is tested that the samples/subsamples are clearly identified.

5.3 Test methods

5.3.1 Slump-flow

Origin and principle

The slump-flow test[2, 4, 6] was developed from the common slump test[1, 2, 7], widely used for TVC. It was first used for highly workable, self-compacting underwater concretes, which showed a collapsed slump, and then for all SCC. Instead of the vertical settlement of the truncated cone of concrete after removal of the mould when TVC is tested, the test allows a sample of concrete to flow out in all directions. The test therefore measures an unconfined horizontal spread of the sample.

A significant degree of confusion has arisen in a few countries (such as Germany) where the otherwise ubiquitous, traditional, 'Abrams' slump cone[7, 8] has not been used to determine the workability of TVC. Such countries used instead a 'flow-test' devised by Graf[1, 2, 9] based on a 'remoulding' of a sample of concrete initially contained within a conical mould shorter that that for the slump test. This mould is only 200 mm high, which is therefore too low to generate an adequate gravity driven outflow or spread of the sample. It is therefore not sensitive enough to measure the filling ability of a mix. The flow-test was developed specifically for TVC mixes of medium workability. The flow or, more appropriately, the spread, of such mixes was determined only after a controlled amount of jolting of the sample, a procedure, which simulated the compaction process but makes it entirely inapplicable to SCC.

There have been attempts to modify the slump-flow test by using the standard conical mould in an inverted position with the smaller opening facing down. The inverted cone does not require manual pressing-down or standing on the side brackets when it is placed on the baseplate and filled with concrete. A single operator can then carry out the basic slump-flow test. This test arrangement was examined during the systematic prenormative evaluation of tests[4], which highlighted its inherent disadvantages. First, unlike in the standard slump-flow test, the inverted cone is lifted up whilst the sample is flowing out. The magnitude of the spread then depends very strongly on the speed with which the cone (with most of the concrete still in it) is lifted and its height above the baseplate. The test would lose its simplicity if the apparatus had to be modified to ensure that the height of the lift of the cone remained constant and that the height limit was always reached within a very narrow time interval. Secondly, results of the inverted cone slump-flow test could not be compared with any existing slump-flow test results, which form the major part of the bulk of existing evidence about the behaviour of fresh SCC in practice. It was concluded that the difficulties outlined above clearly outweighed the advantages and the test was not included in the final prenormative assessment of the performance of the test methods suitable for the first stage of European standardisation.

Sample

A volume of 5.5 l is required to fill the conical mould. In practice, a sample of concrete should have a volume of at least 6 l or approximately a mass of 14 kg, provided that a mix of ordinary density is used. It is usual to determine the value of the slump-flow from two test results and a composite sample of at least 12 l or approximately 30 kg would be required. The maximum size of aggregate is 40 mm.

Equipment required

- Standard mould (Figure 5.1) for traditional slump test (height 300 mm, base diameter 200 mm, top diameter 100 mm), as used for TVC. The mould is fitted at its base with brackets to stand on.
- Metal baseplate, minimun dimensions 900 mm × 900 mm. The baseplate should be made of metal which is resistant to attack by cement paste and corrosion. It must be smooth and rigid enough to remain adequately flat when in use. The maximum permissible deviation from total flatness, measured at any point with straight edges placed diagonally across the baseplate, should not exceed 2 mm (3 mm is allowed in prEN[6]). The top surface of the baseplate is marked with a concentric circle which is 210 mm in diameter. As the equipment is usually also used for measurement of the speed of flow, the baseplate is normally marked with an additional circle which is 500 mm in diameter.
- Ruler 1 m long, graduated in millimetres.
- Metal straight edge 1 m long.

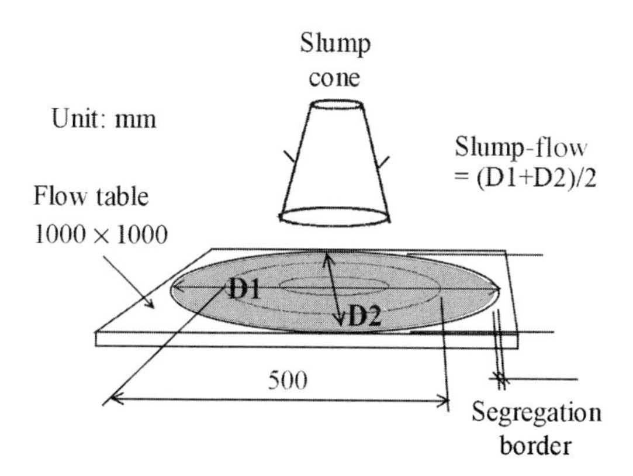

Figure 5.1 *Slump-flow test arrangement.*

- Scoop, preferably with a rounded mouth not more than 100 mm wide.
- Spirit level or a similar device.
- Cleaning cloth/rags.
- Optional: a 'holding-down' collar having a mass of 10 ± 1 kg. The collar should fit tightly onto the top of the conical mould, allowing the top edge of the mould to protrude.

Operating instructions

Two operators are required unless the optional 'holding-down' collar is used.

- Place the baseplate on a firm surface and check that it is in a horizontal position.
- Wet the surface of the baseplate and wipe off the moisture with a cloth.
- Place the conical mould onto the central circle (210 mm diameter).
- Press the mould firmly down by standing on its foot-holds or fit the holding-down collar.
- Fill the conical mould with concrete using the scoop, without any tamping or other compaction, ending with a minimum of surplus material above the top edge of the mould.
- Strike off any surplus concrete from the top off the cone and remove any spilled concrete from the baseplate.
- Lift the conical mould vertically in a single and smooth action within approximately 30 s of filling the mould (Figure 5.2).
- Allow the concrete to spread and once its movement has stopped, use the ruler to measure the largest diameter of the spread and record its value d_m to nearest

Figure 5.2 *Slump-flow test.*

10 mm (Figure 5.3). Follow this by measuring the diameter of the spread in a direction perpendicular to the first one. Record its value as d_r (to nearest 10 mm).

- Carry out a visual check of the spread concrete. Note any indications of potential segregation such as a zone (a 'halo') of paste or mortar without coarse aggregate around the edge of the spread concrete (Figure 5.4) and any 'pile-up' of coarse aggregate in the centre.

Test results

- Determine the difference between the values of d_r and d_m. If the difference exceeds 50 mm the result of the test is rejected as unreliable. Repeat the test procedure and if the difference between the diameters remains excessive, consider either the slump-flow test in principle or the equipment used as unsuitable for the concrete tested.
- Calculate the mean value of d_r and d_m and record it at the resulting slump-flow spread SF (mm) to the nearest 10 mm.
- Record any significant visual observations
- Record complementary test data such as: date, time and location; time lapse between completion of mixing of the concrete tested and the commencement of the slump-flow test; ambient temperature and temperature of concrete.

Interpretation of test results

The higher the slump-flow spread (SF), the greater is the filling ability (or flowability) of the fresh mix. Most SCCs require a slump-flow > 600 mm to achieve an adequate filling ability.

Figure 5.3 *Measurement of the 'slump-flow spread'. Note the uniform distribution of coarse aggregate.*

Figure 5.4 *A rim ('halo') of paste/laitance at the edge of the concrete sample at the end of the slump-flow test.*

The occurrence of any rim or halo of fine mortar or paste or laitance noted in the visual observation (Figure 5.4) is sometimes referred to as a 'segregation border'. Such a halo indicates, that some of the paste or fine mortar was not held strongly enough within the bulk of the concrete sample and has thus segregated. However, full-scale trials[4, 10] failed to support the widespread assumption that the use of SCC, which showed the segregation border invariably led to in situ segregation of the concrete. Samples of hardened concrete extracted from full-scale reinforced concrete trial elements made of SCC, which showed an edge 'halo' in the slump-flow test showed no indications of segregation. It is very likely that the effects of the segregation border may be confined to horizontal surfaces and may be unimportant except when such concrete is used in structures with very large horizontal surfaces, such as pavements or slabs.

Drawbacks and limitations

- The baseplate must be set up on firm, level ground, to prevent deformations, rocking or sloping during the test. A minor slope will not affect the result significantly, provided no concrete has flowed over the edge. The shape will be an oval rather than a circle; it is then essential that the perpendicular measurements are carried out along the longest and shortest length of the spread.
- The sides of the square baseplate must be at least 900 mm, which excludes the smaller (usually 750 mm × 750 mm), hinged baseplates used for the standard flow table test[9] for TVC.
- The results are significantly affected by the condition of the surface of the baseplate, particularly by the presence of moisture. It is recommended to wet the surface and wipe off surplus water immediately before a test. The surface should be protected from corrosion, which would roughen it and affect the results.

5.3.2 Flow-rate, determined during the slump-flow test

Origin and principle

The rate (or speed) of flow of the fresh concrete had been of interest from the very first attempts to measure slump-flow[2, 4, 6]. The time it took for the concrete sample to remould itself from the original shape of a truncated cone into a largely flat 'pizza-like' layer can be determined, usually in addition to the measurement of the slump-flow spread. The flow-rate is expressed as the time taken from the moment the mould is being lifted to the flowing concrete reaching a circle of 500 mm (50 cm) in diameter, concentrical with the position of the mould. It has often been referred to as time t_{50} (seconds), but an alternative expression, as the numerically equal time t_{500}[6] based on the diameter of the target circle being in millimetres (S.I. unit) is now preferred. The timing is difficult; both the start and especially the end of the timing are not easily determined.

Sample

The test for flow-rate is almost always carried out as part of the slump-flow test. There is therefore no need to test another sample; the values of both the slump-flow spread and t_{500} can be obtained from the same sample. The maximum size of the aggregate is 40 mm.

Equipment required

A stopwatch or another timer capable of measuring time to 0.1 s is required in addition to the basic equipment for the slump-flow test. The baseplate must be marked with a concentric ring of a 500 mm diameter, be flat and horizontal.

Operating instructions

Two operators are required. One carries out the slump-flow test, the other does the timing. The timing starts at the moment the cone is lifted (loses contact with the baseplate) and finishes when the spreading concrete first reaches the circle. The result will be unreliable if the spreading of the concrete was not concentric because of a distorted or excessively inclined baseplate (Figure 5.5).

Test results

- Record the flow-time t_{500} to the nearest 0.5 s.
- Record complementary test data such as: date, time and location; time lapse between completion of mixing of the concrete tested and start of the test; ambient temperature and temperature of concrete.

Interpretation of test result

The time t_{500} is related to the viscosity of the fresh mix but the test does not measure directly this fundamental rheological characteristic. The shorter the time t_{500}, the higher will be the flow-rate of SCC during placement. There are practical applications, such as very large pours and in precasting, where the flow-rate will significantly affect the completion times and productivity.

Drawbacks and limitations

- The test procedure requires accurate timing of a very short duration. It is important that the operator is trained and has a good (short/rapid) reaction time.

Figure 5.5 *An insufficiently rigid baseplate. The irregular spread of concrete makes the test result invalid.*

- Determination of the end of the flow is inherently difficult and requires good hand/eye coordination. Determination of the instant when the inscribed circle is touched by the advancing concrete and a slow reaction when stopping the timer are common sources of inaccuracies.
- The condition of the baseplate (smoothness, presence of moisture) is important. Dry and/or rough baseplates will slow down the flow-rate.

5.3.3 Orimet

Origin and principle

The test[1, 2, 4] was developed in early 1970s by Bartos[11] as a practical method specifically for a rapid assessment on construction sites of very highly workable, flowing fresh concrete mixes. The test is based on the principle of an orifice rheometer and measures filling ability (fluidity) of fresh concrete and its flow-rate as the Orimet flow time t_o (sec), which is the time it takes for a concrete sample to flow out through the orifice. The apparatus is also applicable to mortars and grouts, using orifices of appropriately reduced diameters.

The Orimet is also used as a go/no-go device to eliminate mixes of inadequate filling or passing ability for a given size of the orifice. In this mode it becomes a tool for a rapid on-site acceptance test.

Sample

At least 7.5 l or approximately 17 kg of concrete is required for a single test. The concrete is then re-tested twice, using the original sample collected in the bucket. Maximum size of aggregate is 25 mm for the 'standard' Orimet with a casting pipe of 120 mm inside diameter and an orifice of 80 mm in diameter.

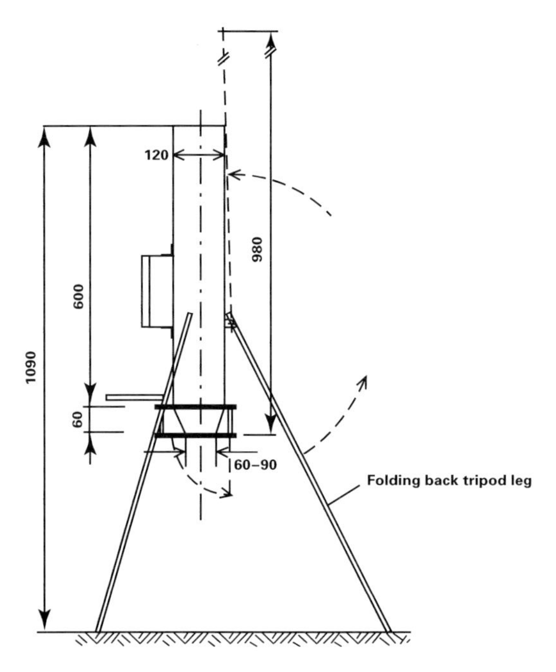

Figure 5.6 *Orimet: basic dimensions (mm). Orifice diameters vary from 60 mm for mortars to 90 mm (25 mm coarse aggregate); 80 mm external diameter is the most common.*

Equipment required

- Orimet apparatus, consisting of a vertical casting pipe with an attachment for an interchangeable orifice and a quick-release trap door at the bottom is required. The apparatus is supported by an integral tripod, which folds back for transport. The dimensions and the general arrangement are shown in Figure 5.6.
- Bucket having a minimum volume of 10 l for collection of concrete discharged from the Orimet.
- Stopwatch or a timer with an accuracy of 0.2 s.
- Cleaning brush.
- Orifice of an appropriate diameter (basic opening has a diameter of 80 mm).

Optional items:

- hopper fitting onto the top of the casting pipe for an easier filling
- scoop with a rounded mouth not wider than 120 mm
- additional orifices (e.g. 70 and 90 mm diameter)
- steel rod (approximately 16 mm diameter, minimum 750 mm long, with one end rounded)

Operating instructions

One or two operators are normally required. They need to:

- Extend and lock in position the supporting tripod and set the apparatus on firm ground.

- Wet the cleaning brush and dampen the inside of the casting pipe and the orifice. Ensure that the trap door is open to allow any surplus water to drain.
- Close the trap door and place the bucket under the orifice.
- Fill the apparatus with concrete without any compaction. Check that the casting pipe is adequately vertical; the surface of the concrete should not be more than 6 mm below the rim of the casting pipe at any point.
- Prepare the timer/stopwatch. Within one minute of filling the apparatus, quickly open the trap door and start timing simultaneously. Avoid tapping, shaking or disturbing the apparatus during the test.
- Look into the casting pipe (Figure 5.7) and stop timing when light is showing through. Record the flow time t_{o1} to the nearest 0.2 s. If the concrete does not discharge within 60 s, record a 'no-flow' result.
- Repeat the procedure twice more, using the sample collected in the bucket, without cleaning or washing out the apparatus and record times t_{o2} and t_{o3}. If there was one 'no-flow' recorded, carry out an additional test.
- Use the cleaning rod and brush to remove concrete from the apparatus following a 'no-flow' result and clean the apparatus.

Test results

- Discard the first flow time. Calculate the average flow time t_o from times t_{o2} and t_{o3} and record it to the nearest 0.2 s. If two tests produced a 'no-flow' result there is no flow time to be recorded.

Figure 5.7 *Orimet test in progress.*

- No timing is required when the Orimet is used as an 'acceptance' test. Either a go (complete flow-through) or a no-go (partial or no flow-through) is recorded.
- Record complementary test data: date, time and location; time lapse between completion of mixing of the concrete tested and the start of the test; ambient temperature and temperature of the concrete.

Interpretation of test results

The flow time t_0 indicates the filling ability of fresh SCC and its flow-rate. A higher flow time indicates a lower filling ability and a lower flow-rate, which, in turn, indirectly indicates a higher plastic viscosity of the mix. If two 'no-flows' were recorded and no flow time is established, the concrete is not suitable for testing using the Orimet. This mostly indicates a too low filling ability, however, it can also indicate cases of very low passing ability (e.g. an oversize aggregate or an excessive content of coarse aggregate block the orifice). A mix of a very low static segregation resistance may cause the coarse aggregate to settle within the orifice and greatly increase the flow time or cause a 'no-flow' to be recorded by blockage or a partial/intermittent flow.

Drawbacks and limitations

If carried out manually by a single operator, the synchronisation between the beginning of the timing, and the opening of the trap door can lead to an operator error.

5.3.4 V-funnel

Origin and principle

The funnel tests[2,12] follow the principle of the Orimet, differing in the shape and size of the test device. It was used first to assess SCC by Ozawa *et al.* in Japan[13]. Unlike the arrangement of the Orimet, where the sample of fresh concrete is held in a length of a straight casting pipe, the funnels use a container, which tapers all the way from the top down, ending in a short length of a pipe of rectangular (V-funnel) or circular (O-funnel) cross-section above the bottom opening (Figure 5.8). The shape of the bottom opening is fixed and is either rectangular, typical for the more common V-funnel test, or circular, which is typical of the less common O-funnel test (e.g. a Lafarge Co. proprietary test). V-funnels of different size and shape, particularly with regard to the size and shape of the rectangular bottom opening have been used.

Sample

A volume of approximately 10.5 l is required to fill the V-funnel described below. A practical sample of at least 12 l or approximately 30 kg of concrete is used. A single sample can be easily retested. The maximum size of aggregate is 20 mm.

Equipment required

- Complete apparatus made of steel, comprising a tapering, funnel-like, container fitted at the bottom with a short length of a straight pipe and a quick-release opening device. The apparatus is attached to a supporting metal frame. The basic dimensions are shown in Figure 5.8.

- Bucket with a capacity of approximately 5 l for filling of the funnel.
- Container with a volume of at least 12 l to collect the discharged concrete.
- Stopwatch or a timer accurate to 0.1 s.
- Straight edge at least 300 mm long.

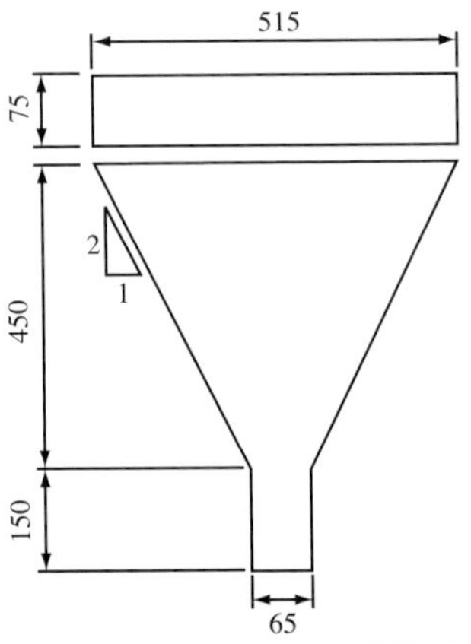

Figure 5.8 *Basic arrangement and dimensions of a V-funnel apparatus (version selected for European standardisation).*

Operating instructions

One or two operators are required. They need to:

- Set up the V-funnel apparatus on a firm ground.
- Place the receiving container below the V-funnel.
- Dampen the inside of the apparatus, ensuring that any surplus water drains out of the funnel.
- Close the gate and fill the apparatus with concrete without any compaction or assistance. Immediately strike-off surplus concrete with the straight edge. Prepare for timing.
- Wait approximately 10 s from the completion of filling of the apparatus. Quickly open the gate and simultaneously start timing the flow.
- Look down inside the funnel (Figure 5.9) and stop timing once daylight shows through the orifice.
- Record the V-funnel flow time t_v to nearest 0.1 s.
- Record complementary test data: date, time and location; time lapse between completion of mixing of the concrete tested and the start of the test; ambient temperature and temperature of the concrete.

Test results

A direct test result (V-funnel flow time t_v) is obtained. There is no need for further processing.

Interpretation of test results

The V-funnel flow time t_v measures the filling ability and flow-rate of a fresh mix. The flow-rate is strongly related to plastic viscosity. The test is also capable of

Figure 5.9 *Measurement of the V-funnel flow time.*

detecting a very low resistance of a mix to static segregation (settlement). This requires additional testing in which the short (10 s) normal waiting time is increased to 5 min or more and the results are compared. A substantial increase of t_V may indicate a partial blocking caused by increased concentration of coarse aggregate at the bottom of the funnel.

Drawbacks and limitations

- If carried out manually by a single operator, the synchronisation between the beginning of the timing, and the opening of the trap door can lead to an operator error.
- The apparatus cannot be adjusted for mixes with small coarse aggregate or mortars.
- The apparatus is bulky and difficult to transport (Figure 5.9).

5.3.5 J-ring

Origin and principle

The precise origin of this idea is unknown, however, the practical form of test, now widely used, was developed by Bartos *et al.* at the ACM Centre in Paisley[2, 10, 14] as a test for passing ability (primary aim) and filling ability and flow-rate (secondary aim). A sample of concrete is allowed to flow/spread in all directions while being confined by a circular arrangement of bars, which simulate reinforcement (Figure 5.10). In the basic arrangement, such as that proposed for European standardisation[15] the flow is generated purely by gravity, with the starting shape of the sample determined by the conical mould used for the slump-flow test.

It is easy to combine the J-ring with the Orimet, in which case the flow of the concrete through the J-ring will relate more closely to flows encountered in the practical placing of SCC.

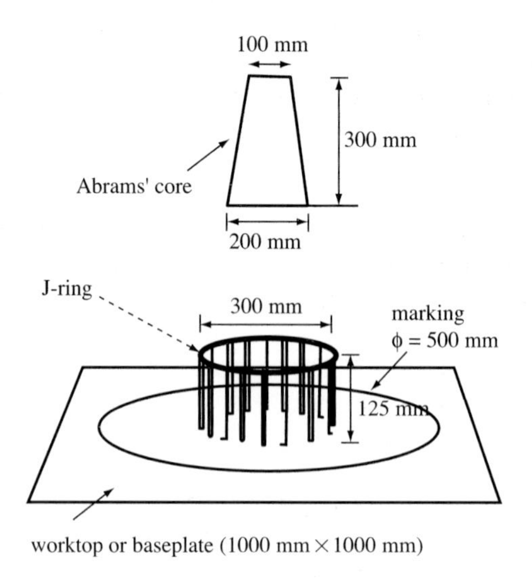

Figure 5.10 *Basic dimensions and arrangement of the J-ring apparatus.*

Sample

The sample is the same as that for the slump-flow test (not less than 6 l) or for the Orimet (not less than 7.5 l).

Equipment required additional to the equipment necessary for the slump-flow (or Orimet) test

- J-ring apparatus, comprising a metal ring of a 300 mm diameter (centreline), which can accommodate the conical mould for the slump-flow test but without its holding-down brackets. Regularly spaced smooth steel bars (18 mm diameter, 125 mm long) are attached (screwed-in) to the underside of the ring in a regular spacing/arrangement. The arrangement selected for European standardisation[15] has either a wide gap setting (12 bars at approximately 59 mm clear spacing) or a narrow gap setting (18 bars at approximately 41 mm clear spacing) of the bars.
- Measuring tape, graduated in mm.
- Straight edge (rigid, minimum length 400 mm long).

Operating instructions

One operator is required for the assessment of the passing ability (measurement of the step) and filling ability (measurement of the spread as in the slump-flow test) if a holding-down collar is used as in the slump-flow test. Otherwise, an additional operator is required, and two operators are always required when the flow-rate is assessed (measurement of the t_{J500} flow time). The operators need to:

- Set-up the baseplate as for the slump-flow test.
- Place the J-ring and the standard conical slump-test mould concentrically onto the baseplate.
- Firmly press down (manually or using the holding down collar) the mould and fill it completely with fresh mix. Avoid any compaction or agitation. Remove any concrete which has spilt onto the baseplate.
- Lift the conical mould vertically (Figure 5.11) in a single and smooth action within approximately 30 s of filling the mould.
- Allow the concrete to flow through the bars of the ring and to spread (Figure 5.12).
- Once the movement of concrete has stopped (Figure 5.13), place a straight edge across the top of the ring. Use a measuring tape to determine vertical distances between the underside of the straight edge and the following points:
  - centre of the ring (h_0)
  - two opposite points (line x) at the outside edge of the ring (h_{x1}; h_{x2})
  - two opposite points (line y, perpendicular to line x) at the outside edge of the ring (h_{y1}; h_{y2})
- Use the ruler to measure the largest diameter of the spread and record its value d_{jm} to nearest 10 mm (Figure 5.14). Follow this by measurement of the diameter of the spread in a direction perpendicular to the first one. Record its value as d_{jr} (to nearest 10 mm).

Figure 5.11 *Lifting of the mould at the start of the J-ring test.*

- Carry out a visual check of the spread concrete. Note any indications of potential segregation such as a zone (a halo) of paste or mortar without coarse aggregate around the edge of the spread concrete.
- Record the complementary test data: date, time and location; time lapse between completion of mixing of the concrete tested and the start of the test; ambient temperature and temperature of concrete.

Figure 5.12 *Concrete flowing through the ring. Deformed bars are used instead of the standard plain bars. The J-ring is in a 'narrow-gap' standard arrangement.*

Figure 5.13 *End of a J-ring test for passing ability. Note the higher level of mix retained within the ring compared with the level of concrete outside which passed through. This created a 'blocking step'* B_J*, indicating the passing ability of the mix. Note that the J-ring spread* SF_J *is approximately 670 mm.*

Optional, for assessment of the flow-rate:

- Start timing at the moment of the cone being lifted (loses contact with the baseplate) and finish it, when the target circle (500 mm in diameter) is reached by the spreading concrete.
- Record the J-ring flow time t_{j500} to 0.1 s

Test results

The J-ring test produces up to three results:

- Blocking step B_J, expressed to the nearest 1 mm, indicating the passing ability:

$$B_J = \frac{(h_{x1} + h_{x2} + h_{y1} + h_{y2}) - h_0}{4} \tag{5.1}$$

- Slump-flow SF_J, (spread) expressed to the nearest 10 mm, indicating the filling ability, constrained by reinforcement:

$$SF_J = \frac{(d_{jm} + d_{jr})}{2} \tag{5.2}$$

- Time t_{j50} (equal to t_{j500}), indicating the speed of flow, constrained by the J-ring reinforcement, expressed to the nearest 0.5 s.

Figure 5.14 *Measurement of the blocking step B_J using a straight edge and tape at the outer edge of the ring. Note the irregular spreads which were caused by too flexible (not rigid enough) baseplates, invalidating the results.*

Alternatively, the Orimet apparatus can be positioned within the J-ring (Figure 5.15 a, b). In addition to the J-ring results, the Orimet flow-time t_o is then determined instead of the time t_{j500}.

The complementary test data: date, time and location; time lapse between completion of mixing of the concrete tested and the start of the test; ambient temperature and temperature of the concrete should be recorded.

Interpretation of test results

Different types of reinforcement at different spacings can be set up, if the J-ring was to be used for simulation of specific project conditions rather than as a test verifying basic passing ability of a fresh SCC. It is therefore essential to verify what arrangements (spacing) and type of bars were used in the test. Results from the J-ring test can be compared only if the same arrangement of bars was used. The greater the spread SF_J, the greater is the filling ability, including the effect of the constraint by the bars of the J-ring. However, the filling ability alone does not guarantee that the mix is self-compacting. The value of SF_J must be complemented by an adequate value of B_J, indicating the necessary level of passing ability. The J-ring test indicates the passing ability by the size of the blocking 'step' B_J (if any) between the levels of the concrete inside and outside of the ring at the end of its flow.

Drawbacks and limitations

- The same drawbacks and limitations as for the slump-flow test apply for the J-ring.
- In addition, care must be taken to record correct readings to the nearest millimetre when the distances between the edge of the ring and concrete within

Figure 5.15 *J-ring used in conjunction with the Orimet test.*

and at the perimeter of the ring are measured. This is not easy when the baseplate is on the ground. However, placing the baseplate onto a firm and stable elevated support substantially facilitates both the basic test procedure and the determination of the blocking step B_J.

5.3.6 L-box

Origin and principle

The test was originally developed to assess the fluidity of a fresh underwater concrete mix by measuring the distance it could flow under its own weight, without passing through any reinforcement. It aimed to measure the 'consistence' of concrete during a 'partially confined' (trough) flow. Later, it was developed further to include the passing ability of SCC[2, 4, 10, 16], which has now become its primary application. It is also possible to assess the speed of flow of the test sample by measuring the time which has elapsed between the opening of the separating gate and for the moving

Figure 5.16 *L-box with four vertical bars during a test. The optimal time, t_L, is also being measured.*

'front' of the flowing sample to reach either the end of the trough or specific distance (e.g. 400 mm) along its length.

The L-shaped apparatus consists of a vertical hollow column of a rectangular cross-section with a horizontal trough attached at the bottom. A vertical sliding door separates the content of the column from the horizontal trough. A set of vertical reinforcing bars is placed in the trough immediately behind the separating gate (see Figure 5.16). The closed door contains the concrete in the column. Once it opens, the mix has to pass through the reinforcing bars before flowing from the column into the trough. The flow is generated by the static weight of the fresh concrete in the vertical column at the beginning of the test. The process simulates semi-static placing conditions, compared with the usually more dynamic practical placing conditions. The primary test result is the passing ratio, PR, which reflects the passing ability.

Many different geometries of the L-box and different bar arrangements/positions have been used in the past. The arrangement of vertical steel bars can be adjusted according to aggregate maximum size, the density and diameters of the reinforcement and practical casting methods. It is therefore essential to ensure that the results are compared on the basis of tests using identical L-boxes. The dimensions given in Figure 5.17 refer to the L-box arrangement adopted for EN standardisation[16].

Sample

A typical L-box requires a sample of at least 14 l of concrete.

Equipment required

- L-box apparatus made of dimensionally stable and rigid material such as steel (resistant to corrosion) or plywood (12 mm thick) with a suitable, durable, surface coating (Figure 5.18). A demountable assembly for easier cleaning is an advantage. The basic (inner) dimensions are:

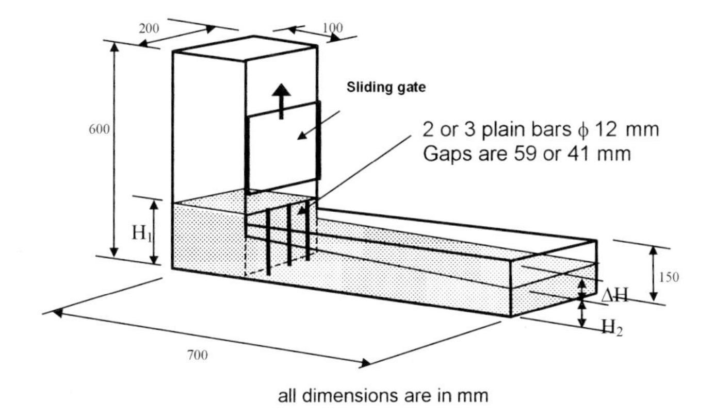

Figure 5.17 *Basic dimensions of the L-box apparatus.*

- vertical column is 100 mm × 200 mm (rectangular) cross-section, 600 mm high. The attached sliding door must be easy to operate, resistant to clogging by cement paste
- trough of rectangular cross-section, 200 m wide, 150 mm deep, 700 mm long
- set of smooth steel bars of 12 mm diameter, either two bars (a gap of 59 mm) or three bars (a gap of 41 mm) in an equidistant spacing

- Stiff measuring tape or a ruler at least 500 mm long, graduated in millimetres from zero at one end.
- Spirit level (unless a levelling indicator is built-in).
- Straight edge, at least 300 mm long.
- Scoop with a rounded mouth not wider than 150 mm or another suitable container for filling the column of the L-box.

Figure 5.18 *L-box made of coated plywood. A typical three-bar arrangement, with an optional, removable filling hopper.*

Operating instructions

One or two operators are required. They need to carry out the following tasks:

- Place the L-box apparatus on firm level ground. Use the spirit level to check that its base is in horizontal position and adjust it, if required.
- Dampen the inside of the L-box, removing any excess moisture. Close the column gate.
- Fill the column with concrete. Take care not to shake the apparatus and avoid any compaction or agitation of the concrete in it.
- Immediately strike off any surplus concrete from the top of the column using the straight edge and wait for one minute.
- Open the gate in one smooth movement. Ensure that the apparatus is not jolted or shaken during the opening.
- Once the flow has stopped, measure the vertical distance between the surface of the concrete sample remaining inside and the top edge of the column at three points equally spaced across the wider side of the column. Record the measurements as H_{1a}, H_{1b} H_{1c} to the nearest millimetre.
- Measure the vertical distance between the surfaces of the concrete sample at the far end of the trough at three points equally spaced across the width of its top inside edge. Record the measurements as H_{2a}, H_{2b} H_{2c} to the nearest millimetre.
- Check the concrete visually and record any indications of segregation or bleeding.
- Clean the apparatus immediately after a test, with a particular attention on ensuring continued easy operation of the sliding separation door.
- It is optional to start timing when the separating door begins to be lifted and end the timing when the moving front of the concrete reaches either one of the marks along the length of the trough or its end. Record the times t_{L20} , t_{L40} or t_L to the nearest 0.1 s.

Test results

The basic result of the L-box test is the passing ratio $PR = H_2 / H_1$, calculated the nearest 0.05. The PR is the ratio between the height of the concrete surface remaining within the vertical, column of the apparatus H_1, obtained from

$$H_1 = \frac{H_{1a} + H_{1b} + H_{1c}}{3} \tag{5.3}$$

and the height H_2, of the concrete surface at the far end of the trough, after its passage through the reinforcing bars. H_2 is obtained from

$$H_2 = \frac{H_{2a} + H_{2b} + H_{2c}}{3} \tag{5.4}$$

The horizontal trough of the L-box may have two additional marks at 200 mm and 400 mm from the sliding door. These permit an assessment of the speed of flow of the fresh concrete. The time which elapses between the opening of the sliding door and the advancing 'front' of the concrete in the trough, reaching each of the marks, produces two additional results: times t_{L20} and t_{L40} (in seconds).

The operators need to report the complementary test data: the bar arrangement used; date, time and location; time lapse between completion of mixing of the concrete tested and the start of the test; ambient temperature and temperature of concrete.

An improvement in the test procedure has been proposed[4], which substantially simplifies the measurements. Considering that the initial volume of the concrete sample in the vertical column is always the same (12 l), the height of the concrete at the end of the trough after a test in which the resulting PR = 1.00 cannot be more than 86 mm (maximum passing ability is reached, concrete surfaces in the column and trough are completely level). The distance from the surface of the concrete to the top edge of the trough is then 150 – 86 = 64 mm. It is therefore necessary to measure only the final height H_e at the end of the trough, in order to establish the PR, which indicates the level of the passing ability of the concrete tested.

Interpretation of test results

Poor performance of a fresh mix in the L-box test shows primarily as blocking, which may be caused by an excessive content of oversize coarse aggregate, which becomes wedged between the reinforcement bars (see Figure 5.19), as well as by a moderate to severe dynamic segregation. It is possible to obtain a limited visual indication of the dynamic segregation resistance of the mix tested, if the particles of coarse aggregate were visibly uniformly distributed on the concrete surface along the full length of the trough.

Figure 5.19 *Blocking in the L-box caused by inadequate passing ability. A three-bar arrangement.*

Drawbacks and limitations

- In rare instances, the calculated value of the PR ratio may be slightly higher than 1.0 ($H_1 < H_2$!). Such a result may occur for a fresh mix with a very high passing and filling ability, low plastic viscosity, which is also highly thixotropic. The fast moving, leading, 'front' of the flowing concrete can gain sufficient momentum to make it pile upwards when it hits the end of the horizontal trough. Due to high thixotropy, it may 'freeze' when its motion stops, and remain at a level higher than the original H_1.
- Readings may become inaccurate when some of the sample tested is lost during the test, if a mix of very high filling ability and flow-rate 'splashes' out of the horizontal trough when it reaches its end.
- Any slight deviation from horizontal of the trough of the apparatus significantly affects the test results.
- Accurate timing of the flow to defined distances, or to the end of the trough, is difficult. The moving front of the mix is unlikely to be uniform across the width of the trough, which makes the end of the timing very difficult to judge. Timing of the L-box flow is not recommended.

5.3.7 Sieve segregation test

Origin and principle

The test[2, 4, 10] was developed in France in the late 1990s by Cussigh[17] as a simple on-site method of quantitatively measuring the resistance of a fresh SCC to 'static' segregation. It is also known as the 'wet sieve segregation test' or 'screen stability test'. The test is based on the principle of determining how much separation occurs between the coarse aggregate and mortar in a sample of concrete[18]. The purpose is to quantify the risk of segregation by settlement/separation of coarse aggregate in an SCC mix. A simple, manual, version (Figure 5.20) and a less operator-sensitive, automated version (Figure 5.21) of the test are available.

Sample

A sample of 10 ± 0.5 l of fresh concrete is required.

Equipment required (manual version)

- A perforated plate sieve with 5 mm square holes with a matching receiver. The frame of the sieve should be at least 30 mm high and at least 300 mm in diameter in order to accommodate the test sample.
- Platform balance with a minimum capacity of 10 kg and accuracy at least 20 g. The platform should be large enough to accommodate the sieve receiver.
- Container for the sample, diameter approximately 250 mm, minimum volume of 11 l.
- Stopwatch or a timer accurate to 0.1 s.
- Thermometer graduated in 1°C.

Operating instructions

One or two operators are required. They need to:

Figure 5.20 *Sieve segregation test. The sample is poured from the sample container onto the sieve and receiver in the prescribed manner, maintaining a height of approximately 0.5 m.*

- Place the container on firm level ground and fill it with the sample of concrete.
- Measure the temperature of concrete in the container and cover the top to prevent evaporation.
- Wait for 15 ± 0.5 min. Ensure that the container is not shaken or the concrete disturbed.

Figure 5.21 *Automated version of the sieve segregation test.*

- Place the balance onto a firm and level support. Place the dry sieve with the receiver onto the platform balance and record the mass m_p (g).
  Place the receiver onto the balance and record its mass.
- Remove the cover from the sample container after the waiting time has elapsed. Record any notable bleeding or other features observed.
- Place the sieve with the receiver onto the platform balance. Pour (decant) steadily and carefully a mass of 4.8–5 kg of concrete from the sample container into the 5 mm sieve. Take care to pour it from a height of approximately 500 mm straight into the centre of the sieve. Record the mass of concrete actually poured, m_c, in grams (see Figure 5.22).
- Allow the concrete to remain in the sieve undisturbed for 2 min before lifting the sieve vertically, without any agitation. Record the mass of concrete which passed through the sieve and that of the receiver as m_{ps}, in grams.

Test results

The degree of static segregation is assessed by calculation of the Segregation index (SI) as the percentage proportion of the mass of concrete, which passed the sieve from the total initial mass poured into the sieve.

$$SI = \frac{(m_{ps} - m_p) \times 100\%}{m_c} \tag{5.5}$$

Interpretation of test results

The results indicate the potential for 'static' segregation (settlement). It is possible for concrete to show potential for some static segregation while remaining self-compacting. In such cases the segregation may not show as 'honeycombed' concrete but it can still affect the construction process. In addition, bleed water should not be detected during the test.

Figure 5.22 *Weighing of the sieve and receiver with the decanted concrete.*

Drawbacks and limitations

The test procedure requires the use of a platform balance, which needs careful maintenance if it is used on a concrete construction site.

5.3.8 Settlement column

Origin and principle

The settlement column segregation test was developed independently by Poppe and De Schutter at the Magnel Centre of the University of Ghent, Belgium[19] and by Rooney at the ACM Centre in the University of Paisley, Scotland[20] in the late 1990s. The settlement column test is the only test in which resistance to both static and dynamic segregation can be assessed.

A concrete sample in a vertical hollow column is either kept undisturbed over a period of time (when static segregation is to be assessed), or it is subjected to a controlled cycle of jolting to promote segregation. Subsamples of concrete are then taken from different vertical positions within the hollow column and the concrete removed is analysed for the coarse aggregate content. The difference between the aggregate content at the top and bottom samples produces a segregation column ratio (SCR) which provides an indicator of the susceptibility of the mix to segregation[2, 20, 21].

Sample

A sample of at least 8 l of concrete is required.

Equipment required

- Complete settlement column apparatus (Figure 5.23). This comprises a column with internal dimensions of 500 mm × 150 mm × 100 mm with 3 hinged doors to facilitate the collection of subsamples from the top, middle and bottom of the column
- Flow table jolting device[22] (see Figure 5.23).
- Two small G-clamps to secure the settlement column apparatus to the flow table apparatus.
- Sample bucket with a capacity of at least 8 l.
- Scoop.
- Timer accurate to 1 s.
- Two small collection trays with a capacity of at least 1.8 l each.
- Large collection tray with a capacity of at least 3.3 l.
- 300 mm diameter, or larger, 5 mm sieve with a matching receiving pan.
- Drying oven (optional).
- Balance with a capacity of 10 kg, accurate to 1 g.

Operating instructions

One or two operators are required.

- The settlement column apparatus is secured to the jolting table by means of two small G-clamps.
- The interior surfaces of the settlement column apparatus and the small collection trays should be dampened but free from excess moisture. The hinged doors of the apparatus are secured in closed positions.

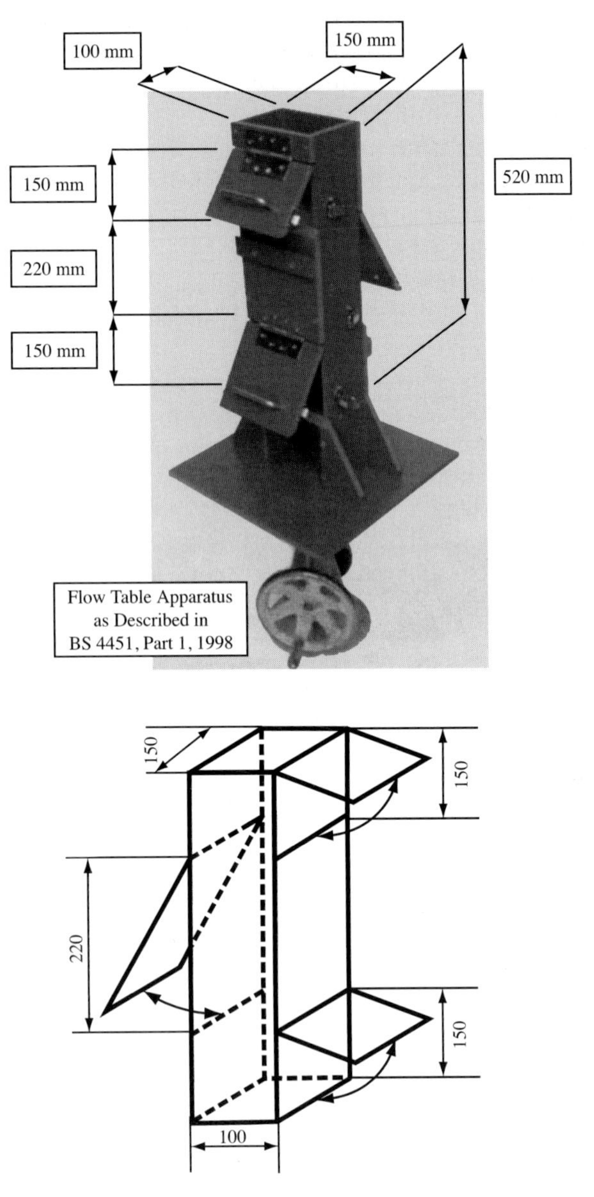

Figure 5.23 *Settlement column test apparatus* [21].

- Concrete is poured from the sample bucket into the column. When completely filled with concrete, it should be allowed to rest undisturbed for a period of 1 min.
- The settlement column apparatus is jolted 20 times within a 1 min period.
- The apparatus is allowed to rest undisturbed and the concrete inside allowed to settle for a period of 5 min.

- Top door of the apparatus is opened and the top sample is allowed to flow into the first small collection tray. If required, the scoop can be used to assist the concrete to flow out. Any excess concrete within the collection tray should be struck off by means of the scoop.
- The middle door of the apparatus is opened and the sample allowed to flow into the large collection tray, to be discarded.
- The bottom door of the apparatus is opened and the bottom sample is allowed to flow into the second small collection tray.
- The top sample should be transferred into the 5 mm sieve and the mortar content should be completely washed out by water until only the clean coarse aggregate remains.
- The sieve is cleaned, the bottom sample is transferred into it and the mortar content is completely washed out by water supply so that only the coarse aggregate remains.
- Coarse aggregate in both samples is surface-dried, preferably in an oven or by other artificial means in order that a result can be obtained in the minimum period of time. Alternatively, the samples may be allowed to surface-dry naturally. Natural drying may result in a higher moisture content being retained within the aggregate than when oven dried. It is important that the top and bottom samples are dried in an identical manner and for approximately the same period of time.
- The mass of the samples of coarse aggregate in both samples is determined by weighing; the amounts are recorded to the nearest 1 g.

Test results

The settlement column segregation ratio SCR is calculated as:

SCR = mass of top sample/mass of bottom sample

The value of the SCR is expressed to two decimal places.

Interpretation of test results

A limited number of full-scale experiments have been carried out in order to provide meaningful relations between the results of the settlement column test and the performance of the same mix when placed in conditions which induce certain degrees of potential segregation[20]. Four levels were identified and are shown in Table 5.1.

Table 5.1 *Interpretation of results from the settlement column test* [20].

Level of segregation	*SCR*
1. No segregation	0.96 and above
2. Mild segregation	0.95–0.88
3. Notable segregation	0.87–0.72
4. Severe segregation	0.71 and below

A simplified approach was proposed[4] for the separation of mixes, which were or were not likely to segregate:

SCR < 0.90 suggested that the mix was likely to segregate
SCR > 0.95 indicated good resistance to segregation during placing

The results of the settlement column test differentiate between different degrees of a more 'dynamic' segregation resistance than that measured by the sieve segregation and penetration tests. However, it is less sensitive in detecting static segregation than the other two tests[4].

Drawbacks and limitations

- It is a recently developed test method. There has only been very limited experience in practical applications.
- The test procedure is complex and requires the use of a platform balance, which needs careful maintenance if it is used on a concrete construction site.

5.3.9 Penetration test for segregation

Origin and principle.

The test was developed by Van *et al.*[23]. An overview of the test method has been given elsewhere[2, 24]. It exploits previously developed tests for measuring the viscosity and density of fresh concrete. It detects a change in resistance against penetration of a suitable tool into fresh concrete, which is caused by the change in the concentration of the coarse aggregate in the upper layer of the concrete. Such a change in concentration of aggregate reflects its settlement; it is a case of static segregation. The test is suitable for a rapid in situ evaluation of the 'static' segregation resistance of fresh SCC.

Sample

The penetration test does not require the removal of samples from the bulk of the freshly placed concrete, provided that adequate access is available and depending on the method of placing the concrete.

Instead of being carried out on concrete within formwork, the test can be carried out on the surface of the fresh mix in a container. The minimum size of the surface of concrete in an appropriate container is 200 mm × 200 mm, the concrete in it being at least 600 mm deep (minimum volume of the test sample is 7.5 l).

Equipment required

- A complete penetration test apparatus, consisting of a frame, which supports a sliding cylindrical 'penetration head' (usually having a mass of 54 g) and indicates its movement against a linear scale graduated in millimetres.
- A container with a sufficient volume and surface area of concrete and on which the penetration test apparatus can be positioned (Figure 5.24). A container, which allows the top surface of the concrete to be levelled by a straight edge, is preferred.
- Timer accurate to 1 s.
- A straight edge (optional).

Figure 5.24 *Setting up of the penetration apparatus before a test.*

Operating instructions

One operator is required, who needs to:

- Fill the container with concrete ensuring that it remains a representative sample (avoid shaking or moving the container, once it has been filled with concrete).
- Immediately position the penetration test apparatus on top of the concrete and adjust the penetration head (cylinder) until it just touches the surface of the concrete.
- Allow two minutes to elapse after placing of the apparatus and then release the penetration head.
- Wait for 45 s before recording the depth of penetration P_d (mm).

Test results

No calculations are required, the penetration depth P_d (mm) is read directly off the scale within the test apparatus.

Interpretation of test results

There is limited experience with practical applications of this test, however, a $P_d \leq 8$ mm is considered to indicate concrete with good resistance against static (settlement) segregation when a 54 g penetration head is used.

The test was evaluated as a potential EN standard test[4] because it is based on a simple method, it is easy to carry out and a direct test result is obtained.

Drawbacks and limitations

The difference in the depths of penetration may be influenced by factors other than the static segregation alone.

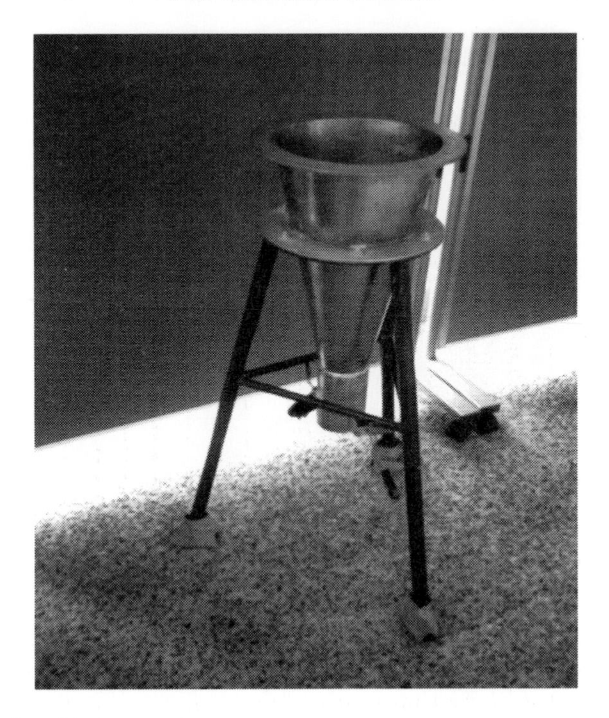

Figure 5.25 *O-funnel test apparatus.*

5.3.10 Other tests

O-funnel test

The test (Figure 5.25) is very similar to the V-funnel and closely related to the Orimet test. It was originally developed in Japan[2], however, it is now rarely used except as a 'proprietary' product-linked test.

U-test

This was originally developed as an in-house test by a large contractor in Japan[2]. Two versions exist, one with a rounded and another with a rectangular/flat bottom part (Figure 5.26). A sample of concrete has to pass through an arrangement of vertical reinforcing bars when flowing from one vertical column into another. It measures the passing ability and it is related to the L-box test. The result is a ratio comparing the levels/heights of concrete in the adjoining columns. If there were no blocking, the passing ability would be highest when the levels in both columns are the same.

Filling test

This test, sometimes known as the 'simulated filling test'[2, 24], and very closely related to the 'simulated soffit test'[2] had been used by Ozawa *et al.* in Japan during the original development of SCC. It simulates the filling of a space containing an array of horizontal reinforcing bars (Figure 5.27). The procedure and the calculation of the filling ratio are complicated. The test apparatus is also difficult to use.

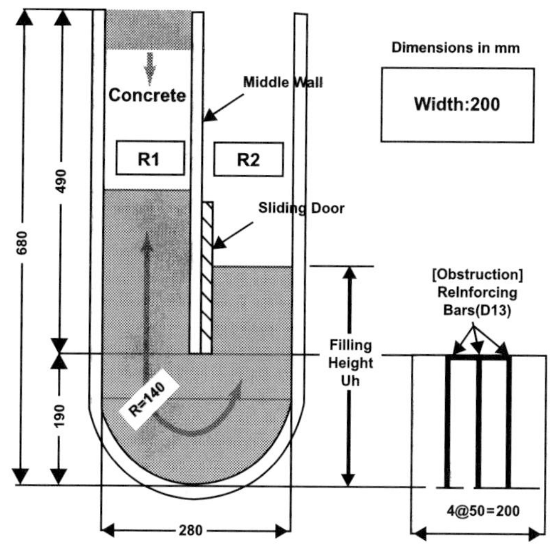

Figure 5.26 *Typical arrangements of the U-test.*

Automatic acceptance test

The acceptance at job site test was developed in the mid 1990s by researchers from the concrete laboratory of the University of Tokyo, Japan as a method of providing a yes or no answer to whether or not the SCC delivered to the site is of an acceptable standard[25, 26]. The acceptance at job site test involves concrete being discharged from a site-mixer or a truck-mixer directly into the apparatus. Once inside, it flows through and around a series of closely spaced horizontal and vertical steel bars, and

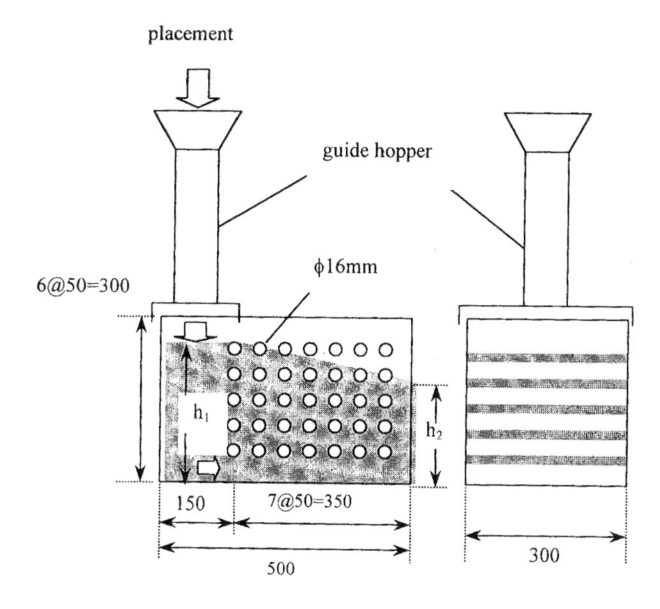

Figure 5.27 *Filling test.*

then out through an opening at the other end directly into the conventional concrete pumping or placing equipment to which it is attached (Figure 5.28).

If the concrete passes through the apparatus it is deemed to be adequately self-compacting and therefore suitable to be used. If the concrete does not pass, its placing is automatically stopped. The device attempts to provide a continuous go/no-go assessment of the passing, and to a lesser degree, the filling ability of the concrete placed (pumped). Segregation resistance is not detected, except, perhaps a very severe case, showing an extremely poor passing ability and blocking.

The practical advantage of the test is that it assesses all of the concrete placed, not just individual samples. However, it is not possible to identify the reason(s) why the concrete is rejected (a no-go result is obtained). This can be obtained by combining this test with another method for a rapid on-site assessment of one or more of the key properties of fresh SCC.

5.4 Review of performance of tests for properties of fresh self-compacting concrete

5.4.1 Basic performance parameters

The most significant parameters, which reflect performance of a test, are its reproducibility (*R*) and repeatability (*r*), which together make up its precision.

Reproducibility

This parameter[27] is related to two different operators. It is a statistical expression of the 'random error' of the test. It is defined as: the difference between two single test results on identical material that should be exceeded only once in 20 tests.

If two operators do the test, the difference between their results should in 95% of cases be less than the value of the reproducibility, such as the 39 mm for 'moderate' filling ability measured by the slump-flow.

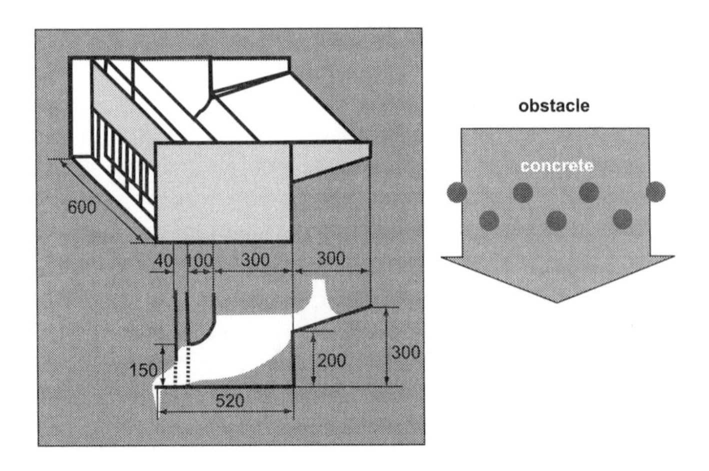

Figure 5.28 *Acceptance at job site test apparatus*[25, 26].

The reproducibilities at 'moderate/typical' levels of fresh properties indicated by the current EN European standards[28] for TVC are:

- The slump test (traditional 'Abrams' cone 300 mm high)
  - reproducibility: 25 mm
  - class limits for moderate consistence (slump): ±20 mm
  - permissible tolerance on a single test result: –30/+40 mm
- The flow table test (not the slump-flow test)
  - reproducibility: 91 mm
  - class limits: ±30 mm
  - permissible tolerance on a single test result: –50/+60 mm

Repeatability

This parameter[27] is related to a single operator. It is another statistical expression of the 'random error' of the test. It is defined as: the difference between two consecutive single test results on identical material which should be exceeded only once in 20 tests (5% probability).

5.4.2 Variations in test results

Despite using identical materials and mix designs, differences were observed in the results of even relatively simple tests, such as the slump-flow, during a large systematic evaluation of the tests in the European Testing-SCC project[4].

The performance of some of the tests was not simply a universal function of the principle or design of the test itself. It was also dependent on the mix and materials used.

Additional and unexplored effects of the mixer type and concrete temperature appeared to be potentially significant, with implications for the specification of the SCC characteristics. It was concluded[4] that: a given test method may be very useful (e.g. sensitive enough) and appropriate for assessment of some mixes, but not for others. The 'intervals' over which a property could be reliably measured varied from test to test. One test may differentiate well between mixes of, say, moderate filling ability, but less so for the same mix at higher levels of filling ability, where another test will perform better. An illustration of variations of performance (R, r) over such 'intervals' is given in Section 5.5.

As the performance of the tests depended on the properties of the mix tested, it was sensible to examine it when the test was applied to a mix, which had been (independently) proved to be self-compacting (in practical applications). Moreover if a meaningful comparison between performances of different test methods was to be made, it was useful to restrict the 'testing domain' further. Table 5.2 shows the performance parameters obtained when the mixes assessed were not only self-compacting but had 'typical' values of the characteristics measured.

Performance at 'typical' values does not, however, take into account the dependence of the observed reproducibility value on the magnitude of the test results. It does not indicate how 'wide' is the spread of values of the property over which the test maintains its best performance, how wide is the region over which its results are acceptably reliable.

Table 5.2 *Performance of selected tests for key properties of fresh SCC, interpreted in accordance with ISO 5725:1994* [4, 27].

Test		*Typical value*	*R, at typical value*	*R, as % of typical value*	*r, at typical value*
Slump-flow	spread *SF*	675 mm	43 mm	6.3	42 mm
J-ring	spread SF_J	675 mm	46 mm	6.8	46 mm
Orimet	flow time t_o	4.5 s	1.8 s	40.0	1.4 s
V-funnel	flow time t_v	8 s	3.1 s	39.0	2.1 s
J-ring	blocking step B_J	15 mm	4.9 mm	32.6	4.6 mm
L-box	H_2/H_1 ratio	0.90	0.12	13.3	0.11
Sieve segregation	*SI*	10%	3.7%	37.0	3.7%
Penetration test	*Pd*	13 mm	9.4 mm	73.0	9.4 mm
Flow-rate (slump-flow or J-ring) t_{500} or t_{J500}		2.5 s	1.2 s	47.2	1.2 s

5.4.3 Basic rheology and tests for key properties of fresh self-compacting concrete

It should be possible, and it is very desirable, to establish correlations, or at least linkages, between the results of the more 'empirical' tests used for practical assessment of properties of fresh SCC and the basic rheological characteristics (yield value, plastic viscosity) obtained from concrete rheometers. However, there are serious difficulties in establishing such relationships, the possibilities are largely theoretical. A fundamental obstacle is in single values being obtained as results of the 'empirical' tests; each result therefore embodies effects of both the basic rheological characteristics, and of other factors. Another difficulty in establishing correlations is due to very different rates of shear and their variations to which a sample of concrete is subjected during a test. The shear rate during an Orimet or V-funnel test was estimated to be two to three times higher than during the slump-flow test[4].

An evaluation of the relationships between the tests with potential for standardisation and rheological characteristics obtained from BML-type concrete rheometers was carried out by Wallevik *et al.* as part of the European Testing-SCC project[4]. The task proved more complex than expected. Over the wide domain of fresh concretes, which were self-compacting, viscosity played a major role and varied over a greater range than the yield stress, which tended towards zero. Significant correlations were established between the results of the tests proposed for standardisation and rheological parameters when assessment of the filling ability and flow-rate was concerned. To a lesser extent this included the passing ability, which, for some mixes and over certain range of values also correlated with the filling ability.

Difficulties arose when segregation resistance was considered. It was impossible to provide a reliable prediction of segregation resistance purely from the two basic rheological parameters (plastic viscosity, yield stress)[4]. The main reason for this appeared to be the inability of current concrete rheometers based on the coaxial cylinders (such as the BML rheometer) to reliably determine the basic rheological characteristics on mixes, which had low resistance to segregation, as measured by the 'empirical' SCC segregation tests.

5.5 Review of applications of tests for fresh self-compacting concrete

In order not to hinder wider acceptance of SCC, and to help engineers inexperienced in SCC better understand what they are dealing with, perhaps for the first time, it is essential, to set up 'limiting values' for the key properties of fresh mixes in identification of 'good' SCC. However, the values used for such for initial benchmarks, as shown below, remain largely arbitrary. This is because although there is now a very substantial body of practical experience in building with SCC[e.g. 29, 30], assessment of the key properties has been carried out using nonstandard test methods and equipment. This has made it difficult, and sometimes impossible, to compare and evaluate the results from different sites, countries and materials.

The limits and intervals, as well as the SCC 'classes' proposed will be therefore subject to revisions, sometimes quite radical, as more data from full-scale practical construction is obtained, based on the use of what will now be standardised test methods.

The preferred method of specifying SCC for a given application is to use the classifications given in Chapter 11, although, in some cases, target values with appropriate tolerances may be used instead. This would apply when the specification requirements are more precise. In addition to any of the tests currently available, visual assessment and judgement gained from practical experience will always remain useful, as they were for TVC.

Following the existing European standard practice, the European guidelines on SCC[31] have proposed classes for some of the key properties. These and others are outlined in the next sections, and may be used for specification (see Chapter 11). The precision data have been derived from the results of the Testing-SCC European project[4].

5.5.1 Tests for filling ability

Slump-flow

Most information about the behaviour of fresh SCC has come from slump-flow tests. The main reason has been the easy availability, low cost, simplicity and familiarity of the basic equipment because it was widely used for the slump test on fresh TVC. The test is easy to carry out and produces a direct numerical result. Slump-flow is often the only test used on-site to check the compliance with the specification of the fresh concrete which has been delivered. However, the validity of results (as for the normal slump test) is critically dependent on the correct use of the correct equipment, correctly maintained, by trained operators. The precision of the slump-flow test is given in Table 5.3. The test is suitable both for site and laboratory use.

Table 5.3 *Precision of the slump-flow test* [4, 6].

Slump-flow spread	*SF (mm)*	*< 600*	*600–750*	*> 750*
Repeatability	r (mm)	n/a	42	22
Reproducibility	R (mm)	n/a	43	28

It is likely that the 'plain' slump-flow test will be gradually replaced by the J-ring test, which offers important additional information in the fresh mix at a minimum extra cost/effort.

It should be noted that care should be taken that the conical mould used has the correct dimensions – those of the original Abrams truncated cone (300 mm tall) and not the smaller (200 mm tall) cone used for the flow table test[9] on TVC.

The typical range of slump-flow spread SF is 600–800 mm.

The classes of filling ability based on slump-flow spread[31] are:

- SF1 550–650 mm low filling ability: The minimum level, which provides self-compaction of the mix lies within this range. The exact value of any minimum level depends on individual mix design. Fresh mixes with values of slump-flow spread SF < 600 are not usually self-compacting in practical applications.
- SF2 660–750 mm good filling ability: Concrete is suitable for most practical applications.
- SF3 760–850 mm high filling ability: Mixes with filling ability in this range usually flow very easily (low or zero yield value), rapidly, and long distances. Such performance may be required in casting of very complex shapes or heavily reinforced concrete elements. It may reduce the production times, in precasting. Because of the interactions of the key properties, particular attention must be paid to the passing ability and segregation resistance of concrete in this range to ensure that it remains self-compacting and produces homogeneous concrete when placed.

J-ring

The test was developed primarily as a tool for a realistic assessment of passing ability, however, it has shown[4] to be as capable of measuring the filling ability as is the slump-flow test itself. It is possible that the J-ring test will eventually make the slump-flow test redundant. The test is suitable both for site and laboratory use. The precision of the J-ring test is given in Table 5.4. The typical practical range of the SF_J

Table 5.4 *Precision of the J-ring test (spread SF_J).*

J-ring spread	*SF_J (mm)*	*< 600*	*600–750*	*> 750*
Repeatability	r (mm)	59	46	25
Reproducibility	R (mm)	67	46	31

spread is the same as that for 'plain' slump-flow spread SF, 600–800 mm without the confining ring.

Classes of filling ability based on J-ring spread SF_J are identical[31] to the classes SF1–3 as described above.

- SF_J 1 550–650 mm low filling ability.
- SF_J 2 660–750 mm good filling ability.
- SF_J 3 760–850 mm high filling ability.

5.5.2 Tests for passing ability

J-ring: Blocking step B_J

The test was introduced in the late 1990s and less data from practical applications is therefore available. However, the simplicity of the test, inexpensive, robust and very easily handled equipment, combined with easily interpreted and meaningful results are making it very popular. The test result provides an indication of the degree of risk of the mix blocking when passing through reinforcement. The test is suitable both for site and laboratory use. The precision of the blocking step is given in Table 5.5.

The typical range of the (narrow gap) blocking step B_J is 3–20 mm

Suggested classes of passing ability based on the blocking step B_J:

- B_J $1 \leq 10$ mm zero to low risk of blocking[4]. Suitable for structural elements with dense reinforcement
- B_J $2 > 10 \leq 20$ mm moderate to high risk of blocking. Widely spaced or no reinforcement, few obstacles to flow

No data is available for performance of the J-ring test with 'wide gaps'.

L-box: passing ratio PR

Primarily, the test gives an indication of passing ability, but it is significantly interrelated with filling ability (flow). Mixes with low filling ability are more likely to block, even if their aggregate sizes and grading are the same. The performance (sensitivity) of the L-box to detect blocking is much reduced when the filling ability (expressed as slump-flow) is high. The L-box test results therefore reflect, to a variable extent, both the filling ability (flow) and passing ability (blocking). For some fresh mixes, but not all, the PR reading would indicate whether or not the mix is self-compacting. The precision of the passing ration is given in Table 5.6. The test is suitable for both site and laboratory use.

Table 5.5 *Precision of the J-ring test (blocking step-narrow gap only).*

J-ring blocking step	B_J	*≤ 20 mm*	*> 20 mm*
Repeatability	*r*	4.6 mm	7.8 mm
Reproducibility	*R*	4.9 mm	7.8 mm

Table 5.6 *Precision of the L-box test (PR).*

L-box	*PR*	*≥ 0.8*	*< 0.8*
Repeatability	*r* (mm)	0.11	0.13
Reproducibility	*R* (mm)	0.12	0.16

The typical range of passing ratio PR measured by the L-box is 0.85–0.95.

The classes of passing ability based on the L-box[31] are:

- PR1 ≥ 0.80 with 2 bars adequate passing ability for general-purpose applications with light or no reinforcement
- PR2 ≥ 0.80 with 3 bars mix suitable for placing into formwork with more closely spaced, denser, reinforcement

The minimum value of the passing ratio is usually shown as 0.80[2–4,10, 27, 29–31], but it has been largely 'quoted over' and there is little evidence linking it to performance of such SCC in practical applications. However, results giving PR < 0.75 become unreliable[2] and PR values of SCC mixes tend to be around 0.90, safely above the 0.80 limit. The mix is likely to either block severely or it has an extremely low filling ability (low workability) when the PR is at or below 0.80, both being indications of a mix that cannot be reliably deemed to be self-compacting.

5.5.3 Tests for segregation resistance

Segregation resistance is more significant in fresh SCC than in TVC. This has led to the development of entirely new tests (see Sections 5.3.7–5.3.9). Experience in hand from application of the new tests in construction[4] indicates that segregation resistance is unlikely to be carried out on-site. Tests for segregation resistance will be used mostly in laboratory work during development of SCC mix designs.

Once a mix design with an adequate SR for its application is specified, the on-site testing can be limited to simple tests for filling ability and passing ability, used primarily as indicators of uniformity of supply of the specified mix. It will be an advantage to use simple site-tests, which can also detect concrete with an extremely low segregation resistance (e.g. the Orimet).

Sieve segregation

The test is a relatively simple one to conduct and one person can carry it out. A single test takes around 25 min to complete. An 'automated', albeit more expensive, version of the test (see Figure 5.21) is recommended instead of the 'manual' one. The precision of the sieve segregation test is given in Table 5.7. The test is more suitable for laboratory than for site use.

The typical range of SI ratios is 10–20%.

Table 5.7 *Precision of the sieve segregation test (SI).*

SI %		*≤ 20%*	*> 20%*
Repeatability	*r*	3.7%	10.9%
Reproducibility	*R*	3.7%	10.9%

Segregation resistance can be classed (European guidelines)[31] according to the level of the segregation index, SI

- SI 1 ≤ 20   adequate resistance to static segregation (settlement)
- SI 2 ≤ 15   good resistance to static segregation (settlement)

The original French classification[17] for casting of vertical reinforced concrete elements recommended:

- 0 < SI< 15   satisfactory segregation resistance (stability)
- 15< SI< 30   segregation resistance questionable; an in situ trial is recommended
- 30 < SI   Inadequate segregation resistance, mix unusable for placing as a SCC

It also suggested that a very high resistance to segregation, indicated by a segregation index of less than 5% could be associated with filling ability inadequate for self-compaction of the mix. The paste/mortar may be too viscous. Mixes with a segregation ratio greater than 30% tend to be susceptible to severe segregation (very poor stability).

5.5.4 Tests for flow-rate

The flow-rate (rate of flow) is closely linked to, but it does not measure, a value of (plastic) viscosity of the fresh mix. Very fast flowing SCC indicates a very low plastic viscosity, usually linked to low or zero yield value. Mixes with a high flow-rate, which have a low flow-time (e.g. t_{500} < 1 s) are likely to require above zero yield value in order to prevent severe static and dynamic segregation.

Mixes with a low flow-rate, moving very slowly, show high plastic viscosity. This is often a feature of an underwater SCC with a degree of washout-resistance, which requires a very cohesive but still self-compacting mix. Four tests are used for the flow-rate:

- Slump-flow flow-time t_{500}, and the J-ring t_{j500} flow-time. Both tests are more suitable for use in a laboratory than on site.
- Orimet and V-funnel. These tests measure both filling ability and flow-rate. In practice it is the flow-rate, which is more often assessed, hence their inclusion as additional/alternative methods to those of slump-flow and J-ring.

Typical ranges are: slump-flow flow-time t_{500} is 2–5 s and J-ring flow-time t_{J500} is 2–6 s. The precision of the flow-rate tests is given in Table 5.8.

Table 5.8 *Precision of the flow-rate tests.*

Slump-flow	*flow-time t_{500}*	*< 3.5 s*	*3.5–6.0 s*	*> 6.0 s*
Repeatability	*r*	0.66 s	1.18 s	n/a
Reproducibility	*R*	0.88 s	1.18 s	n/a
J-ring	*flow-time t_{J500}*	*≤ 3.5 s*	*3.5–6 s*	*> 6 s*
Repeatability	*r*	0.70 s	1.23 s	4.34 s
Reproducibility	*R*	0.90 s	1.32 s	4.34 s

Relevant classes, which have been recently proposed[31] are still shown as 'viscosity' classes, based on the flow-times mentioned above.

- VS1 flow-times t_{J500} or t_{500} ≤ 2 s
- VS2 flow-times t_{J500} or t_{500} > 2 s

V-funnel: flow-time t_v

The test was adopted in Japan and its use spread with that of SCC, mainly through Japanese contractors. It is an easy test to carry out, with direct test results, but it is bulky and difficult to transport. The precision of the V-funnel test is given in Table 5.9. The test is suitable both for site and laboratory use.

The typical range of flow-time t_v, is 5–12 s.

The proposed classes of filling ability based on V-funnel time t_v (s) are described as 'viscosity' classes in the European guidelines[31]. They are, equivalent to the classes based on the slump-flow flow-rate time $t_{500.}$

- VF1 $t_v \leq 8$ good filling ability and moderate to high flow-rate, equivalent to VS1 measured by t_{50} or t_{J50}
- VF2 $9 \leq t_v \leq 25$ moderate to low filling ability, low flow-rate

Orimet: flow time t_o

The test uses a simple, rugged, durable, easily maintained and portable apparatus, with good simulation of movement of fresh flowing concrete during actual placing. No power supply is limited and no calibrations are necessary. The testing procedure is very quick: a set of three tests can be completed in less than three minutes. No

Table 5.9 *Precision of the V-funnel test (flow time t_v).*

V-funnel flow time	*t_v s*	*3.0*	*5.0*	*8.0*	*12.0*	*> 15.0*
Repeatability	*r* s	0.4	1.1	2.1	3.4	4.4
Reproducibility	*R* s	0.6	1.6	3.1	5.1	6.6

Table 5.10 *Precision of the Orimet test (flow time t_o).*

Orimet flow time t_o s		*< 5*	*5–10*	*10–20*	*> 20*
Repeatability	r s	0.2	0.9	1.9	2.2
Reproducibility	R s	0.4	1.3	1.9	2.8

specially prepared ground is required for setting-up of the Orimet; small deviations from verticality do not affect the test results. The test is suitable both for site and laboratory use. The precision of the Orimet test is given in Table 5.10.

The typical range of flow times t_0 (80 mm diameter orifice) is 1.5–5 s.

No Orimet classes have been proposed. A target value of $t_0 \leq 5$ s has been proposed for 'ordinary' SCC[32].

5.6 Conclusions

SCC is designed to be both fluid and cohesive; it has to possess both properties in correct proportions. Fluidity is the filling ability of the mix. Cohesion is its ability to maintain homogeneity, the basis of segregation resistance. However, no single test can be expected to provide separate values of both the fluidity and cohesiveness. In addition, when the fresh mix has to flow through constricted spaces and pass between reinforcing bars, it requires passing ability to fill the space completely with homogeneous, compacted concrete.

The J-ring test is the nearest to the ultimate aim of a single test method for assessment of more than one of the key properties of fresh SCC, namely:

- filling ability by measuring the spread SF_J as in the slump-flow test, but with the added reality of concrete being confined by reinforcement
- passing ability by measuring the blocking step B_J
- flow-rate by measuring the time t_{J500}

When more 'dynamic' placing conditions are expected, the J-ring can be used with the Orimet, providing a measure of filling ability and flow-rate by Orimet time t_0 and the blocking step B_{JO}.

Performance data and practical convenience[4,10], suggest the following ranking of the best performing final group of tests:

(1) J-ring: spread SF_J, blocking step and flow-time t_{J500}
(2) Slump-flow: spread SF, flow-time t_{500}
(3) Sieve segregation: segregation index SI

Complementary tests providing additional characterisation are:

(4) L-box
(5) and (6) Orimet and V-funnel
(7) Penetration test for segregation (laboratory use only)

It should be noted that none of the tests is without drawbacks, including those tests most commonly used to-date, such as the slump-flow test.

References

[1] Bartos P. *Fresh Concrete: Properties and Tests*, Elsevier Science, Amsterdam, the Netherlands, 1992.

[2] Bartos P.J.M., Sonebi M. and Tamimi A.K. *Workability and Rheology of Fresh Concrete: Compendium of Tests*, RILEM Publications, Cachan, France, 2002.

[3] Ouchi M. and Ozawa K. (Eds.) *Proceedings of the International Workshop on Self-compacting Concrete*, Kochi, 1998, Japan Society of Civil Engineers.

[4] Bartos P.J.M, Gibbs J.C. *et al. Testing SCC*, EC Growth contract GRD2-2000-30024, 2001–2004, http://www.civeng.ucl.ac.uk/research/concrete/Testing-SCC, Final Report, 2005.

[5] EN 12350-1, Testing fresh concrete, Part 1. Sampling fresh concrete, 2000.

[6] prEN 12350, Testing fresh concrete, Part 8. Self-compacting concrete – Slump-flow test, 2007.

[7] EN 12350-2, Testing fresh concrete, Part 1. Slump test, 2000.

[8] ASTM C143: Standard test method for slump of hydraulic cement concrete.

[9] EN 12350-5, Testing fresh concrete, Part 5. Flow table test, 2000.

[10] Grauers M. *et al.* Rational production and improved working environment through using self-compacting concrete. EC Brite-EuRam Contract No. BRPR-CT96-0366, 1997–2000.

[11] Bartos P. Workability of flowing concrete: assessment by a free orifice rheometer, *Concrete* 1978, 12, 28–30.

[12] prEN 12350-9. V-funnel test, 2007.

[13] Ozawa K., Sakata N. and Okamura H. Evaluation of self-compactability of fresh concrete using the funnel test. *Proceedings of the Japan Society of Civil Engineering*, 1994, 23, 59–75.

[14] Bartos P.J.M. An appraisal of the Orimet test as a method for on-site assessment of fresh SCC. In: Ozawa K. and Ouchi M. (Eds.) *Proceedings of the Intl. Workshop on Self-compacting Concrete*, Japan Society of Civil Engineers. Tokyo, March 1999, pp 121–135.

[15] prEN 12350: Testing fresh concrete, Part 12. Self-compacting concrete-J-ring test, 2007.

[16] prEN 12350 Testing fresh concrete, Part 10, Self-compacting concrete-L-box test, 2007.

[17] Cussigh F. (Ed.) Self-compacting concrete – Provisional recommendations, Association Française de Génie Civil Documents scientifiques et techniques (in French) July 2000.

[18] prEN 12350 Testing fresh concrete, Part 11, Self-compacting concrete – sieve segregation test, 2007.

[19] De Schutter G., Poppe A.-M., Audenaert K. and Boel V. Self-compacting concrete: fundamental and applied research (in Dutch) *Bouwkroniek*, October 2001, pp 28–33.

[20] Rooney M.J. Assessment of properties of fresh self-compacting concrete with reference to aggregate segregation. Ph.D. thesis, University of Paisley, Scotland, 2002.

[21] Rooney M.J. and Bartos P.J.M. Development of the settlement column test for fresh self-compacting concrete. In: Ouchi M. and Ozawa K. (Eds.) *Proceedings of the Second Intlernational Symposium on Self-compacting Concrete*, COMOS Engineering Corporation, Tokyo, Japan, October 2001, pp 109–116.

[22] EN1015-3 Methods of test for mortar for masonry, 1999.

[23] Van B.K., Montgomery D.G., Hinczak I. and Turner K. Rapid testing methods for segregation resistance and filling ability of self-compacting concrete. In: Malhotra V.M. (Ed.) *Proceedings of the fourth International CANMET/ACI/JCI Conference on Recent Advances in Concrete Technology*, Tokushima, Japan, 1998, ACI-SP 179-6, pp 85–103.

[24] Skarendahl Å. and Petersson Ö. (Eds.) *Self-compacting Concrete*, RILEM Publications, Cachan, France.

[25] Ouchi M. Self-compactability evaluation for mix proportioning and inspection. In: Ouchi M. and Ozawa K. (Eds.) *Proceedings of the International Workshop on Self-compacting Concrete*, Japan Society of Civil Engineers, Tokyo, March 1999, pp 111–120.

[26] Watanabe T., Nakajinma Y., Ouchi M. and Yamamoto K. Improvement of the automatic testing apparatus for self-compacting concrete at jobsite. In: Wallevik O.H. and Nielsson I. (Eds.) *Self-compacting concrete*, RILEM Publication, Cachan, France, 2003, pp 895–903.

[27] ISO 5725. Accuracy (trueness and precision) of measurement methods and results, 1994.

[28] EN 206-1 Concrete – Part 1: Definitions, specifications and quality control, 2000.

[29] Bartos P.J.M. Self-compacting concrete in bridge construction – guide for design and construction, Concrete Bridge Development Group, Technical Guide 7, Camberley, UK, 2005.

[30] Day R.T.U. and Holton I.X. (Eds.) Self-compacting concrete: a review, Concrete Society Technical Report 62, Camberley, UK, 2005.

[31] European guidelines for self-compacting concrete, Joint Project Group, EFNARC, Brussels, Belgium, 2005, www.efnarc.org.

[32] Concrete Institute of Australia Recommended practice, super-workable concrete, September 2005.

Chapter 6
Mix design

6.1. Introduction

Combining the constituent materials in the optimum proportions to give concrete with the required combination of fresh and hardened properties for a particular application is clearly an essential part of the SCC production process.

Mix design (or mix proportioning) must start with the definition or selection of the required properties. The fresh properties are normally defined by values of measurements from appropriate tests selected from those described in Chapter 5 and so this chapter starts by considering the level and ranges of properties that may be required and how they have been classified.

We then discuss some general considerations of the mix types and proportions that experience has shown will achieve these properties. It is possible to develop a mix or mixes for a specific application by starting with a potential mix based on these previous experiences and then to proceed on a trial-and-error basis. However, it is normally more beneficial to use a more rigorous method based on greater knowledge of the effect of the various parameters on the mix properties; two such methods, the so-called general method, which originated from the early work on SCC in Japan, and the CBI method, developed in Sweden, and their development and extensions by others are then described in some detail. How these methods have led to a greater understanding of the effect of the mix proportions on SCC behaviour is then considered, following which a number of other methods with different approaches are briefly summarised (space precludes a greater consideration of these).

Finally a method which is considered suitable for those new to SCC but with some knowledge of concrete technology and with access to reasonably well-equipped laboratory facilities is described.

6.2 Concrete requirements

We have seen in previous chapters that successful SCC needs a combination of three key distinct properties

- Filling ability (also called unconfined flowability): the ability to flow into and completely fill all spaces within the formwork under its own weight. This property is a combination of total flow capacity and flow rate.
- Passing ability (also called confined flowability): the ability to flow through and around confined spaces between steel reinforcing bars without segregation or blocking.

- Segregation resistance (also called stability): the ability to remain homogeneous both during transport and placing (i.e. in dynamic conditions) and after placing (i.e. in static conditions).

In rheological terms the first and third properties require a combination of a low yield stress and a moderate plastic viscosity.

When considering production, supply and quality control issues we should add the properties of robustness, which is the capacity of a mix to tolerate the changes and variations in materials and procedures that are inevitable with production on any significant scale, and consistence retention (sometimes called open-time) which is the ability of the mix to retain its self-compacting properties until placing.

Different levels and combinations of these properties are required for different applications, for example:

- The required passing ability will depend on the bar spacing in the member being filled.
- The required degree of segregation resistance will depend on the transport and placing methods (pump, skip etc.) and the size (particularly the height) of the space being filled.

An important first step in mix design and development is therefore the definition of the required properties for the specific application in terms of values measured by the tests described in Chapter 5. Figure 6.1 shows examples of different combinations of flow capacity (as measured by the slump-flow test) and flow-rate (as measured by the V-funnel test) for various applications that have been given by Walraven[1].

Standards and specifications will be covered in detail in Chapter 11, but it is useful here to illustrate the possible requirements by giving two examples of the classification of SCC, one produced by the Japan Society of Civil Engineers (JSCE) in 1999[2], and one by the EFNARC European project group in 2005[3].

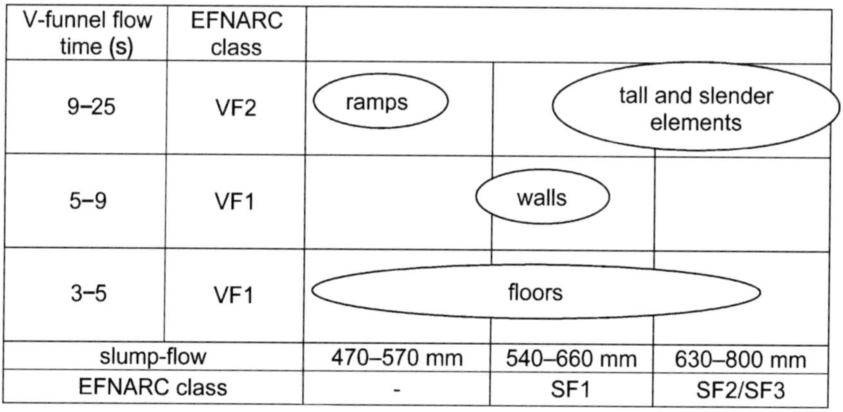

V-funnel flow time (s)	EFNARC class			
9–25	VF2	ramps	tall and slender elements	
5–9	VF1		walls	
3–5	VF1		floors	
slump-flow		470–570 mm	540–660 mm	630–800 mm
EFNARC class		-	SF1	SF2/SF3

Figure 6.1 *Applications of SCC of different properties (adapted from Walraven [1]).*

Table 6.1 *JSCE ranking of SCC mixes according to passing ability.*

Rank	*Reinforcing bar spacing (mm)*	*Normal steel content (kg/m³)*	*U-box requirement*
1	< 60	< 350	Filling height >300 mm with 5 bar gate
2	60–200	350–100	Filling height >300 mm with 3 bar gate
3	> 200	< 100	Filling with no gate

The JSCE recommends ranking mixes according to their suitability for applications with different levels of bar spacing and reinforcement content (Table 6.1). The ranking is defined by the U-box filling height (see Chapter 4). No levels of slump-flow are specified and so it can be inferred that the U-box test is thought sufficient to characterise both the passing and filling ability.

The more recent EFNARC classification (Table 6.2) gives sets of classes for slump-flow, viscosity, passing ability and segregation resistance (or stability) based on the values obtained with the tests as specified in the European standards[4–7]. The accompanying guidelines state that:

- slump-flow will normally be specified for all SCC
- the viscosity may be important where good surface finish is required or the reinforcement is very congested but need not be a requirement in other cases
- passing ability is not a requirement if there is little or no reinforcement
- in most cases class SR1 segregation resistance will be adequate

Table 6.2 *SCC classes proposed by EFNARC.*

Slump-flow		*Viscosity*			*Passing ability*			*Segregation resistance*	
Class	Slump-flow (mm)	Class	t_{500} (s)	V-funnel (s)	Class	L-box ht ratio (mm)	J-ring step height* (mm)	Class	Sieve segregation (%)
SF1	550–650	VS1/ VF1	≤ 2	≤ 8	PA1	≥ 0.80 with 2 rebars	≤15 with 59 mm bar spacing	SR1	≤ 20
SF2	660–750	VS2/ VF2	> 2	9–25	PA2	≥ 0.80 with 3 rebars	≤15 with 41 mm bar spacing	SR2	≤ 15
SF3	760–850								

*J-ring step height criteria not specified by EFNARC – values have been estimated from results of Testing-SCC project

The EFNARC passing ability classes are based on L-box measurements. However, the alternative J-ring test is also included in the European standards[8] and results from the Testing-SCC project of the step-heights corresponding to the L-box height ratios have been added to Table 6.2.

The corresponding EFNARC classes for the suggested properties have been included in Figure 6.1 and it is interesting to note that mixes with a flow capacity lower than that of the minimum EFNARC slump-flow class (SF1) are appropriate for some situations. Indeed, the American Concrete Institute[9] defines three slump-flow targets: less than 550 mm, 550–650 mm and greater than 650 mm, but they do note that mixes with a slump-flow of less than 550 mm may require minor vibration. However, an analysis of 68 case studies of worldwide uses of SCC in the period 1993–2003[10] showed that nearly 50% of the applications used concrete with slump flows in EFNARC class SF1, and a further 35% with slump flows in EFNARC class SF2 (see Figure 6.2)

6.3 General considerations for mix proportioning

The early development of SCC in Japan established the essential criteria for achieving the three key properties[11]:

- a low water/cement (or water/powder) ratio with high doses of superplasticiser to achieve high flow capacity without instability or bleeding
- a paste content sufficient to overfill all the voids in the aggregate skeleton to the extent that each particle is surrounded by an adequate lubricating layer of paste, thus reducing the frequency of contact and collision between the aggregate particles during flow
- a sufficiently low coarse aggregate content to avoid particle bridging and hence blocking of flow when the concrete passes through confined spaces

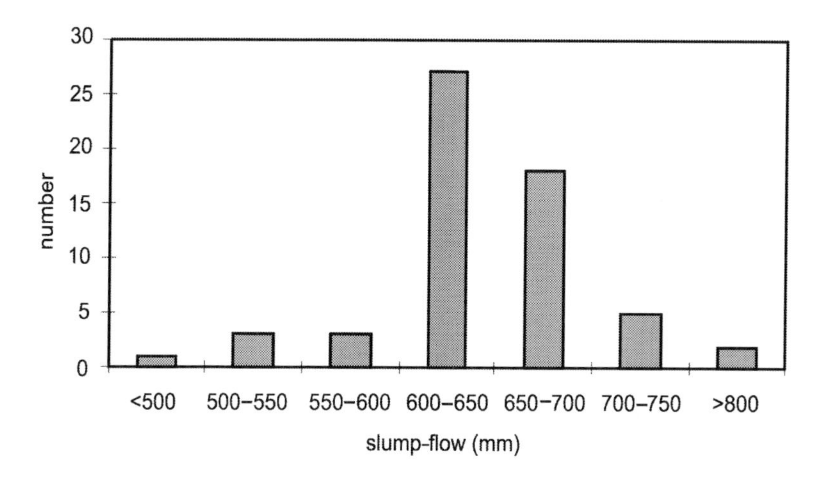

Figure 6.2 *Slump-flow values from an analysis of 68 worldwide case studies of applications of SCC (from Domone[10]).*

Such mixes are sometimes called powder-type mixes. An early alternative is to achieve sufficient stability at higher water/powder ratios by incorporating a VMA, thus producing a viscosity-agent type SCC, analogous to underwater concrete. In practice, such mixes have not proved popular and an intermediate combined type SCC is more often used. In this, a superplasticiser provides fluidity and a VMA is used to improve the robustness of the mix, and in particular tolerance to variations in material supply such as the grading or moisture content of the fine aggregate.

There are an enormous number of publications on laboratory studies on SCC which contain details of the mix proportions, but it is more instructive when describing the possible ranges of the proportions to consider mixes which have been used in practice. The analysis of case studies of successful mixes in the period 1993–2003[10] has given the medians and ranges of key mix proportions shown in Table 6.3. As discussed in Chapter 3, the analysis showed no effect of aggregate type and size on the mix proportions, and there was also no correlation between powder content or water/powder ratio on slump-flow or other fresh properties. This therefore illustrates that there is no unique mix of a given set of materials to give a particular set of fresh and hardened properties.

Table 6.3 also includes:

- Typical ranges of mix proportions as given in the EFNARC guidelines, which broadly correspond with those from the case study analysis, though in some

Table 6.3 *Medians and ranges of key proportions of SCC mixes.*

		From analysis of 68 worldwide reported case studies of SCC from 1993–2003[10]			*Typical ranges given by EFNARC*[3]	*Typical mid-range normally vibrated mix**
		Median	10th centile	90th centile		
Coarse aggregate (>4 or 5 mm)	% by vol	31.2	29.1	34.8	27–36	46
Fine aggregate	% by vol	30.5	22.9	40		25
	% by wt total agg	49.5	44	54	48–55	35
Paste (water + powders)	% by vol	34.8	32.3	39.0	30–38	29
Powder	kg/m^3	500	445	605	380–600	355
Free water	kg/m^3	176	161	200	150–210	160
Water/powder ratio	by wt	0.34	0.28	0.42		0.45
	by vol	1.03	0.83	1.28	0.85–1.10	1.41

* 40MPa char. cube strength, 75 mm slump, 20 mm gravel aggregate, CEM I, typical superplasticiser dose

cases they are a little wider. This is partly the result of the case study analysis being given in terms of the 10th and 90th centiles to avoid including exceptional mixes.
- The proportions of a medium quality traditionally vibrated mix, from which the lower coarse aggregate content and higher powder and paste content of SCC is apparent.

6.4 Mix design procedures

There are many examples of SCC mixes being developed by starting with mix proportions somewhere in the ranges given in Table 6.3, and then modifying these either systematically or more randomly in a programme of tests until a satisfactory mix is obtained. Indeed, both the JSCE and the ACI have recommended this approach. The JSCE guidelines[2] are summarised in Table 6.4, for the proportions of mixes of the three types and rankings discussed above. It is clear from this that there is greater flexibility with viscosity agent or combination-type mixes than with powder mixes. Nawa *et al.*[12] have extended this with suggestions for proportions of viscosity-agent type mixes for various types of viscosity agent.

The ACI guidelines[9], summarised in Table 6.5, are broadly similar, although a slightly different set of parameters are specified. Such an approach is satisfactory in the sense that a suitable mix or mixes is eventually produced, but the result may not necessarily be either the most efficient or the most robust mix possible. A more rigorous or methodical approach is therefore often preferred.

A number of such methods of varying complexity have been proposed. Two interesting and useful methods (and some developments of these) are outlined in this chapter: the so-called 'general' method which arose from the early studies on

Table 6.4 *Limits for initial mix proportions as recommended by JSCE (for 20 or 25 mm coarse aggregate)*[2].

	Powder mixes	*Viscosity agent type mixes*	*Combination type mixes*
Coarse aggregate (volume %)	Rank 1: 28–30 Rank 2: 30–33 Rank 3: 32–35	Rank 1: 28–31 Rank 2: 30–33 Rank 3: 30–36	Rank 1: 28–30 Rank 2: 30–33 Rank 3: 30–35
Water content (kg/m^3)	155–175	Minimum for required fresh properties	Sufficient for specified deformability and segregation resistance
Water/powder ratio	0.28–0.37 by wt 0.85–1.15 by vol	Maximum for required hardened properties	
Powder content (volume %)	16–19	Sufficient for self-compactability	> 13

– The powder composition should be selected from consideration of the performance requirement of the fresh, hardening and hardened concrete.
– Initial superplasticiser and VMA dosages are selected from previous experience or manufacturer's recommendations.

Table 6.5 *Suggested ranges for initial mix proportions as recommended by ACI (for coarse aggregate > 12.5 mm)*[9].

	Slump-flow < 550 mm	*Slump-flow 550–650 mm*	*Slump-flow > 650 mm*
Powder content (kg/m^3)	355–385	385–445	445+
Coarse aggregate (volume %)		28–32	
Paste (volume %)		34–40	
Mortar (volume %)		68–72	
Water/powder ratio		0.32–0.45	

SCC in Japan and the method based on blocking (or passing ability) developed in Sweden. We will see that these also have the advantage of giving an understanding of the influence of some important factors on the performance of the concrete. Following this some comments on some other methods are given, and finally a method is described which is suitable for those with typical, but not advanced, laboratory facilities at their disposal, based on work at UCL.

Mix design methods for SCC have two important differences from those for TVC:

- The dominant consideration is to produce satisfactory self-compacting properties, with less initial attention being given to the subsequent hardened properties. In particular, the self-compacting criteria will normally govern the coarse aggregate content, the paste content, the water/powder ratio and the admixture dose. Strength and other early age and long-term properties are then controlled by an appropriate combination of different powder materials (Portland cement, fly ash, ground granulated blast furnace slag, limestone powder etc.). This is discussed in more detail in Chapters 8 and 9.
- The number of possible combinations of the component materials for SCC means that a large number of variables need to be reconciled (e.g. powders, superplasticiser and viscosity agent). Since their interaction is often difficult to predict with any certainty some testing, often extensive, of potential combinations is necessary. This is often carried out on the mortar component. Not only is this convenient in the laboratory but the high mortar content of SCC results in good correlations between mortar and concrete properties and the required mortar properties are sufficiently well defined that, if achieved, the subsequent tests on the concrete are minimised[13–14].

A feature of all the methods is that they consider the volumetric composition of the mix, with subsequent conversion to proportions by weight for batching.

6.4.1 General method

This relatively simple step-by-step method arose from the extensive early work on SCC at the University of Tokyo by Okamura, Ozawa and co-workers[11, 15, 16]. It was developed for powder-type mixes with a limited range of materials, including coarse aggregate with a size range of 5–20 mm, fine aggregate with a maximum size of 5 mm, a low heat, high belite content Portland cement. Air entrainment at a level of 4–7% is usually incorporated, primarily to provide freeze–thaw resistance, but this also has the advantage of reducing the water and cement contents. The main features are:

- The air content is that required for the exposure condition.
- The coarse aggregate content is set as 50% of the dry-rodded weight in the concrete volume less the air content. This allows for the effect of the particle shape and grading.
- The fine aggregate content is set at 40% of the resulting mortar volume. For this purpose, all particles larger than 0.09 mm are considered as aggregate, and all less than 0.09 mm as powder. This content is considered critical: if the content is too high, the fine aggregate particles would interfere with each other during flow and cause blocking; if too low, the resulting higher cement and water content would be detrimental to the hardened concrete properties. It depends to some extent on the particle shape and size distribution of the aggregate and the properties of the cement.
- The water/powder ratio and superplasticiser dosage are determined by tests first on the water/powder paste and then on the mortar. Both use a spread test carried out on a glass plate (Figure 6.3), the mortar tests also use a V-funnel (Figure 6.4). These are smaller scale versions of the slump-flow and V-funnel test used for concrete and measurements are made in the same way.
    - The paste is tested at increasing water/powder ratios to determine the effect on the spread. Expressing this as the ratio of the change in the area to the initial area (called the relative flow area R_a), a linear relationship with the water/powder ratio is obtained. The intercept on the water/powder ratio

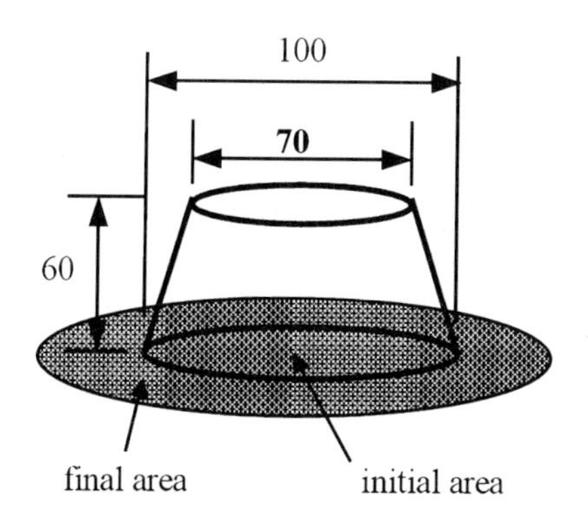

Figure 6.3 *Dimensions (mm) of truncated cone for paste and mortar spread test.*

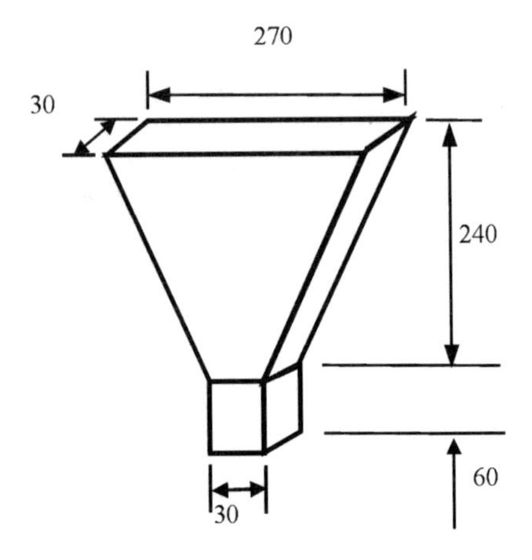

Figure 6.4 *Dimensions of V-funnel test for mortar.*

axis is the point at which the water content is just sufficient to initiate flow, and is called the retained water ratio of the powder (β_p) (see Figure 6.5).

- Spread and V-funnel tests on the mortar are then carried out with an initial water/powder ratio of $0.85\beta_p$ and varying superplasticiser dosage. The superplasticiser dosage and water/powder ratio of the mortar are adjusted until a flow time of 9–11 s and relative flow area of 5 (a spread of 245 mm) are obtained. These are then the starting point for tests on concrete. The mortar tests can also be used to assess the time-dependent properties.

- For the resulting concrete, a slump-flow of 650 mm is considered adequate and, if necessary, the superplasticiser dosage is adjusted until this is obtained. If the V-funnel flow time is in the range of 10–20 s then the self-compactability is considered satisfactory. A U-box test (see Chapter 5) is used to assess the passing ability and a final height of at least 300 mm is required.
- The time dependent properties of the concrete are also assessed.

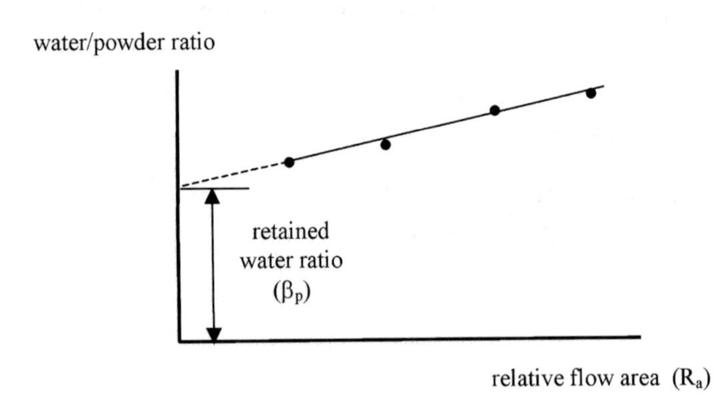

Figure 6.5 *Determination of retained water/powder ratio of powder from the spread test on paste.*

The resulting quantities of coarse and fine aggregate were considered to be on the safe side for SCC, and therefore the method was appropriate for SCC for general use. There is no inclusion of a concrete strength requirement in the mix design process, but some control over strength and other hardened properties was achieved with suitable powder compositions, for example the use of a low-heat type cement with fly ash and blast furnace slag in combination with a limestone powder.

The target slump-flow and V-funnel flow times for the concrete are at the low and high ends of their respective ranges of the EFNARC classes (Table 6.1), which was a distinctive feature of the mixes used in Japan in the 1990s. There has been a general trend towards using mixes with increasing flow and lower viscosity as SCC technology and practice has spread to other countries, but this does not mean that the method cannot be used for such mixes. As will be seen later in this chapter, adjusting the target values of the mortar properties will result in suitable concrete properties.

6.4.2 Modifications and developments of the general method

There have been a number of modifications and developments of the general method aimed at producing mixes which are more efficient in terms of paste content, and which are applicable to a wider range of materials. Edamatsu *et al.*[17] have developed a method of determining the required fine aggregate content of the mortar for fine aggregates with different particle shapes and size distribution and different powders. The mortar/aggregate interaction was evaluated on mixes using 10 mm glass beads as the aggregate.

Ouchi *et al.*[18, 19] have proposed a more rational method of determining a satisfactory combination of water/powder ratio and superplasticiser dosage than that used in the general method. Tests on mortar are used to determine each parameter independently, thus minimising the number of tests required. The method makes use of the forms of the relationships between relative flow area and V-funnel flow time which were obtained from extensive tests.

Working with Dutch materials, with in particular a maximum aggregate particle size of 16 mm, Pelova *et al.*[20] and Takada *et al.*[21] have found that the quantity of coarse aggregate can be successfully increased to 60% of the dry-rodded bulk density. This corresponds to the point at which the maximum degree of packing of the aggregate mixture was obtained. At this aggregate content, the paste content required was about 10% less than that obtained using the general approach, and therefore a more efficient mix was obtained. In other respects, the general approach was followed, and was found to be applicable to the materials used.

6.4.3 CBI method

This mix design method, developed at the CBI, Sweden by Billberg and Petersson[22–24], considers the concrete as a solid aggregate phase in a liquid paste phase formed by the powder, water and admixtures. The paste fills the void in the aggregate matrix and provides a lubricating layer around each aggregate particle.

There are two principal stages:

- Selection of the maximum aggregate content, and hence minimum paste content, based on a blocking criterion likely to ensure sufficient passing ability.

- Determination of the required composition of the paste to give adequate hardened properties and sufficient fluidity of the fresh concrete.

Blocking analysis

The blocking analysis is based on earlier work in Japan and Thailand[25–27] which showed that an overall blocking risk can be obtained by considering the contribution to blocking of each aggregate fraction. A satisfactory mix will be obtained if the overall risk is less than 1.

The advantage of this approach is that it considers the overall grading of the combined aggregate, thus avoiding the difficulty of different sized specifications of aggregates in different countries, particularly those of the division between fine and coarse aggregate and the maximum coarse aggregate size. It also takes into account the placing conditions, including the ratio of the aggregate size to the minimum gap through which the concrete has to pass.

The basis of the analysis is:

(1) For any aggregate particle size fraction there is an associated volume proportion or ratio in an aggregate/paste mixture that will be just sufficient to cause blocking when the mixture tries to pass through a gap. This blocking volume ratio is dependent on:
- the ratio of the gap size to particle diameter (the dominant factor)
- the particle shape
- the ratio of particle size to reinforcing bar diameter

but it is assumed independent of the paste properties provided there is no segregation whilst the mix is standing.

(2) In a concrete mix, the sum of the volume ratio of each fraction of the aggregate (i.e. the volume of aggregate per unit volume of concrete) divided by its blocking volume ratio gives the risk of blocking, which should be less than or equal to 1

$$\text{i.e. risk of blocking} = \Sigma(n_{ai}/n_{abi}) \leq 1 \tag{6.1}$$

where n_{ai} is the volume ratio of the single size group i of the aggregate, and n_{abi} is the blocking volume ratio of this group.

Values of n_{abi} were initially determined experimentally by Ozawa *et al.*[25] and Tangtermsirikul and Van[27] to produce the relationships shown in Figure 6.6. The parameters in this figure are: n_{abi} which is the blocking volume ratio of aggregate group $i = V_{abi}/V_t$, where V_{abi} is the blocking volume of aggregate group i and, V_t is the total concrete volume. c is the clear spacing between reinforcement or other obstacles. D_{af} is the aggregate fraction diameter. It equals $M_{i-1} + 0.75(M_i - M_{i-1})$ where M_i and M_{i-1} are the upper and lower sieve sizes of aggregate size fraction, respectively. K is the ratio of the reinforcing bar diameter to the maximum aggregate size

For any combination of aggregate, the volumes of aggregate in each of the fractional groups can be calculated from the combined aggregate grading curves and the specific gravities of the aggregate particles. These can be then expressed as

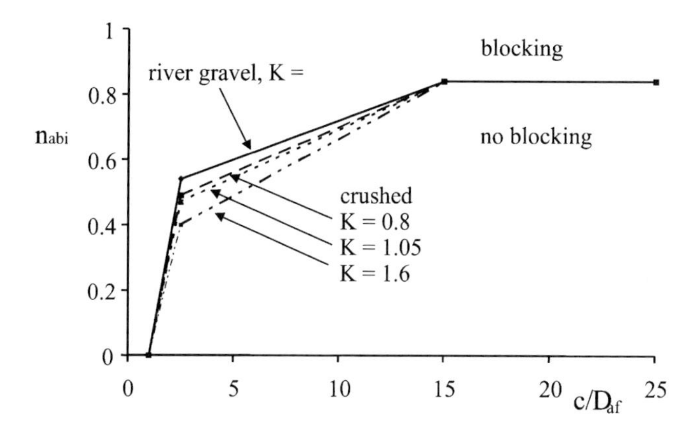

Figure 6.6 *Relationship between the blocking volume ratio, n_{abi}, and the ratio of clear spacing, c, to aggregate fraction diameter, D_{af} (K = bar diameter/maximum aggregate size) (from Tangtermsirikul and Van[27]).*

proportions of the concrete volume, n_{ai} in terms of the paste content which is then determined from equation (6.1) (a spreadsheet analysis is convenient).

If the combination of coarse and fine aggregate is defined by the volume of coarse/total aggregate (N_{ga}), then carrying out this calculation for the whole range of N_{ga} from zero (no coarse aggregate) to 1 (no fine aggregate) results in curves for the minimum required paste content (V_{pwmin}) such as those in Figure 6.7 for a typical crushed and a rounded gravel coarse aggregate. The extra paste required with the crushed aggregate is readily apparent. The curve for the voids content of the aggregate mixtures is also shown. Up to an N_{ga} value of about 0.2 (i.e. 20% coarse aggregate) the blocking analysis gives a required paste content below the voids content of the aggregate system, which shows that the analysis is not valid for such mixes. However, very few, if any, concretes will have such low values, and the approach has been verified for N_{ga} values is the range 0.3–0.6. N_{ga} is normally chosen to be that for minimum voids content i.e. 0.45–0.5 in the example shown.

Paste composition

The water/cement ratio and the type of cement are defined by the required hardened properties of the concrete, and assuming that the cement content is also set then the required volume of paste as calculated above is obtained by adding a combination of air entrainment (if required) and inert crushed dolomite fine filler. If the cement and water contents are too low and more filler is required, then the cost is increased by the increased amount of superplasticiser required, and the concrete can have a high viscosity and a low flow-rate. No mention is made of the use of fly ash or ground granulated blast furnace slag.

The water/powder ratio as well as the paste volume is important for control of stability. If this is too high the mix will segregate, and if it too low the plastic viscosity becomes too high for practical purposes.

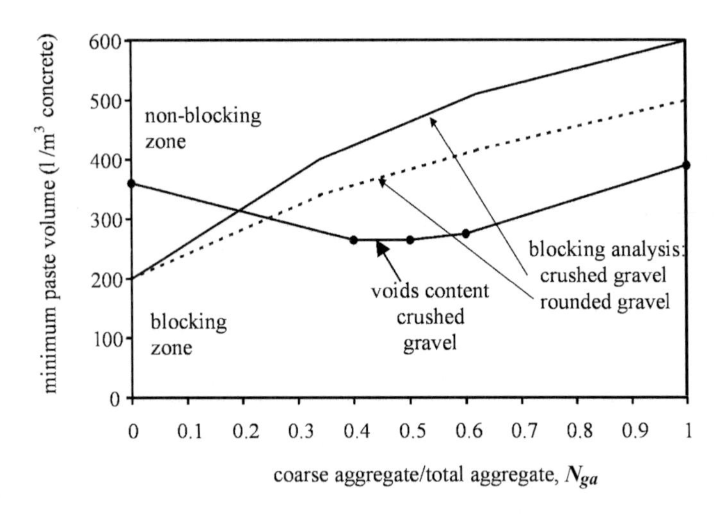

Figure 6.7 *Typical results of required minimum paste volumes from blocking analysis with two coarse aggregate types (from Billberg[24]).*

The most suitable filler and the type and dosage of superplasticiser are obtained from rheological tests in a concentric cylinder viscometer on mortar incorporating fine aggregate with a maximum particle size of 0.25 mm. The stability of the mortar is indicated by the form of the flow curves. The dosage of the superplasticiser which reduces the yield stress to near zero (i.e. the saturation dosage) is determined.

The compatibility of the air entraining agent with the superplasticiser can also be assessed at this stage, but the satisfactory performance of the combination of admixtures still needs to be confirmed by tests on the concrete.

Concrete properties

A slump-flow value of about 700 mm is considered appropriate, with the stability evaluated by observing any tendency to segregation at the outer rim. The t_{500} time is also measured, but no specific recommendations for values of this are given. However, a very low value will indicate a segregating mix, and a very high value will lead to slow casting rates.

The blocking behaviour of the concrete is measured with the L-box (see Chapter 5), which also assesses the flowability and segregation. The blocking ratio should be greater than 0.8. Blocking can be seen by coarse aggregate being held up by the bars, and the mix is considered stable if the coarse aggregate is evenly distributed all the way up to the end of the box.

Although not part of the mix design procedure, determining the rheological constants (yield stress and plastic viscosity) with the BML viscometer provides useful confirmation of the properties. Results from a series of trial mixes have shown that satisfactory SCC had a yield stress below 12 Pa and a plastic viscosity of about 150–250 Pas.

The method gives no recommendations for the design of mixes containing a viscosity agent.

6.4.4 Extensions of the CBI method

Van and Montgomery[29] have proposed adding an additional criterion to the calculation of the minimum cement paste volume to the solid blocking criterion of the CBI method. Called the liquid phase criterion, it involves ensuring that there is sufficient paste volume to provide a required minimum spacing between the aggregate particles.

The average particle spacing (D_{ss}) is calculated from:

- the average particle size, calculated from grading curves
- the (measured) volume of voids in the compacted aggregate
- the volume of paste, in particular the excess over that required to fill the voids in the aggregate system

Assuming spherical particles, D_{ss} is given by:

$$D_{ss} = D_{av}\left(\left[\frac{V_{pw}-V_{void}}{V_t-V_{pw}}+1\right]^{\frac{1}{3}}-1\right) \quad (6.2)$$

where V_{pw} is the paste volume, V_{void} is the volume of voids in the densely compacted aggregate, V_t is the total concrete volume, and D_{av} is the average particle diameter, which is given by:

$$D_{av} = \frac{\sum_{i=1}^{n} D_i M_i}{\sum_{i=1}^{n} M_i} \quad (6.3)$$

where D_i is the average size of aggregate fraction i, M_i is the percentage of aggregate in fraction i (obtained from sieve analysis) and n is the number of aggregate fractions.

The minimum required values of aggregate spacing (D_{ssmin}) were obtained from a programme of tests on concrete containing:

- five types of cement: ordinary Portland, shrinkage compensated Portland (two types), blast furnace slag cement (two types with 30% and 65% slag)
- three types of limestone powder of fineness 380, 870 and 1680 m^2/kg
- fly ash
- five sources of coarse aggregate, two each of maximum sizes 20 mm, 14 mm and 10 mm
- two sources of river sand
- a naphthalene-based superplasticiser

The mixes did not contain a viscosity agent.

It was found that D_{ssmin} was dependent on:

- the properties of the aggregates, powder materials and the superplasticiser
- the water/powder ratio
- the void content of the fine/coarse aggregate mixture
- the interaction between the components of the paste

In particular, for mixes with a sufficient slump-flow (650 mm), to achieve sufficient segregation resistance, higher values of D_{ssmin} were required with higher water/powder ratios and higher aggregate average particle diameters. Typical results for a maximum aggregate size of 20 mm are shown in Figure 6.8. The value of D_{ssmin} increases significantly when D_{av} exceeds 6.5 mm. The coarse to total aggregate ratio (N_{ga}) should therefore be chosen so that D_{av} is below 6.5 mm.

In the mix design method, having obtained the required D_{ssmin}, the required minimum paste, V_{pdmin}, is then obtained from Equation (6.2) rearranged:

$$V_{pd\,min} = V_t - \frac{V_t - V_{void}}{\left(\frac{D_{ss\,min}}{D_{av}} + 1\right)^3} \tag{6.4}$$

with the symbols having the same definitions as before.

Figure 6.9 shows typical variations of minimum paste volumes with coarse/total aggregate ratio (N_{ga}) when calculated according to the liquid and blocking criteria, the latter for two gap widths. The liquid criterion dominates at low N_{ga} values, thus overcoming the previous problem of low paste contents calculated with the blocking criterion. In this example, with an N_{ga} value of 0.5 (not unreasonable) the blocking criterion dominates with a gap of 41 mm, and the liquid criterion if the gap is increased to 58 mm.

The stepwise mix design procedure for the CBI method and its extension is therefore as follows:

(1) Obtain the construction criteria of:
   - spacing between and diameter of reinforcing bars
   - compressive strength and any other hardened concrete properties.

(2) Determine the material characteristics of:
   - specific gravity of the powders
   - specific gravity and gradings of the aggregate fractions (from which the average diameters, D_{av}, for various coarse total aggregate ratios, N_{ga}, can be calculated)
   - voids content for N_{ga} values between 0.4 and 0.6 with D_{av} ≤6.5 mm.

(3) Calculate the minimum paste volume, V_{pwmin}, with respect to N_{ga} according to the blocking criterion.

(4) Select the powder composition and water/powder ratio to give the required compressive strength (and other early age and hardened properties as required).

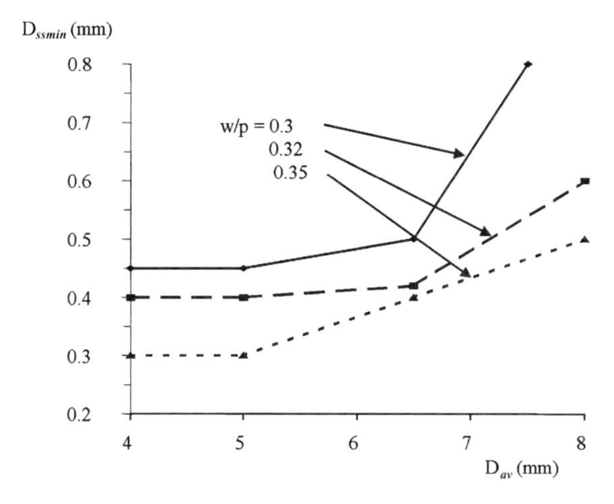

Figure 6.8 *Typical variation of minimum average spacing between aggregate particles, D_{ssmin}, with particle size D_{av} for 20 mm aggregate (from Van and Montgomery[29]).*

(5) Calculate the minimum required values of D_{ssmin} for different D_{av} with respect to N_{ga} and hence the minimum required paste volumes, V_{pdmin}, according to the liquid phase criterion.

(6) The optimum N_{ga} is that which requires the lowest paste volume according to the liquid phase criterion, provided that its D_{av} is not larger than 6.5 mm; otherwise choose the N_{ga} for D_{av} = 6.5 mm.

(7) Select the paste volume as the higher of V_{pwmin} and V_{pdmin}, but if this is above 420 l/m^3, which may be the case with very narrow gaps between the reinforcing bars, then consider using a coarse aggregate with a smaller maximum size.

(8) Estimate the superplasticiser dosage and carry out trial mixing and testing. If necessary, adjust the water/powder ratio and superplasticiser dosage until the required combination of properties is achieved.

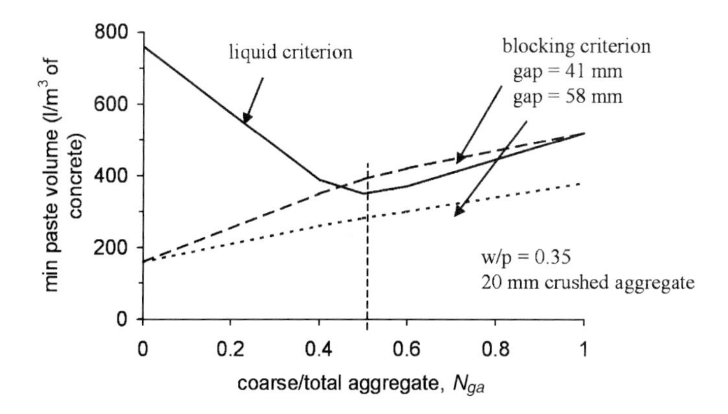

Figure 6.9 *Typical minimum required paste values for liquid phase and blocking criteria (from Van and Montgomery[29]).*

When used together, as above, the CBI method and its extension enable mixes to be designed for a specific bar spacing (the blocking criterion) and for achieving sufficient lubrication between the aggregate particles (the liquid phase criterion). They are not in themselves sufficient: adequate paste properties have also to be obtained through appropriate combinations of powder types, water/powder ratio and admixtures (e.g. by the fine mortar rheology tests described above). They do, however, also lead to an understanding of the importance of the two criteria in SCC behaviour, for example by using them in reverse for the analysis of successful SCC mixes that have been produced by other methods.

6.4.5 Analysis of mixes

The approaches described above give the basis for the analysis of mixes whose proportions and properties are known.

First, the risk of blocking for any particular bar spacing can be calculated from the proportions of each aggregate particle size fraction. As an example Figure 6.10 has been obtained from a series of four mixes with different proportions of 20 mm gravel aggregates but with the same proportion of fine aggregate in the mortar and the same water/powder ratio and superplasticiser dosage i.e. constant paste properties. Each mix was tested with a set of three J-rings with bar spacings of 21 mm, 41 mm and 60 mm. Although not wholly conclusive, the data do show an increasing step height with risk of blocking and hence the value of the risk of blocking approach to mix analysis.

Secondly, calculating the average particle spacing using Equation (6.2) for the same set of mixes as above, i.e. with different coarse aggregate contents but the same mortar composition, gives the relationship with slump-flow shown in Figure 6.11. The importance of this factor in influencing slump-flow is clear.

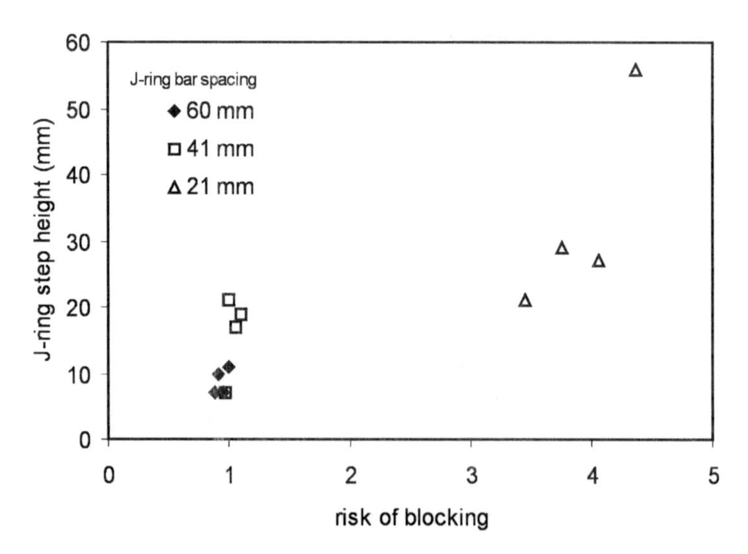

Figure 6.10 *Relationship between risk of blocking and J-ring step height for mixes with varying aggregate content and constant paste properties (UCL data).*

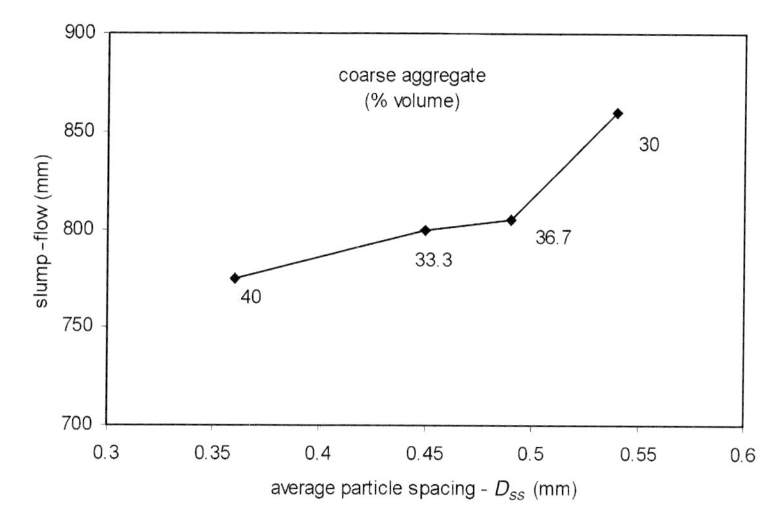

Figure 6.11 *Effect of average particle spacing on the slump-flow of mixes with varying coarse aggregate content and constant mortar composition and properties (UCL data).*

6.4.6 Other approaches

A considerable number of other approaches to mix design have been reported that are of varying complexity. Five of those with distinct differences from those already described are briefly discussed here.

LCPC method

The mix design method developed by Sedran and de Larrard at LCPC, France[30] aimed to produce an optimal (low paste volume) mix as efficiently and simply as possible, i.e. with the minimum of laboratory testing. However, a number of preliminary tests on concrete mixes are required. Its central feature is the use of a mathematical model called the 'compressible packing model' for which a software package was produced; this was an extension of a previous method called the 'solid suspension model' which was developed for a wider range of high performance concrete[31, 32].

Any size ranges of material can be considered, but it differs from other methods in that the cement and other powders as well as at the aggregates are included in the packing analysis and an optimised overall particle size distribution is obtained. Jonasson *et al.*[33] have found that the model is very useful for predicting the effect of aggregate type (natural gravel or crushed rock) on SCC properties, and for studying the robustness of the concrete to variation in the mixed parameters.

Space precludes a full summary of this method, and the reader is referred to the references cited above for a detailed description.

Icelandic Building Research Institute

Wallevik and Nielsson[34] have considered the rheological properties to be the dominant requirement of SCC, and have included rheological testing as an essential and integral part of the mix development process. Dispersing agents and superplasticisers are assessed by measuring the Bingham constants of mortars, and the effect of admixture dosage and the change of the properties with time are determined.

Tests on concrete mixes using the BML viscometer suggested that successful SCC has:

- a yield stress in the range 50–70 Pa (and this should not exceed 80 Pa before placing)
- a plastic viscosity in the range 20–30 Pas

Successful mixes were developed with moderate powder contents of 450–460 kg/m^3, and water/powder ratios of 0.41–0.42. Limestone powder could be used to reduce the cement content, and relatively small amounts of a viscosity agent (welan gum at 0.02–0.04% by weight of mix water) was incorporated to compensate for the lack of fine material in the sand.

However, Wallevik and Nielsson[34] have pointed out that these mixes used Icelandic materials which have some different characteristics from those used elsewhere: in particular the cement has a high alkali content, a high water demand and high workability loss and also contains 7% microsilica, and the aggregates are a relatively porous basalt.

Optimisation of mixes with factorial design

Khayat and co-workers at the University of Sherbrooke[35, 36] and Sonebi and co-workers[37, 38] at the University of Paisley have used statistical factorial methods to design sets of experiments to determine the relative magnitude of the influence of mix parameters and their interactions on the properties of SCC with a minimum of testing. Key parameters, such as powder content, powder composition, water/powder ratio and superplasticiser and VMA dosage were chosen for investigation. Three or four values of each parameter covering the likely range in SCC were selected, and the factorial method was used to design a set of mixes with appropriate combinations of these parameters. Each mix was then tested for the required properties (slump-flow, V-funnel time, L-box height ratio, consistence loss, compressive strength etc.) and the factorial analysis gives a set of models of the effect of the individual and combined parameters on each property. The resulting understanding of the interaction between the key parameters can be used for both mix optimisation and quality control.

Segregation controlled mix design

This method, proposed by Saak *et al.*[39], assumes that for a given aggregate particle size distribution and volume fraction both the fluidity and segregation resistance of the concrete is controlled by the rheology and density of the cement paste. The approach suggests that:

- the concrete will have its greatest fluidity at the lowest paste yield stress and viscosity at which segregation is avoided
- the segregation resistance is optimised at the highest yield stress and viscosity that still produce self-flowing

There is therefore a self-flow zone (SFZ) in which the concrete has both high consistence and segregation resistance. The extent of the SFZ can be defined in terms of values of the ratios of the yield stress and apparent viscosity of the paste divided by the difference in density between the aggregate particles and the paste. SFZ boundaries are suggested for paste with cement alone, cement plus silica fume, and cement plus silica fume plus a cellulose VMA. The concept has not been taken to the stage of incorporation into a complete mix design procedure.

Aggregate packing approach

Su *et al.*[40] have developed a method which they claim is simpler and less time-consuming than some other methods (particularly the Japanese general method) and results in smaller binder quantities, and hence reduced cost.

The method is based on the loose bulk density of the coarse and fine aggregate fractions, assuming that in the concrete these are more densely packed with a packing factor in the range 1.12–1.18. The coarse/fine aggregate ratio is preselected, and the combined packing or grading of the aggregate is not considered. The binder composition (typically, the combination on cement, fly ash and ground granulated blast furnace slag) and the water/binder ratio are chosen from a combination of strength requirements and measured flow characteristics of the paste. The superplasticiser dosage is initially chosen from experience, and adjusted if necessary in trial mixes. In the examples of mixes published, the resulting concrete generally has a coarse aggregate volume and binder contents at the lower end of the typical ranges given in Table 6.3 (28% of the concrete volume and 420–450 kg/m^3, respectively) and a fine aggregate content towards the upper end (36% of the concrete volume).

6.5 UCL method of mix design

The experience and understanding of the behaviour of SCC gained from other methods of mix design has enabled a relatively simple but effective mix design method to be developed at University College London (UCL). This has been developed as suitable for those new to SCC but with some experience of mix design of TVC and with access to a normally equipped concrete laboratory.

As with most methods for TVC, the method involves making the best estimates of the mix proportions for a given set of required properties, and then carrying out trial mixes to prove and if necessary adjust and refine the mix. However, an important feature of the first part of the process is tests on the mortar fraction of the concrete, using the spread and V-funnel tests described earlier, to determine the water/powder ratio and admixture dosage for the optimal mortar properties. Testing mortar is far quicker and more convenient than testing concrete, and therefore a greater range of variables can be readily examined. Sufficient confidence has been obtained in the relationship between the water and concrete properties for this to be an effective step in the mix design process.

The method has been produced for use for concrete with a coarse aggregate of 20 mm or 16 mm maximum size, crushed or uncrushed. Some comments are included if smaller (8 mm or 10 mm) aggregate is being used. The coarse/fine aggregate division is either 4 mm or 5 mm.

6.5.1 Equipment

The equipment available should include:

- spread and V-funnel tests for mortar
- slump-flow and V-funnel test, for concrete, although if viscosity is not critical for the application, the t_{500} time from the slump-flow test can be used as an alternative to the V-funnel flow time. The t_{500} time is approximately 3.5 times smaller than the V-funnel flow time[41]
- J-ring test, if passing ability is a requirement. This is the preferred test for passing ability in this method. If the L-box test is to be used, then the equivalent height ratio given in Table 6.2 can be used
- segregation resistance test
- mortar and concrete mixers, scales etc.

6.5.2 Procedure

Specified concrete properties

The properties required should always include slump-flow; depending on the application a viscosity or passing ability or both may also be required, as discussed in Section 6.1. Adequate segregation resistance will also be necessary, although assessment of the visual appearance and behaviour during, for example, the slump-flow test may be thought sufficient. Specification in terms of the EFNARC classes (Table 6.2), or similar, may be thought appropriate.

Some caution should, however, be exercised when specifying passing ability in addition to slump-flow and viscosity to ensure that it is feasible to obtain a mix meeting all the class criteria. As discussed in Chapter 5, the passing ability is dependent on the slump-flow and viscosity as well as the aggregate bridging. The effect of the first two properties is illustrated in Figure 6.12 which shows the variation in J-ring step height with the slump-flow and V-funnel time of mixes containing a low volume (29%) of 10 mm maximum size aggregate and in which aggregate blocking therefore did not occur. The two graphs show that:

- with a 59 mm bar spacing, to achieve a step height of less than 15 mm (EFNARC class PA1) a slump-flow of above 700 mm and a V-funnel flow time of less than 7 s is required;
- with a 41 mm bar spacing, to achieve a step height of less than 15 mm (EFNARC class PA2) a slump-flow of above 750 mm and a V-funnel flow time of less than 4 s are required.

Materials data

Details of the materials to be used are required including:

- coarse and fine aggregate type and grading
- cement, additions and their likely combination
- superplasticiser type
- VMA, if proposed
- relative density of all materials

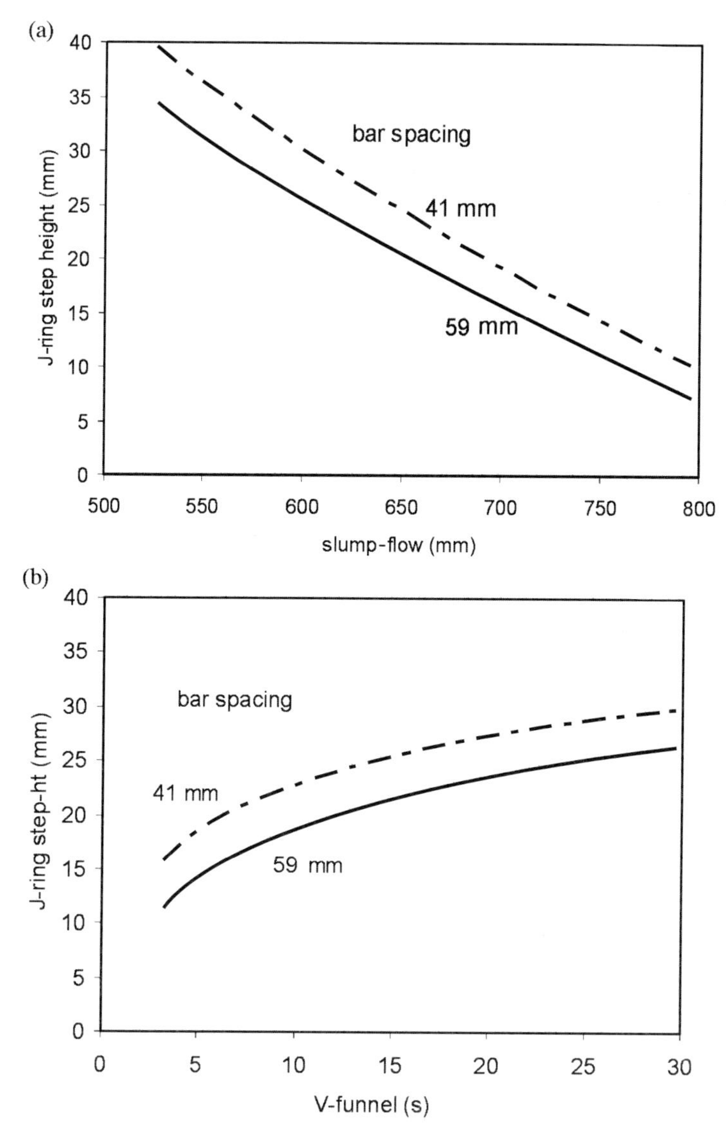

Figure 6.12 *Effect of (a) slump-flow and (b) V-funnel flow time on the J-ring step height for mixes with no aggregate bridging (UCL data).*

As was discussed in Chapter 3, all the materials should conform to the relevant standards for use in concrete. The aggregates, cement and additions may simply be those locally available for TVC production. The types and relative proportions of cement and additions will affect the strength of the concrete and/or other factors such as a heat of hydration, alkali content etc. and may therefore be chosen from previous knowledge of the materials' behaviour and the required early age and long-term properties.

Coarse aggregate content

All the key properties of SCC are influenced by the coarse aggregate content, and so the initial content selected will depend on the specified concrete properties. Recommend values are given in Table 6.6 The EFNARC class limits have been added to the table for convenience.

Fine aggregate content

The volume of fine aggregate (V_{fa}) is set at 45% of the resulting mortar volume, i.e.

$$V_{fa}\ (\%) = 0.45\ (100 - V_{ca})$$

Any fine aggregate particles smaller than 0.125 mm should be considered to be part of the powder fraction, and therefore will not contribute to this volume. It is however, reasonable to ignore these if they constitute less than 2% of the fine aggregate.

For an air-entrained concrete, the air content can be considered to be part of the mortar volume.

Paste volume and composition

The paste volume (V_{pa}) is then simply calculated from

$$V_{pa}\ (\%) = 100 - V_{ca} - V_{fa}$$

Table 6.6 *Recommend values of coarse aggregate content for initial mixes in UCL mix design method.*

Specified properties (EFNARC class)				Initial coarse aggregate % by volume (V_{ca})
Slump-flow	Viscosity: V-funnel flow time	Passing ability: J-ring step height 59 mm bar spacing	41 mm bar spacing	
Any (SF1, SF2, SF3)	Not specified	Not specified		38
Any (SF1, SF2, SF3)	≤ 8 s *(VF1)*	Not specified		30
	>8 and ≤15 *(VF2)*			35
	> 15 s *(VF2)*			38
< 700 mm (SF1/SF2)	≤ 8 s *(VF2)**	< 15 mm *(PA1)*		No mix possible
700–750 mm (SF2)				34
> 750 (SF3)				38
< 700 (SF1 /SF2)	≤ 4 s *(VF1)**		<15 mm *(PA2)*	No mix possible
700–800 mm (SF2/SF3)				32
>800 mm (SF3)				35

*max recommended values for mixes with a PA1 or PA2 passing ability requirement

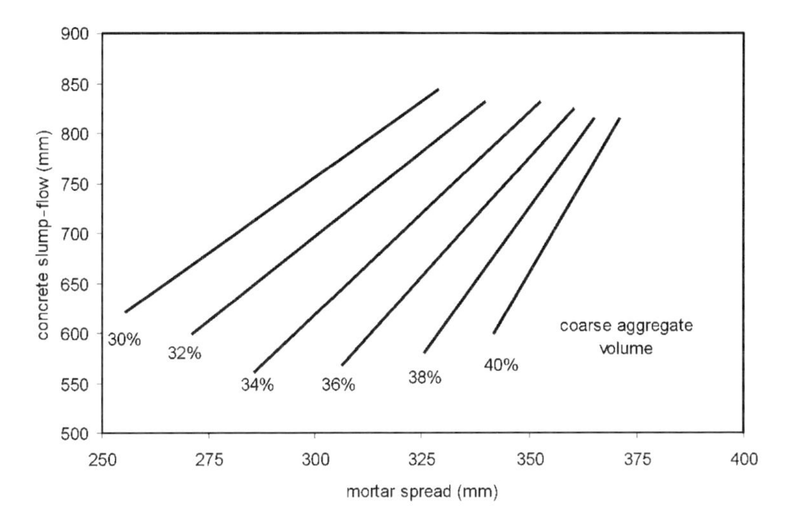

Figure 6.13 *Mortar spread values for concrete slump-flow and coarse aggregate contents (from UCL data).*

The paste composition, i.e. the water/powder ratio and the admixture dosage, is obtained from tests on mortar using the spread and V-funnel tests shown in Figures 6.3. and 6.4. The required or target mortar properties are first obtained from the required concrete properties and the estimated coarse aggregate content using Figures 6.13 and 6.14.

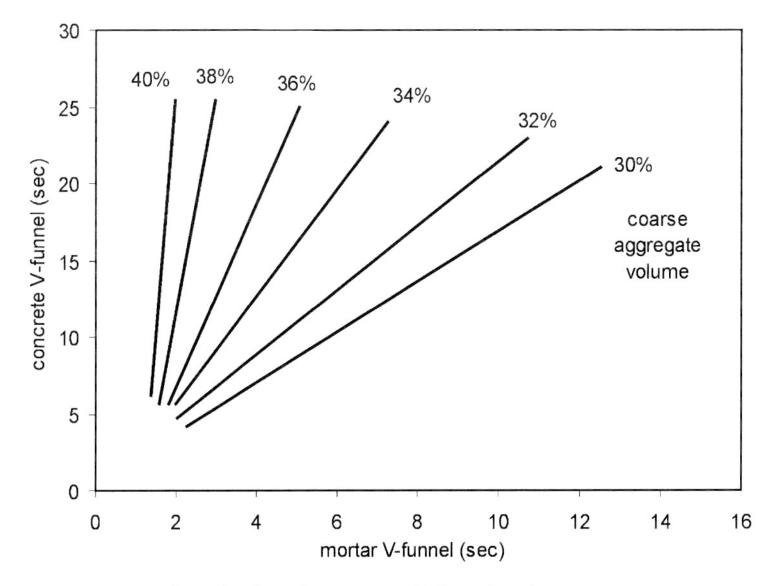

Figure 6.14 *Mortar V-funnel values for concrete V-funnel and coarse aggregate contents (from UCL data).*

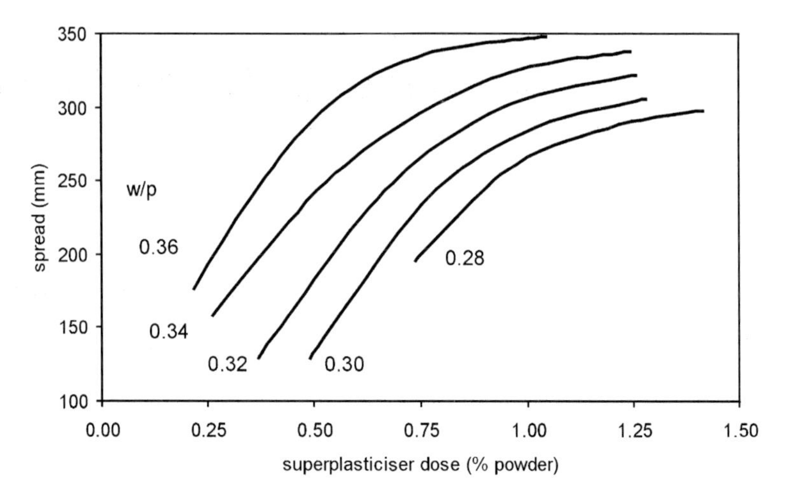

Figure 6.15 *Typical relationships between superplasticiser dose, w/p ratio and mortar spread (adapted from Domone*[42]*).*

(If a T_{500} time has been specified for the concrete, then this should be multiplied by 3.5 to obtain the equivalent V-funnel flow time. If no concrete V-funnel flow time has been specified, then a value of 8 s can be used.)

Mortars are then tested until a mix that meets the required combinations of spread and flow time is obtained. Testing a comprehensive set of mixes with varying water/powder ratios and superplasticiser contents will produce relationships such as those shown in Figures 6.15 and 6.16 (but with different values and axis scales depending on the constituent materials). Transferring these onto a spread vs. V-funnel flow time chart and

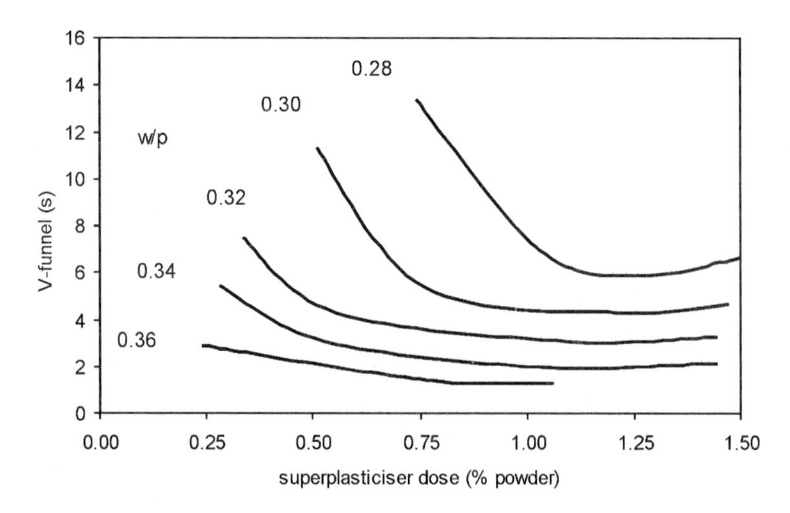

Figure 6.16 *Typical relationships between superplasticiser dose, w/p ratio and mortar V-funnel flow time (adapted from Domone*[42]*).*

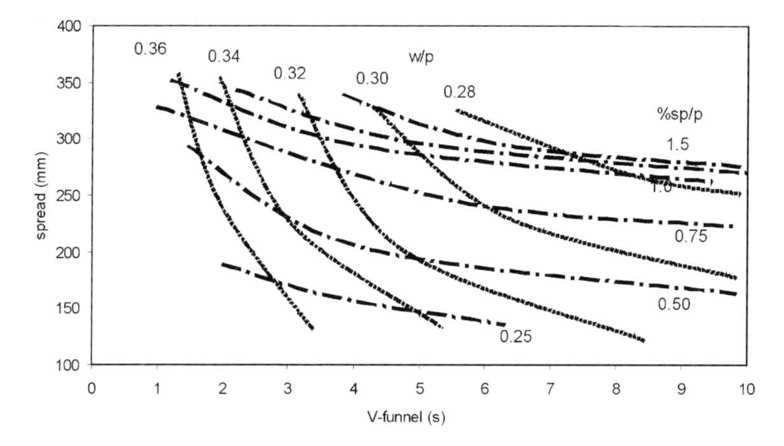

Figure 6.17 *A typical mortar spread. V-funnel flow time diagram showing the effect of w/p ratio and superplasticiser dose (adapted from Domone[42]).*

plotting contours of the water/powder ratio and superplasticiser dose gives a diagram such as Figure 6.17. From this the water/powder ratio and superplasticiser dose for the required combination of spread and V-funnel time can be estimated.

In practice, after some experience it is not necessary to carry out tests to define the whole of this diagram. It is sufficient to define the area of the diagram that is of particular interest. Variations in spread and V-funnel time for changes in mortar composition around the selected value can also be estimated from the plot, giving an indication of the likely robustness of the mix. Further explanation of the background and details of this approach can be found in Domone[42].

The effect of VMA can also be investigated with the mortar tests.

Trial mixes of concrete

All of the mix proportions of the concrete have now been estimated and so a trial mix can be batched and tested. Since all of the relationships used to produce the estimates are typical or average it is almost inevitable that not all of the concrete's target properties will be obtained with the first mix; adjustments and further testing will be therefore be required. The relationships used for the initial estimates of the mix properties can be used to make the adjustments.

6.6 Conclusions

We have seen that SCC can be formulated to produce a wide range of properties, thereby enabling the user of the concrete to select the properties required for a particular application. The principal differences in the mix proportions of SCC compared to TVC are lower coarse aggregate contents, higher powder contents, lower water/powder ratios and significant superplasticiser doses. The powder normally comprises Portland cement with significant quantities of one or more additions, and the powder composition is often chosen to give satisfactory early age and long-term properties.

The need to obtain a number of distinct fresh properties and the potential combinations of materials that could be used for this means that mix design procedures are necessarily more complicated than those for TVC, and many mix design methods of varying complexity have been developed. Some of these have the advantage of leading to a greater understanding of the concrete behaviour, as illustrated by the two methods considered in this chapter.

Sufficient experience has now been obtained with self-compacting mixes that it is possible to produce an initial estimate of mix proportions by considering the ranges of the key proportions of typical previously successful mixes. However, this may lead to a significant number of time-consuming trial mixes before an optimum is reached, and a method has been proposed which involves mortar tests before the concrete trials which is both relatively simple and efficient.

References

[1] Walraven J. Structural applications of self compacting concrete. In: Wallevik O. and Nielsson I. (Eds.) *Proceedings of the Third RILEM International Symposium on Self-compacting Concrete*, Reykjavik, Iceland, RILEM Publications, Bagneux, France, 2003, pp 15–22.

[2] Japan Society of Civil Engineers Recommendations for self-compacting concrete, Concrete Engineering Series 31, JSCE, Tokyo, 1999.

[3] EFNARC European Project Group The European guidelines for Self-compacting concrete: specification, production and use, May 2005, http://www.efnarc.org/pdf/SCCGuidelinesMay2005.pdf

[4] prEN 12350-8 Testing fresh concrete – Part 8: Self-compacting concrete – Slump-flow test, 2007.

[5] prEN 12350-9 Testing fresh concrete – Part 9: Self-compacting concrete – V-funnel test, 2007.

[6] prEN 12350-10 Testing fresh concrete – Part 10: Self-compacting concrete – L-box test, 2007.

[7] prEN 12350-11 Testing fresh concrete – Part 11: Self-compacting concrete – Sieve segregation test, 2007.

[8] prEN 12350-12 Testing fresh concrete – Part 12: Self-compacting concrete – J-ring test, 2007.

[9] ACI Committee 237 Self-consolidating concrete ACI Report 237R-07, American Concrete Institute, Detroit, MI, USA, 2007.

[10] Domone P.L. Self-compacting concrete; An analysis of 11 years of case studies. *Cement and Concrete Composites* 2006, 28(2) 197–208.

[11] Okamura H. and Ozawa K. Mix design method for self-compacting concrete, Concrete Library of Japan Society of Civil Engineers No 25, June 1995 pp 107–120.

[12] Nawa T., Izumi T. and Edamatsu Y. State-of -the-art report on materials and design of self-compacting concrete. In: *Proceedings of the International Workshop on Self-compacting Concrete*. 1998, Kochi, Japan. pp 160–190.

[13] Domone P.L. and Jin J. Properties of mortar for self-compacting concrete. In: Skarendal Å. and Petersson O. (Eds.) *Proceedings of RILEM International Symposium on Self-compacting Concrete*, Stockholm, September 1999, RILEM Proceedings, Cachan, France, pp 109–120.

[14] Jin J. and Domone P.L. Relationships between the fresh properties of SCC and its mortar component. In: *Proceedings of the First North American Conference on the Design and Use of Self-consolidating Concrete*, Chicago, IL, USA, November 2002, pp 33–38.

[15] Okamura H., Maekawa K. and Ozawa K. *High performance concrete* (in Japanese, translation of relevant parts by Takada T.). Gihouda Publishing Co, Tokyo, Japan 1993.
[16] Okamura H. and Ozawa K. Self-compactable high performance concrete. In: *International Workshop on High Performance Concrete*, American Concrete Institute, Detroit, MI, 1994, pp 31–44.
[17] Edamatsu Y., Nishida N. and Ouchi M. A rational mix-design method for self-compacting concrete considering interaction between coarse aggregate and mortar particles. In: *Proceedings of First International RILEM Symposium on Self-compacting Concrete*, Stockholm, Sweden, September 1999, RILEM Publications, Cachan, France, pp 309–320.
[18] Ouchi M., Hibino M., Ozawa K. and Okamura H. A rational mix-design method for mortar in self-compacting concrete. In: *Proceedings of the Sixth South-East Asia Pacific Conference of Structural Engineering and Construction*, Taipei, Taiwan, 1998, pp 1307–1312.
[19] Ouchi M., Hibino M. and Okamura H. Effect of superplasticizer on self-compactability of fresh concrete. *Transportation Research Record* (1574), 1997, pp 37–40.
[20] Pelova G., Takada K. and Walraven J. Aspects of the development of self-compacting concrete in the Netherlands, applying the Japanese mix design system. In: Andreikiv O.Y and Luchko J.J. (Eds.) *Fracture Mechanics and Physics of Construction Materials and Structures*, 3rd edn., The National Academy of Sciences of Ukraine, Kamaniar, 1998.
[21] Takada K., Pelova G. and Walraven J. The first trial of self-compacting concrete in the Netherlands according to the Japanese design method. *Proceedings of FIP Congress*, Amsterdam, the Netherlands, 1998, pp 113–115.
[22] Petersson Ö., Billberg P. and Van B.K. A model for self-compacting concrete. In: Bartos P.J.M., Marrs D.L. and Cleland D.J. (Eds.) In: *Proceedings of RILEM International Conference on Production Methods and Workability of Fresh Concrete*, Paisley, June 1996, E&FN Spon, London, UK, pp 484–492.
[23] Petersson Ö. and Billberg P. Investigation on blocking of self-compacting concrete with different maximum aggregate size and use of viscosity agent instead of filler. In: *Proceedings of First International RILEM Symposium on Self-compacting Concrete*, Stockholm, Sweden. September 1999, RILEM Cachan, France, pp 333–344.
[24] Billberg P. Self-compacting concrete for civil engineering structures – the Swedish experience. Report 2:99. Swedish Cement and Concrete Research Institute, Stockholm, 1999.
[25] Ozawa K., Tangtermsirikul S. and Maekawa K. Role of powder materials on the filling capacity of fresh concrete. In: *Proceedings of the 4th CANMET/ACI Conference on Fly Ash, Silica Fume, Slag and Natural Pozzolans in Concrete*, Istanbul, May 1992, ACI, Detroit, MI, USA, pp 121–137.
[26] Van B.K. A method for the optimum proportioning of the aggregate phase of highly durable vibration-free concrete. MSc thesis, Asian Institute of Technology, Bangkok, Thailand, 1994.
[27] Tangtermsirikul S. and Van B.K. Blocking criteria for aggregate phase of self-compacting high-performance concrete. In: *Proceedings of Regional Symposium on Infrastructure Development in Civil Engineering*, Bangkok, Thailand, December 1995, pp 58–69.
[28] Ozawa K., Tangtermsirikul S. and Maekawa K. Role of powder materials on the filling capacity of the fresh concrete. In: *Fourth CANMET/ACI International conference on Fly Ash, Silica Fume, Slag and Natural Pozzolanas in Concrete*, Istanbul, Turkey, 1992, supplementary papers, pp 121–137.

[29] Van B.K. and Montgomery D. Mixture proportioning method for self-compacting high performance concrete with minimum paste volume. In: *Proceedings of the First International RILEM Symposium on Self-compacting Concrete*, Stockholm, Sweden, September 1999, RILEM Publications, Cachan, France, pp 373–396.

[30] Sedran T. and de Larrard F. Optimization of self-compacting concrete thanks to packing model. In: *Proceedings of First International RILEM Symposium on Self-compacting Concrete*, Stockholm, Sweden, September 1999, RILEM Publications, Cachan, France, pp 321–332.

[31] Sedran T. and de Larrard F. Rene-LCPC: Software to optimize the mix design of the high-performance concrete. In: *Fourth International Symposium on Utilization of High Strength/High Performance Concrete*, 1996, pp 169–178.

[32] Sedran T., de Larrard F., Hourst F. and Contamines C. Mix design of self-compacting concrete (SCC). In: Bartos P.J.M., Marrs D.L. and Cleland D.J. (Eds.) *Proceedings of RILEM International Conference on Production Methods and Workability of Fresh Concrete*, Paisley, E&FN Spon, London, UK, 1996, pp 439–450.

[33] Jonasson J.-E., Nilsson M., Simonsson P. and Emborg M. Designing robust SCC for industrial construction with cars in place concrete. In: *Proceedings of the Second North American Conference on the Design and Use of Self-consolidating Concrete and the Fourth International RILEM Symposium on Self-compacting Concrete*, November 2005, Hanley Wood, Minneapolis, MN, USA, pp 1251–1257.

[34] Wallevik O.H. and Nielsson I. Self-compacting concrete – a rheological approach. In: *Proceedings of International Workshop on Self-compacting Concrete*, August 1998, Kochi, Japan, pp 136–159.

[35] Khayat K.H., Ghezal A. and Hadriche M.S. Utility of statistical models in proportioning self-consolidating concrete. In: *Proceedings of the First International RILEM Symposium on Self-compacting Concrete*, Stockholm, Sweden, September 1999, RILEM Publications, Cachan, France, pp 345–360.

[36] Ghezal A. and Khayat K.H. Optimising self consolidating concrete with limestone filler by using statistical factorial design methods. *ACI Materials Journal*, 2002, 99(3) 264–272.

[37] Sonebi M., Bahadori-Jahromi A. and Bartos P.J.M. Development and optimisation of medium strength self-compacting concrete by using pulverised fuel ash. In: Wallevik O. and Nielsson I. (Eds.) *Proceedings of the Third RILEM International Symposium on Self-compacting Concrete*, Reykjavik, Iceland, RILEM Publications, Bagneux, France, 2003, pp 514–524.

[38] Sonebi M. Medium strength self-compacting concrete containing fly ash: modelling using factorial experimental plans. *Cement and Concrete Research* 2004, 34 1199–1208.

[39] Saak A.W., Jennings H.M. and Shah S.P. New methodology for designing self-compacting concrete. *ACI Materials Journal* 2001, 98(6) 429–439.

[40] Su N., Hsu K-C. and Chai H-W. A simple mix design method for self-compacting concrete. *Cement and Concrete Research*, 2001, 31, pp 1799–1807.

[41] Bartos P.J.M. (project director) European Union Growth Contract No. G6RD-CT-2001-00580 Testing-SCC: Measurement of properties of fresh self-compacting concrete. Final report, September 2005. http://www.civeng.ucl.ac.uk/research/concrete/Testing-SCC/

[42] Domone P.L. Mortar tests for material selection and mix design of SCC, *Concrete International*, 2006, 28(4) 39–45.

Chapter 7
Construction process

7.1 Batching and mixing

7.1.1 Requirements for the production plant

Reliable and cost-effective production of SCC does not require significant changes to the storage, handling and batching of raw materials compared with that of TVC, provided that existing general good practice is followed. However, SCCs usually require more additions than for TVC mixes. This may include not only the more common additions such as fly ash or ground granulated blast-furnace slag, but also limestone powder or fine-grained inorganic wastes (e.g. industrial dusts). The production plant may therefore need additional silos or other types of storage. The mechanisms for dispensing of the admixture may have to be upgraded to cope with more than one type of admixture being added at different stages of the mixing.

7.1.2 Control of materials

SCC may require greater control over the moisture content during the storage of aggregate and batching of concrete, particularly in cases of 'high-performance' SCC mixes. Such SCCs have extreme levels of filling ability and segregation resistance. They are capable of either flowing very long distances without segregation, or free-fall from substantial heights and through reinforcement. However, in order to achieve this outstanding performance consistently, greater control of the moisture content is required. If an automatic moisture content sensor is used, it is advisable to check its accuracy and calibrate the system more often than for TVC. Control of the moisture/water content of an SCC mix extends to the truck-mixers or agitators being used to mix and deliver fresh SCC in the same manner as it applies for TVC. It is essential to include in the water content of a batch any water, which may remain in the drum of a truck-mixer after it has been washed out.

Often, when producing TVC, the batcher has some discretion in the addition of water. With less robust (more sensitive) SCC mixes, there is a general view that this discretion should be (at least) limited. If the batcher can himself use judgement in adjusting the water contents, he should be aware of how much can be added as a maximum, and this will depend on the pre-assessed robustness of the mix.

7.1.3 Mixers

The concrete mixer itself is a key item in the concrete production process. The significance of the mixing process has tended to be underrated for TVC and practical technical difficulties have restricted research into the performance of concrete mixers[1]. The low consistence of most of the TVC often masked the effects of the mixing process on the concrete when fresh. It was impossible to judge whether the nonuniformity of the hardened concrete was due to inadequate mixing, or to segregation, or to some other site problem, namely poor compaction.

The development of self-compacting mixes for general use has shifted the focus of interest to the properties of the fresh mix and it has drawn new attention to the mixing process. It is now appreciated that SCC mixes of identical proportions and types of constituent materials can have different properties, both in the fresh and the hardened state, simply because they were produced in different mixers, or in the same mixer but using different charging sequences and lengths of mixing time. In addition to the factors outlined above, the size of the batch in relation to the maximum useful batch volume of the mixer has an effect on the properties of the mix produced, both when fresh and hardened. Any such effect tends to be greater for SCC than traditional vibrated mixes[1, 2].

The type of mixer has an effect on the efficiency of production of a well-mixed (adequately homogeneous) fresh SCC. The capability of different types of concrete mixers depends on the target consistence of the mix and the behaviour of the fresh mix also varies, depending on the type of mixer. Information available regarding mixing of SCC is not adequate to meaningfully rank existing types of concrete mixers according to their performance in the production of fresh SCC.

Mixing plant is sometimes fitted with a wattmeter, which (indirectly) measures the 'consistence' of the mix as it 'develops' during the mixing process. Readings from such devices are also used to determine any additional amounts of water required to adjust the consistence of the batch to match as closely as possible the specified one. Such mixer-based devices may prove to be less effective in controlling the production of fresh SCC because its consistence will be near to or even outside their limit of sensitivity. If their use is possible, it may be necessary to 'calibrate' the readings differently for different SCC mixes.

Mixers of the free-fall type, where the mixing action takes place inside a drum fitted with internal blades usually require longer mixing times when producing SCC. Mixing times twice as long as those for TVC have been reported. One of the reasons is the higher amount of additions (fines) in many SCCs, which may cause the initially stiff paste to adhere to some parts of the drum. This considerably reduces the mixing efficiency, and may also result in the mortar actually hardening on the blades, and affecting all subsequent mixes. Altering the charging sequence by adding all of the mixing water and some of the superplasticiser and coarse aggregate before the other constituents may reduce such excessive adhesion of the paste to the drum and blades.

Forced action mixers, in which the ingredients of a mix are moved by paddles or blades in different arrangements, generally produce adequately uniform mixes in very much shorter mixing times than the free-fall mixers. Mixers of this type, pan and trough mixers, are the norm in the precast industry, and are also used in some ready-mixed concrete plants where concrete is mixed in the plant (wet-mixed) and then delivered by truck-mixers acting as agitators.

In some countries, such as the UK, ready-mixed concrete is usually 'dry batched' i.e. preweighed, and then mixed in the (free-fall) truck-mixer itself. With good combinations of materials this can produce satisfactory SCC, though producers may still prefer to supply from central-mix plants. The mixing time in the truck-mixer may have to be extended compared with mixing of TVC of lower consistence. Some specifications permit the addition of all or part of the admixtures after the arrival of the truck-mixer at the site but advice varies regarding the addition of more admixture and a short 'full-speed' remixing of the batch. Many specifications do not permit this. Such retempering can be beneficial in that it can restore the properties of the fresh mix to those measured when the transport commenced, but this may not always be the case. Norwegian guidance for production of SCC[3] cautions against attempts to add more of the copolymer-based SCC admixture at a site, after transport. Such admixtures require more intensive or longer mixing than others to 'activate', which may be more than can be obtained in an ordinary truck-mixer.

Concrete mixes having high levels of consistence such as fresh SCC are particularly suitable for production in a M-Y mixer (Figures 7.1 and 7.2). This is a novel type of a low energy mixer, which may be regarded as a free-fall mixer in principle but where there is no drum-like mixing vessel. The mixing apparatus comprises a set of mixing units, which have no moving parts, assembled into a stationary vertical column. All of the energy for the mixing action is derived from gravity.

Figure 7.1 *M-Y mixer on site in Japan. (Reproduced with permission of Maeda Co.)*

Figure 7.2 *Prototype of the M-Y mixer undergoing mixing efficiency trials at the ACM Centre, University of Paisley, Scotland.*

7.1.4 Quality control of the production process

It is essential that the performance of a satisfactory lab mix be verified at the full-scale production plant which will be used. The efficiency of the mixing is likely to differ and the mixing process found to be suitable in the laboratory, may have to be adjusted for optimum full-scale production at a given plant. It is also possible that the mix itself may have to be adjusted, particularly the admixture dosage, since mixers vary widely in their ability to properly and rapidly disperse the superplasticiser.

Related to this, in countries with large temperature differences between summer and winter temperatures, such variations have an effect on the admixture dosage required. Regardless of the type of mixing process and the arrangement of the concrete production plant, it is necessary to verify the mixing times and check the charging sequences to achieve the specified properties of the concrete, not only in the fresh state but also in the hardened state (see also Chapters 8 and 9). Trial mixing and production of 'mock-up' elements will continue to play an important part in mix design and production of SCC for more significant applications for some time despite advances in the development of mix designs.

Requirements for quality control include:

- As a minimum, normal levels of quality control need to be applied, but testing is likely to be necessary at plant and site, especially at the beginning of a day's production. It is likely that the slump-flow test will be the normal way of assessing the consistence of deliveries, in the same way that slump testing is used. Originally designed as a workability test for TVC, its use is now extended to act as an indicator of the uniformity of supply. It should be recognised that the slump-flow test alone, used on-site for acceptance purposes or identity testing is not a guarantee of self-compactability.
- Greater control of moisture content is necessary, for the reasons described above. Results of tests during full-scale casting trials within the EC Testing-SCC project[4] suggested that variations of ± 10 l/m^3 may be acceptable for TVC, but that the equivalent figure for SCC might be 5 l/m^3.
- Once the mixing process has been established for any SCC mix, in order to consistently produce concrete of the required performance (fresh and hardened), that process must always be strictly followed. This is no longer a difficult task for the types of modern and highly automated concrete production plants used in the ready-mixed and precast concrete industries.
- Greater supervision of the production process is likely to be necessary for SCC. Many national guidelines refer to the need to test every load for slump-flow. Others suggest this requirement at the beginning of a job or a day's production, and then allow the testing frequency to be reduced to every third load. In other cases, the producer will have to establish an appropriate testing regime. Often this will involve having supervision at the plant as well as on-site. When concrete is entirely truck-mixed, an assessment will have to be made of how representative any sample taken at the plant actually is.
- Consideration will have to be given to the necessity for any form of routine testing other than slump-flow testing. This may include J-ring or Orimet tests, which are easy to carry out on-site.
- For the moment the inclusion of SCC in the hierarchy of concrete 'families', on which producers may base their production control systems to ensure conformity, may not be justified. (This involves treating different types of concrete as a single family for purposes of production control and analysis. The different types are linked by proven relationships. There is not normally enough information available about SCC to establish such relationships or to include it in a family.) For the moment, therefore, SCC, and in fact each SCC mix, is likely to require a separate control system.

7.2 Transport

Transport of fresh concrete over longer distances, generally outside the construction site, is mostly by truck-mixers or simpler versions acting as agitators. Tipper trucks are not suitable for SCC. The high consistence level of SCC (filling ability, fluidity) may lead to a need to reduce the size of a batch, to prevent spillages during transport (especially when going steeply uphill) and to maintain the stability of the vehicle.

Conditions during transport, which influence the properties of fresh traditional vibrated mixes of low consistence also apply to fresh SCC. High temperature leads to faster and greater changes and the temperature of the mix may also change. More important changes to properties may include a loss of filling ability and segregation resistance. If the mix is air-entrained, the air content of the mix may decrease by 0.5–1.5%.

Successful transport of fresh SCC, especially over greater distances, requires the retention of an adequate level of self-compactability throughout the period from the end of mixing to the end of placing. Typical fresh SCC for many applications is likely to maintain its fresh properties for up to 45 min. It is not difficult to produce mix designs, which ensure the maintenance of self-compactability over a period of up to 90 min, or longer, in moderate environmental conditions (air temperatures in the range 15–25°C). Transport of fresh SCC for longer periods of time in hotter environments requires special technical measures in addition to a modified mix design. Such measures may include remixing of the batch, with further addition of admixture just prior to placing, cooling of the fresh mix by the addition of crushed ice during the initial mixing etc.

7.3 Placing

One of the key characteristics of fresh SCC is its filling ability, its much-enhanced capacity to flow under its own weight. It is bad practice to allow, or encourage, TVC to flow, but this is an essential characteristic of SCC, which is exploited in the construction process. Depending on a number of factors such as: delivery rate, element and reinforcement geometry, and lift height (the thickness of the layer cast) as well as the filling ability of the concrete itself, fresh SCC is capable of flowing more than 10 m without segregation. A horizontal flow to a distance of 5–6 m from a single discharge point is readily achieved in routine applications. Some of the national guidelines (see Chapter 11) suggest limits for this, but in the current state of knowledge, this may be one of the things that have to be assessed on a job-by-job basis.

When considering the lift height and permissible height from which SCC may be dropped, it should be borne in mind that any entrapped air has to escape on its own, without the assistance of vibration. Typical cases of placing fresh SCC in precast concrete production and in situ are shown in Figures 7.3 and 7.4.

7.3.1 Pumping

Pumping of concrete is a common method for on-site transport and placing. Normally the pumping process is easier with SCC. The high mortar content and high flowability make the mix generally more pumpable: lower pumping pressures are needed, which in turn allows delivery to greater distances or heights. Placing concrete by pumping into the formwork from below is simple when SCC is used. This method is impossible with the much less workable fresh TVC.

A sliding, cut-off valve, as shown in Figure 7.5 is usually fitted to the formwork opening.

The ease of pumping of ordinary SCC permits more efficient pipeline layouts, including ‘splits’ and ‘bifurcations’, which enable much faster placing rates to be achieved (Figures 7.6 and 7.7).

Figure 7.3 *Precasting with lightweight SCC. (Reproduced with permission of R. Hela and M. Hubertova.)*

SCC can be delivered simultaneously to several discharge points (see Figure 7.7) and very large pours may be completed in much shorter times than those required for TVC.

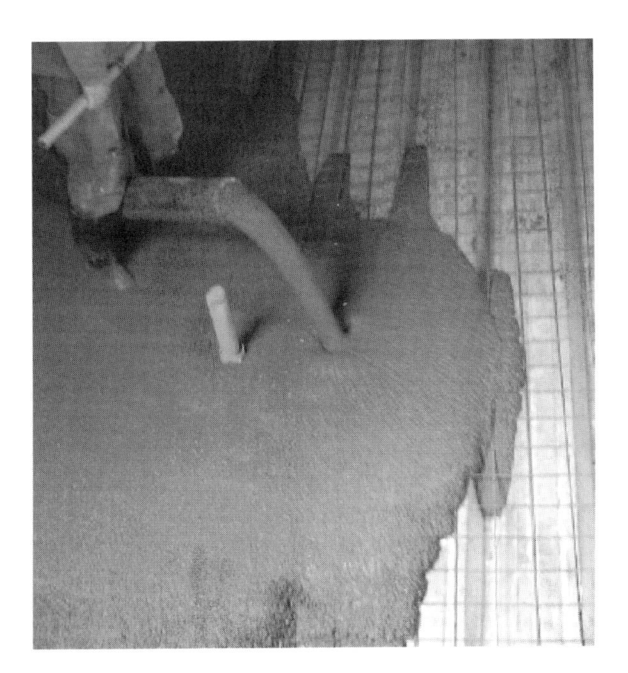

Figure 7.4 *A typical case of one man placing SCC in situ. (Reproduced with permission of R. Hela and M. Hubertova.)*

Figure 7.5 *A sliding 'cut-off' valve attached to column formwork. (Reproduced with permission of John Doyle Co. Ltd.)*

Recent developments have led to the production of high-performance concrete mixes, which include unusual ingredients and are self-compacting. Mixes containing fibres and produced in the traditional manner, often have a consistence so low that they require high compaction efforts and are unsuitable for pumping. Such mixes, when designed to be self-compacting, on the other hand can be pumpable[2], though this cannot be reliably predicted from the mix design alone. When such high-performance mixes have to be pumped to great heights, their resistance to pressure-segregation, or the stability of the mix when moving subject to pressure, are likely to become the

Figure 7.6 *Bifurcated pipeline increases the rate of placing of fresh SCC. (Reproduced with permission of Å Skarendahl.)*

Figure 7.7 *Multiple discharge points when casting SCC for anchorage blocks of the Akashi-Kaikyo bridge, Japan.*

critical parameter for successful delivery and placing. It is therefore important to carry out pumping trials before the concrete is used, namely in a large-scale practical application. The segregation resistance and filling ability of such special SCC mixes becomes even more important than for ordinary SCC.

7.3.2 Skip and crane and other methods

Transport and placing of fresh concrete on a site by a skip and a crane can be carried out equally well for traditional vibrated and self-compacting mixes. For SCC it is common to use a skip which has a length of a flexible hose attached, either to reduce the free-fall of the concrete, or to function as a 'mini-tremie', with the discharge end of the hose submerged to a depth of approximately 0.30 m in the concrete already placed. The hose, especially when used as a tremie, reduces the possibility of a dynamic segregation of the mix during casting.

Skips for placing fresh SCC need to have an easily operated and tightly closing opening system, which should enable slow and controlled discharge, of the SCC.

For applications where horizontal surfaces are specified, SCC can be also designed to be self-levelling. On the other hand, there are applications such as bridge decks (Figure 7.8) and pavements, where a sloping surface is required. SCC can be used for such applications, provided it is designed with reduced filling ability (fluidity). This allows the mix to maintain the profile of the slope, while retaining adequate self-compactability. By reducing the filling ability, fresh SCC can maintain a surface gradient of up to approximately 4%. In these circumstances, any attempt to compact the mix by vibration will make the concrete quickly 'run' down the slope and

Figure 7.8 *Casting of a heavily reinforced deep bridge deck using SCC. (Reproduced with permission of J.L. Vitek, Metrostav a.s.)*

'level-off'. A light vibrating screed may be used, but its purpose is limited to correction of the slope, and finishing – certainly not compaction of the concrete which has been placed, which should anyway be unnecessary for good SCC.

SCC is also suitable for applications where steeply sloping or curved top surfaces are required. In such cases an SCC mix with a high filling ability is cast into spaces with negative (top) formwork, which must be provided with appropriate 'bleeding' or de-airing vents.

7.4 Formwork

7.4.1 General requirements

TVC is temporarily liquefied by vibration, and the joints in the formwork are designed to be completely tight to prevent the escape of grout. In the early development of SCC, there was an initial concern about the high fluidity of the fresh concrete which, it was thought, might lead to potentially greater leakage of paste from the formwork, with a subsequent need for better seals between the parts of the typical modular formwork. However, practical experience and full-scale trials using different types of formwork have shown that there was no need for higher quality formwork when SCC was used. A typical fresh SCC mix possesses sufficient internal cohesion to prevent leakage of the paste or fine mortar through standard formwork joints. Indeed, SCC mixes in general are less likely to leak than TVC. Reduction of leakage in the case of SCC leads to a reduction in unsightly blemishes on surfaces of hardened concrete and to hardened concrete with better-defined, sharper corners and arises.

An additional benefit of using SCC, particularly in precast concrete production is the saving on the strength and rigidity of moulds, which are no longer subject to vibration. A lighter formwork, which is easier to handle, can be used provided its load bearing capacity is sufficient when filled with fresh SCC.

7.4.2 Formwork pressure

The high filling ability and tendency towards a 'liquid-like' behaviour by fresh SCC has automatically raised concerns about the development of unduly high formwork pressures, particularly when deep formwork is used. It was thought reasonable to assume that such a highly fluid material would exert full hydrostatic pressure on the walls of the containing formwork. Such pressures would be markedly higher than those occurring when TVC is cast. Alternative arguments were also proposed – that the thixotropic nature of (many) SCCs would reduce the formwork pressure. It should also be borne in mind that the high fluidity of SCC mixes is not due to their high water contents.

Numerous experiments have been carried out in order to provide evidence as to what the formwork pressures actually were (Figure 7.9), sometimes with contradictory results. The pressures measured varied from full hydrostatic pressure to a pressure which was equal to, or even slightly lower than, when an ordinary vibrated concrete was cast.

Figure 7.9 *Full-scale trials for formwork pressure of SCC. (Reproduced with permission of GTM Vinci.)*[2]

Most of the experiments were carried out on full-scale structural elements. The variables included: different SCC mix designs, rates of placing and placing methods. The method of placing, particularly in combination with the mix design, was found to be the critical factor influencing the magnitude of the formwork pressure[2, 5].

High formwork pressure

In some circumstances a high formwork pressure even higher than the hydrostatic pressure can develop, due to the additional dynamic/impact load at the beginning of the pour. Also, when concrete is pumped from the bottom of the formwork, there is some evidence that the pressures may actually exceed the hydrostatic levels, due to the additional pressure imparted by the pump[6].

Generally, when SCC is being pumped in this way, high pressure is generated when the body of fresh concrete is constantly agitated by the stream of concrete emerging from the inlet of the pipeline at the bottom of the formwork. The agitation maintains the internal shearing in the mix, which then maintains or increases its filling ability and remains in a highly fluid state. Faster filling further increases the agitation within the already placed concrete and contributes to the generation of high, often fully hydrostatic, formwork pressure. High formwork pressures associated with casting from the bottom is an inevitable side effect to offset against the many practical advantages of SCC for this otherwise very convenient method of concrete construction.

Low formwork pressure

Low formwork pressure (in the region of pressures developed with TVC) occurs in the following circumstances:

- When the concrete has low filling ability (flowability).
- When the type of admixture used gives it a measure of thixotropy.
- When it is poured slowly from top down and without significant free-fall. Casting from the top, using a crane and skip or discharging through a flexible pipe/hose/elephant trunk keeps only the top layer of the concrete agitated, subject to an internal movement. Concrete below the top layer remains largely undisturbed. Depending on the degree of thixotropy the freshly placed concrete stiffens very fast, sometimes almost instantaneously.

7.5 Curing

The increased content of fine particles and the consequent closer particle density produces a more cohesive microstructure in SCC than in TVC. This restricts the migration and loss of water during the early stages of the setting and hardening of SCC. Some limited evidence[2] suggests that for this reason the performance of a typical hardened SCC mix may be less sensitive to the quality of the curing than that of TVC, and that the in situ compressive strength of SCC may be less dependent on curing (see Chapter 8). The freshly placed SCC mix may be 'self-curing'. Though this might indicate that there is a lesser need for protection during curing, the evidence was related only to the SCC mix designs examined and it is therefore advisable to apply the established good curing practice for TVC to SCC.

7.6 Finishing

7.6.1 Features and properties of surfaces

The surface finish of SCC can be extremely good, with near-perfect reproduction of the surface characteristics of the formwork, and a clear definition of the profiles and cast-in features. One of the advantages claimed when SCC was introduced into construction practice in Europe, was this better quality surface finish – a significant general advantage, since the negative public image of concrete is often based on its perceived poor appearance. There is no doubt that very high quality finishes have been produced with SCC, undoubtedly better than could be expected from conventionally (and properly) placed TVC. Figures 7.10–7.12 demonstrate how good the surface finish of hardened SCC can be.

More recently, less emphasis has been placed on this claim on behalf of SCC, because it has become apparent that this standard of excellence was not always, automatically, achieved and that there could still be some problems, particularly with blowholes (known as 'bug holes' in the USA and some other countries).

In principle, however, it remains likely that SCC will not suffer from many of the blemishes in the appearance of TVC, simply because these are undoubtedly associated with the process of compaction.

The surface finish of TVC often exhibits features such as:

- hydration and segregation discolouration
- aggregate transparency
- scouring

Figure 7.10 *Precast SCC. Note the blemish-free surfaces and perfect, sharp edges. (Reproduced with permission of Prefa-Dywidag a.s., Czech Republic.)*

Figure 7.11 *Blemish-free surface of a precast SCC element. Note the perfectly formed edges. (Reproduced with permission of Prefa-Dywidag a.s., Czech Republic.)*

- sand textured areas
- aggregate bridging
- gross honeycombing
- problems caused by grout loss

It might be assumed that SCC would eliminate these problems. Indeed, blemishes such as these have not normally been apparent in structures built with SCC. In general, it is still true to say that 'knowledge of the relationship between any blemish and factors causing it … is far from clear'[6, 7] However, it is reasonable to assume that the removal of the need to vibrate, together with the high fines content, greater cohesion and water retention, and the reduced moisture movement in SCC are responsible for the observed improvement in these areas. In practice almost perfect surface finishes can be obtained, better than those observed in even the best conventional concrete.

The surface finish of SCC exhibits properties such as:

- dense, closed surface texture
- perfect definition of corners and angles
- perfect reproduction of textured formwork and other designed formwork features
- fewer colour variations caused by vibration
- elimination of problems caused by grout loss, which is minimised by the thixotropic nature of SCC

In addition, good finishes can be achieved where it would be impossible to do so with TVC – where dense reinforcement makes the use of conventional compaction methods impossible.

Figure 7.12 *Excellent surface finish of large precast T-beams. (Reproduced with permission of C. Richards.)*

7.6.2 Factors affecting surface finish in concrete

There is at present only limited, and unsatisfactory, understanding of the relationship between ordinary concrete and surface finish. Little research had been done for many years and what had been done was limited to largely empirical comparisons of the appearance and 'durability' of selected typical types of surface treatment of concrete.

The surface of any concrete is its vulnerable point. Partly because of the 'wall effect' of the formwork, the surface has a lower aggregate/cement ratio, i.e. more paste, and here the paste is the weak link in the composite chain. The surface is sometimes reported[8] to have a higher water/cement ratio and lower real cement content, which makes it more permeable than the 'inner' mass of the concrete; and it contains voids, bubbles of air trapped against the formwork. The negative influence of the 'wall-effect' tends to be less in SCC than in TVC because SCC has a lower content of coarse aggregate and more paste in the mix. This vulnerability and the implications for long-term durability are well recognised in concrete practice and design codes, where there are long-established requirements governing the cover of the reinforcing steel.

There is much less guidance available regarding the appearance of the concrete surface, which reflects the 'quality' of the cover zone. Problems with the cover zone still bedevil concrete construction, and the layman's antipathy to concrete as a material is rooted in its appearance. Some research was done, mostly many years ago[7–11] on the subject of concrete surface finish (restricted, of course, to TVC), but this has proved to be a very intractable subject. Many of the problems that provoked that research effort still exist. In the UK, these problems are clearly demonstrated in the reference panels to the BS 8110 specification, set up in various parts of the country under the auspices of the British Cement Association and CONSTRUCT, the Concrete Structures Group, with the support of the British government. The full-scale demonstration panels available for inspection in different parts of the UK show the different levels of quality of surface finish in an effort to provide specifiers, contractors and clients alike, with real examples of what can be achieved (or what is permissible) with TVC.

TVC relies on the temporary 'liquefaction' of the fresh mix, by vibration during the compaction process, to drive out entrapped air. However, this process is rarely entirely successful, and most vertical surfaces of hardened concrete have appearance problems which are directly related to the vibration process. These have been categorised in the UK by Monks[12].

The factors affecting surface finish, generally, and the appearance of blowholes in particular, are considered to include:

- mix design (including type and content of admixtures)
- temperature and the related setting time
- dimensions and shape of formwork
- concrete pressure on formwork
- nature of formwork surface (hardness, permeability, texture, wettability)
- type of release agent
- application of release agent
- method of placing
- rate of placing
- coarse aggregate grading
- fineness of sand
- type of addition (fine filler) if any
- uniformity of mixing
- workmanship

Mix design itself is complex. The complexity, and the many variables listed above, are interrelated to varying degrees. This makes systematic research into the production of a good surface finish very difficult, and explains the lack of reliable, integrated advice for practical applications.

In traditional concrete construction the vibration process itself affects the quality of the surface finish. With SCC, additional factors have to be taken into account, particularly the nature and effectiveness of the admixtures used, which critically affect the cohesion and segregation resistance of the concrete.

An investigation, which examined these factors for traditional concrete[8], was unable to pinpoint the reasons for the occurrence of blowholes even with the assistance of an analysis of variance of several of the factors being considered. It is instructive to appreciate that there is not even a general consensus as to whether the blowholes originate as air bubbles or as drops of water, or are linked to the release agent. Other research[10–12] looked specifically at blowholes and concluded that their total avoidance is currently beyond the state-of-the-art, and this continues to be a generally held view today, when most formwork used remains impervious.

7.6.3 Factors affecting surface finish in self-compacting concrete

It is important to repeat that surfaces much better than those of TVC have been produced using SCC, while acknowledging that the mere fact that a concrete is nominally self-compacting does not guarantee such high-quality finishes. Practical applications of SCC, where outstanding surface finishes were obtained were invariably based on systematic 'trial and error' development. Such trials usually examined not only different mix designs but casting techniques, placing processes and different types of moulds/formwork.

Considering the variability of the conditions during placing on a site, it is not surprising to find the best examples of excellent surface finish not on in situ but on precast concrete. Production of concrete for precasting is repeated time after time, batch after identical batch (see Figures 7.10–7.12). The high quality of surface finish achievable with SCC has been one of the key motivators for the precast concrete industry in its wide adoption of SCC technology. The introduction of SCC usually follows a period of experimentation, usually with local materials whose supply is known to be economical, consistent and of guaranteed availability. This period allows the establishment of the best mix design (or a set of mix designs) capable of producing the finishes for different strength grades of concrete. The batching, mixing and casting process are well controlled in precast concrete production, and existing handling/casting processes are easily modified to handle fresh SCC. It is therefore possible to minimise the variations in the production and placing process, and consistently and reliably achieve excellent surface finishes.

In contrast to the situation in precasting, the quality of surface finish obtained during in situ casting of SCC has sometimes been seen to vary significantly even on a single job with a single mix. The reasons for this variation are not entirely clear. There is sometimes no real understanding of why good (or less good) surface finishes are achieved with SCC, and even experienced practitioners may only be able to guess at the reasons: reasons which seem to apply to one job seem irrelevant on others. In principle, as mentioned in Section 7.6.1, SCC surfaces can be perfect, but success depends on the ability of the concrete to flow, retain enough filling ability during placing, and allow air to escape. The expulsion of entrapped air cannot be given a helping hand by mechanical vibration when SCC is placed.

Acceptance that the surface finish was not always perfect led to attempts to establish which factors contributed to a good finish, and which were responsible for a poor finish. In fact, much of what constitutes good practice for TVC was also found to apply to SCC. Some of the factors believed to influence the surface quality of SCC are discussed below – much of this is adapted from the European guidelines on SCC[13]. This represents the state-of-the-art of current knowledge and includes the views of experienced practitioners, but it is largely based on observation rather than on proven, systematic and fundamental research.

The concrete mix itself

The constituent materials and concrete mix design are of paramount importance. The fresh properties of the concrete, in particular the filling ability and flow-rate, must be suitable. The following aspects are important:

- Poor filling and/or passing ability, low flow-rate (high plastic viscosity) or their rapid loss can all contribute to the presence of excessive numbers of blowholes.

- Low viscosity and high water/powder ratio can encourage bleeding, and vertical scouring of the concrete surface adjacent to the form face as the water migrates upwards. The same factors can produce a weak layer of grout on the concrete surface, and consequent scaling. (It should be noted that scaling can also occur because of rapid drying and/or inadequate curing. This raises the point that many of these problems with appearance are not specific to SCC, and can also occur, for similar reasons, in TVC.)
- Very high flow-rate and associated high formwork pressure can lead to movement/displacement of formwork and surface irregularities.
- Admixtures may have an excessive retarding effect and lead to colour variations.
- In the production process, overlong mixing time is likely to introduce excessive amounts of air, which are difficult to lose in the subsequent placing process.
- If the materials/mix design are completely unsuitable, honeycombing can still occur, as in TVC. Insufficient filling ability, or a plastic viscosity too low to prevent segregation, will simply prevent the fresh concrete from filing the mould properly.

It should be emphasised that some of these problems are extreme cases. If the concrete is self-compacting at all, it is unlikely that these gross problems will be encountered.

Formwork

In any concrete, formwork materials have a principal part to play in achieving good surface finish, and this remains true of SCC. It is well recognised that permeable surfaces are better than impermeable ones, not least for the avoidance of blowholes. Some contractors choose to avoid steel formwork when surface finish is important, although experience with SCC in the precast industry has shown that perfect finishes can also be obtained with steel. In some difficult cases, permeable linings can assist, but their effectiveness may depend on the degree to which the movement of moisture will be restricted in a given, fresh SCC mix.

SCC has the advantage of making fewer demands on formwork, which may as a consequence be able to be reused more frequently. The elimination of formwork 'burns' from the use of poker vibrators also helps to obtain good finishes.

Preparation

Preparation of the mould is important, just as it is with TVC. Badly cleaned or heavily worn formwork exhibits greater surface roughness than one with new smooth surfaces, and this is likely to prevent the full escape of air bubbles. Because SCC faithfully reproduces the formwork surface, any defects in that surface will also be reproduced. An uneven formwork surface has been observed to contribute to the formation of blowholes. Because of the potential for formwork pressure to be high when fresh SCC is placed, formwork must be sufficiently strong and stiff and absolutely secure in position, to avoid any movement.

The choice of form release agents has been shown to be important: wax-based agents may be too viscous, particularly when used in conjunction with impermeable formwork. Vegetable-, mineral- or water-based agents are more likely to be successful, always provided that the application is properly carried out. Care with the application of the form release agent is also important. The application of too much release agent has been observed to contribute to the formation of blowholes, apparently because air bubbles adhere more strongly to a thick layer. Uneven application contributes to poor, uneven surfaces. Thus the application should be at the correct rate with proper equipment and a suitable spray nozzle, at the correct pressure.

Mixing

The effect of the mixing process itself on the properties of fresh SCC has already been mentioned. This also applies to the surface finish and reinforces the importance of maintaining a consistent batching procedure.

Temperature

The temperature of the fresh concrete and/or ambient temperatures can have an effect on the surface finish. This may be significant in itself, but its impact is often through one of a number of contributory factors. High temperatures can play a part in the formation of blowholes, and also in the formation and appearance of visible joint planes between different batches of concrete, particularly when there is any delay between deliveries – the temperatures (if above, say, 25°C) encourage rapid stiffening of the concrete and the formation of a 'crust', before the placing of the subsequent delivery.

In a similar way, by encouraging rapid stiffening, high temperatures may contribute to the problem of surface scaling. Conversely, low temperatures may result in colour differences on the surface of the concrete, or between different batches.

Casting

The casting process and technique are particularly important for the avoidance of blowholes. The contractor has to bear in mind that the casting procedure must be such as to allow air to escape freely. Some elements of this include:

- rapid casting rate hinders the escape of air
- conversely, if the casting rate is too slow, blowholes may still result because the necessary degree of constant agitation, to help air escape, is absent
- high lifts obstruct the escape of air
- permitting a large free-fall distance is likely to entrap more air and make it more difficult for all of it to escape
- a minimum flow distance is necessary to encourage the release of air by agitation but, equally, if that distance is too long, excessive blowholes may result
- discontinuous casting can lead to variations in colour
- intermittent deliveries of concrete also contribute to 'pour lines' between deliveries, or between lifts, which can become visible when the formwork is stripped. The best experience has been when casting is continuous, and these conspicuous lines have been avoided when pumping from the bottom of a form, or when using an immersed tremie, so that concrete is not placed on top of still-fresh earlier concrete

There may be a number of contributory factors to many of the issues discussed here. In the case of pour lines, for example, these may include high temperature, settlement (segregation) of coarse aggregate and thixotropy preventing the intimate mingling of concrete from successive deliveries.

It is difficult to discuss surface finish without seeming to concentrate unduly on the problems which may arise. It is therefore worth emphasising, that the highest quality finishes can be achieved with SCC, as experience in the precast industry, in particular, is increasingly demonstrating. That experience also shows that some of the problems of surface finish, which are common to TVC and SCC, can actually be more easily avoided with SCC. (A long detailed tabular presentation on improving the surface finish of SCC is given in Annex C of the European guidelines. Much of the information in this section is based on that document.)

7.6.4 Special case of large slabs and pavements

Two matters arise related to the restriction of movement of moisture in typical fresh SCC. First, this may, in some cases, cause significant difficulties in finishing slabs and other elements with large horizontal surfaces. The absence of water migrating to the upper surface, which normally generates a very thin 'water-rich' layer and 'lubricates' the finishing action, causes problems particularly in the case of mechanical and manual finishing of large slabs and pavements. A freshly placed pavement mix based on cohesive SCC, which flows slowly and retains mix water, may develop an 'elephant skin', a crust of stiffer concrete on its surface. This may be difficult to finish using normal 'power-floating' techniques. The mix design of SCC for such purpose has to be adjusted to ensure that modern finishing methods can be applied reliably. Secondly, attention should also be paid to the risk of plastic cracking. Briefly, this is caused by loss of water from the surface by evaporation, and consequent shrinkage of the concrete mass. In TVC the risk of cracking is somewhat reduced by the replacement of the water lost by bleed water from below. With SCC this water movement by bleeding is much reduced, and it may be inferred that additional precautions should be taken against the risk of this type of early cracking.

7.7 Conclusions

Excellent surface finishes of SCC are clearly possible (Figures 7.8–7.12), but equally clearly, such a high quality of surface is not a universal property of SCCs. There is still insufficient knowledge about the formation of the best quality surfaces of hardened SCC, to allow a reliable guide to how it can be reliably obtained. Neither mechanisms in TVC which produce good finishes are fully understood, nor do we yet fully understand those occurring when SCC is placed.

Examples of outstanding finishes achieved when SCC was used may lead to an erroneous conclusion that any SCC will invariable produce a good to excellent surface finish. SCC covers an extremely wide spectrum of concretes. It is therefore impossible to expect a good surface finish to be obtained from any hardened SCC regardless of its composition, production and placing. It is, however, possible to conclude that 'ordinary' SCCs tend to be based on mix designs (additives/fines, admixtures), which can produce surface finishes of quality comparable to or better than that achieved with good TVC, provided appropriate production methods are adopted.

SCC generally produces better surface finishes than normal concrete, and the expectations are high. In reality, however, it does not solve all problems; nor is it uniformly perfect, and nor, at present, is it reasonable to expect it to be so. The mechanisms in ordinary concrete which produce good finishes are not fully understood, and we do not yet fully understand those in SCC. It is clear that given time and experience extremely good finishes can be consistently produced (e.g. in precast applications, where mixes are repeated time after time, and experience can be gained). Used in situ, there is no quick fix for the production of good finishes, but given good practice, attention to detail and engineering experience, SCC can undoubtedly improve the standards of finish observed in our concrete structures.

References

[1] Bartos P.J.M. and Cleland D.J. (Eds.) *Special Concrete: Workability and Mixing*, E&FN Spon, London, UK, 1993.

[2] Grauers M. *et al.* Rational production and improved working environment through using self-compacting concrete. EC Brite-EuRam Contract No. BRPR-CT96-0366, Project BE96-3801, 1997–2000.

[3] The Norwegian Concrete Association: Guidelines for production and use of self-compacting concrete. Publication 29, 2002.

[4] Bartos P.J.M., Gibbs J.C. *et al.* Testing SCC, EC Growth contract GRD2-2000-30024, 2001–2004, http://www.civeng.ucl.ac.uk/research/ concrete/Testing-SCC.

[5] Leeman A. and Hoffmann C. Pressure of self-compacting concrete on the formwork. In: Wallevik O. and Nielsson I. (Eds.) *Self-compacting Concrete*, RILEM Publications, Bagneux, France, 2003, pp 288–297.

[6] Billberg P. Mechanisms behind reduced form pressure when casting with SCC. In: Yu Z., Shi C., Khayat KH. and Xie Y. (Eds.) *SCC2005-China*, RILEM Publications, Bagneux, France, 2005, pp 589–597.

[7] Murphy W.E. The influence of concrete mix proportions and type of form face on the appearance of concrete. Cement and Concrete Association, Wexham Springs, UK, Technical Report 384, May 1967.

[8] Wilson D. Controlled permeability formwork. *Concrete* 2000, 34(6), 20–21.

[9] Reading T.J. The bughole problem, *ACI Journal,* 1972 March, 165–171.

[10] Blake L.S. Recommendations for the production of high quality concrete surfaces. Cement and Concrete Association, Wexham Springs, UK, 1967.

[11] Kinnear R.G. Concrete surface blemishes. Cement and Concrete Association, Wexham Springs, Technical Report 380, 1964.

[12] Monks W. The Control of Blemishes in Concrete. Cement and Concrete Association, Wexham Springs, UK Publication 47–103, 1981.

[13] BIBM, CEMBUREAU, EFCA, EFNARC and ERMCO, The European Guidelines for self-compacting concrete, Joint Project Group, EFNARC, Brussels 2005. www.efnarc.org

Chapter 8
Hydration and microstructure

8.1 Introduction

As we have seen in Chapter 6, the introduction of SCC resulted in significant changes to mix compositions and in the utilisation of the constituent materials. To obtain the very specific properties of a SCC, fillers and chemical admixtures generally comprise a significant part of the composition. This has, of course, an important influence on the fundamental behaviour and properties of the concrete.

At the onset of the application and examination of SCC the question arose whether the microstructure of SCC and TVC were different, and to what extent. In the first part of this chapter a theoretical overview is given on the development of the microstructure of traditional cement paste and concrete. The calculation of the porosity according to Powers' model is described, as this parameter will be used within the study of transport mechanisms and durability. Based on the difference in composition between SCC and TVC, the influencing factors of the hydration process and the microstructure are discussed. Literature results with respect to the hydration process and the microstructure of SCC are then highlighted.

8.2 Development of the microstructure of concrete

8.2.1 Hydration process of Portland cement

Portland cement

On a centimetre scale concrete can be considered as a composite material consisting of mortar and coarse aggregates. On a millimetre scale, mortar consists of a composition of cement paste and sand. On a micrometre scale, the microstructure of hardened cement paste is composed of unreacted cement, calcium silicate hydrates, calcium hydroxide, capillary pores and other chemical phases. To fully understand the microstructure of concrete, it is better to start at the level of the hardened cement paste. Therefore we will first consider the hydration process of Portland cement[1, 2]. Portland cement consists of angular particles with dimensions of 1–50 μm, and is produced by crushing clinker and a small amount of calcium sulfate ($CaSO_4$). The clinker is composed of several materials which are formed by reactions at a high temperature between calcium oxide (CaO, or C in short notation) and silicium dioxide (SiO_2 or S), aluminium oxide (Al_2O_3 or A) and iron oxide (Fe_2O_3 or F). The chemical composition of the clinker minerals or the components of Portland cement consists mainly of calcium silicates (C_3S and C_2S), calcium aluminate (C_3A) and calcium alumino ferrite (C_4AF). The chemical composition of cement is mostly described by means of oxide contents derived from chemical analysis. Based on Bogue's formula

Table 8.1 *Oxides and components of Portland cement.*

Oxide	*Abbreviation oxide*	*Component*	*Abbreviation of the component*
CaO	C	$3CaO.SiO_2$	C_3S
SiO_2	S	$2CaO.SiO_2$	C_2S
Al_2O_3	A	$3CaO.Al_2O_3$	C_3A
Fe_2O_3	F	$4CaO.Al_2O_3.Fe_2O_3$	C_4AF
MgO	M	$4CaO.3Al_2O_3.SO_3$	$C_4A_3\bar{S}$
SO_3	$\bar{S}$	$3CaO.2SiO_2.3H_2O$	$C_3S_2H_3$ (*)
H_2O	H	$CaSO_4.2H_2O$	$C\bar{S}H_2$

* Only in the case of complete hydration

it is possible to calculate the different components of cement. Table 8.1 gives an overview of the most important oxides and components of Portland cement, together with the common abbreviations used in cement chemistry. The mean values of the most important oxides of Portland cement are, respectively, 65% for C, 22% for S, 5% for A and 3% for F.

Hydration of the clinker minerals

The hydration reactions of the four most important components of Portland cement in contact with water are described in Equations (8.1)–(8.6) ((*) indicates the notations common in cement chemistry). The clinker minerals C_3S and C_2S are not stable when exposed to water and react according to Equations (8.1) and (8.2). The reaction products are calcium silicate hydrates (CSH) and calcium hydroxide (CH). The accurate chemical composition of the calcium silicate hydrates depends on the water/cement ratio, the temperature and the hydration degree. The ultimate reaction products CSH consist of a porous solid substance with gel-like properties. Their total volume is 50–60% of the total volume of solid substances in the completely hydrated Portland cement. As such they are the most important components of the hydrated cement paste. They are crystalline in a very limited way and could be observed as small fibrous crystals. The exact structure of CSH is unknown. Only at full hydration can the composition be approached by $C_3S_2H_3$. During the hydration of C_3S much of the CH develops as hexagonal prismatic crystals. They represent 20–25% of the total volume of solid substances in the cement matrix. Due to the presence of CH the pore solution of hardened concrete exhibits a strong alkaline character (pH about 13). The reaction product CH is easily soluble and favours the transformation into other compounds.

Complete hydration of C_3S*:*

$$2\,(3CaO.SiO_2) + 6H_2O \rightarrow 3CaO.2SiO_2.3H_2O + 3Ca(OH)_2 \qquad (8.1)$$

$$2C_3S + 6H \rightarrow C_3S_2H_3 + 3CH \qquad (*)$$

Complete hydration of C_2S:

$$2(2CaO.SiO_2) + 4H_2O \rightarrow 3CaO.2SiO_2.3H_2O + Ca(OH)_2 \qquad (8.2)$$

$$2\,C_2S + 4H \rightarrow C_3S_2H_3 + CH \qquad (*)$$

Due to the hydration of C_3S, 61% $C_3S_2H_3$ and 39% CH is formed, whereas due to the hydration of C_2S, 82% $C_3S_2H_3$ and 18% CH is formed (mass percentages).

In contact with water the clinker mineral C_3A binds immediately causing an important heat development and ultimately a lower strength. The reaction process is slowed down by the addition of gypsum which diminishes the negative effects. When $CaSO_4$ is present as gypsum reaction (Equation (8.3)) occurs. In the first stage ettringite (crystallising as short prismatic needles) is developed which contributes to the first strength development.

In absence of gypsum reaction (Equation (8.4)) occurs and C_3AH_6 (hydrogarnet) is formed which further reacts with the earlier formed ettringite from reaction (Equation (8.3)). From this reaction monosulfate ($C_3A.CaSO_4.H_{12}$) is formed as thin hexagonal plates (Equation (8.5)).

Hydration of C_3A (when gypsum is available):

$$3CaO.Al_2O_3 + 3CaSO_4 + 32H_2O \rightarrow 3CaO.Al_2O_3.\,3CaSO_4.32\,H_2O \qquad (8.3)$$

$$C_3A + 3\,C\bar{S}H_2 + 26H \rightarrow C_6A\bar{S}_3H_{32} \qquad (*)$$

Hydration of C_3A (in the absence of gypsum):

$$3\,CaO.Al_2O_3 + 6\,H_2O \rightarrow 3CaO.\,Al_2O_3\,.\,6H_2O \qquad (8.4)$$

$$C_3A + 6H \rightarrow C_3AH_6 \qquad (*)$$

Formation of monosulfate:

$$C_6A\bar{S}_3H_{32} + 2\,C_3A + 4H \rightarrow 3\,C_4A\bar{S}H_{12} \qquad (*)\ (8.5)$$

The clinker mineral C_4AF is said to react with water and part of the CH developed from the hydration of C_3S and C_2S. Two compounds are formed (C_3AH_6 and C_3FH_6) from reaction (8.6).

Hydration of C_4AF:

$$4CaO.Al_2O_3.Fe_2O_3 + 2Ca(OH)_2 + 10H_2O \rightarrow 3CaO.\,Al_2O_3.6H_2O + 3CaO.Fe_2\,O_3.6H_2O \qquad (8.6)$$

$$C_4AF + 2CH + 10H \rightarrow C_3AH_6 + C_3FH_6 \quad (*)$$

Figure 8.1 gives two detailed images of the microstructure of hardened cement paste. Both images have been obtained by means of electron microscopy (images from the National Institute of Standards and Technology)[3].

In between the reaction products from the earlier reactions some additional reaction products can occur. These are, however, less important. The reactivity of the hydration reactions not only depends on the content of the different compounds but also on the fineness of the cement. A higher fineness accelerates the reactions. The fineness can be determined by means of the Blaine permeameter (EN 196-6 (1991)).

Physical changes of the cement paste

The hydration process can be divided into three different stages[4], the induction period, the setting and the hardening. Within the first hour after the addition of the mixing water, during the induction period, the hydration products on the cement grains delay the hydration process to some extent. During this period the cement paste is still workable and the paste is said to be fluid or plastic. The setting starts at the age of about 1–3 h. The hydration products formed in the first stage are broken down and new hydration products are formed on the revealed cement grains. The hydration products are long and fibrous. After a while the cement grains become connected and there is a stiffening of the cement paste. It is then possible to develop some stiffness and strength, and the hardening process starts. After a time the reactivity decreases and then almost stops. Figure 8.2 shows the different phases of the hydration process as a function of the hydration degree and time.

At any time after the start of the hydration process it is possible to calculate the hydration degree α ($0 \leq \alpha \leq 1$) as the ratio of the amount of hydrated cement to the initial amount of cement. As it is not easy to determine the amount of hydrated cement, an approximation is made based on the heat development[5–7]. Complete hydration will never occur. For this reason the ultimate hydration degree is defined as the ratio of the maximum hydrated cement amount to the initial amount of cement.

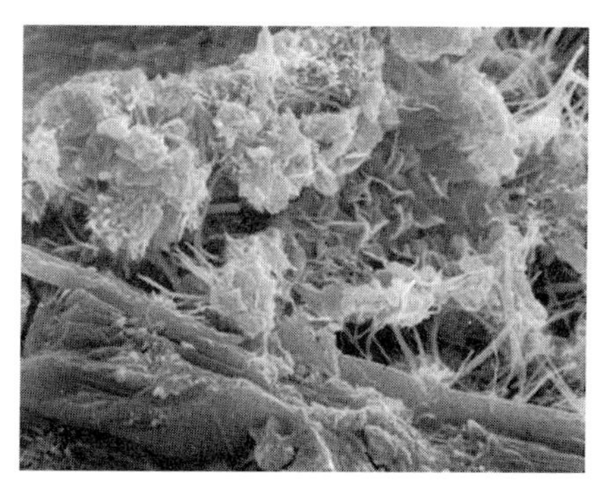

Figure 8.1 *Microstructure of hardened cement paste, images from electron microscopy*[3].

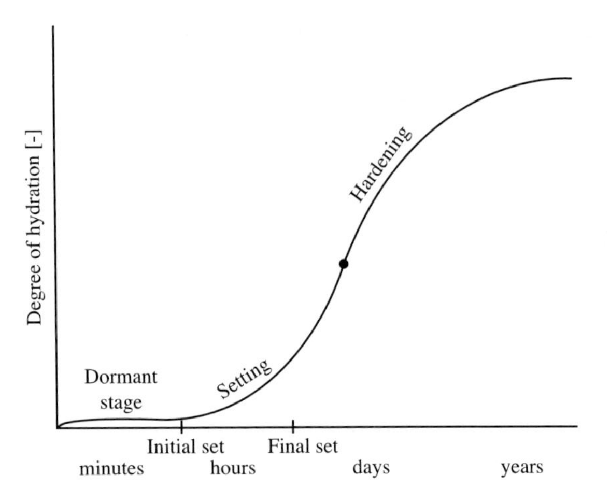

Figure 8.2 *Induction period, setting and hardening.*

The changes of the microstructure during the hydration process are shown schematically in Figure 8.3. Some images obtained by electron microscopy are shown[3] in Figure 8.4. Several phases can be distinguished: hydration products (light grey), unhydrated cement grains (white) and capillary pores (black).

Heat development

The hydration of cement is an exothermic reaction. The instantaneous heat production rate q (J/gh) is measured under isothermal circumstances as a function of time, and the cumulated heat of hydration Q (J/g) can then be calculated by integrating the heat production rate q over time. The total heat of hydration released after complete hydration of the cement is called Q_{max} (J/g). More detailed information on the heat development can be found in the literature[5–7].

8.2.2 Microstructure and pore structure of hardened cement paste

General

In a simplified way, the solid components of hardened cement paste consist mainly of three phases: nonhydrated cement, surface products such as CSH which are

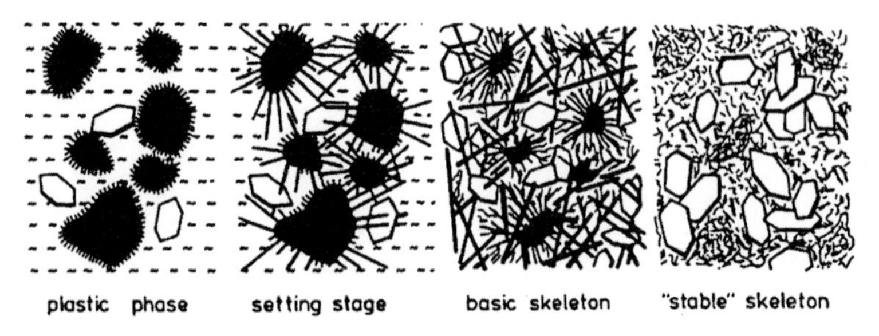

Figure 8.3 *Schematic visualisation of the hydration process*[4, 8].

Figure 8.4 *Images of the hardened cement paste obtained by electron microscopy*[3].

amorphous and nanoporous, and pore products such as CH which form crystals[9]. Besides the solid components, voids or pores are also present in the hardened cement paste. These pores are empty spaces, different from cracks which are present in the cement matrix. The pores are important for the ultimate properties of the cement paste, mortar and concrete[10]. The water which can be present in the voids also influences the properties. An overview of the typical dimensions of the solid components and the voids of hardened cement paste is given in Figure 8.5[1].

Voids in hydrated cement paste

Gel pores

The spaces in between the hydration products (CSH) are known as the gel pores. According to Powers the mean free distance in between the CSH structure is about 1.8 nm, and comprises 28% of the CSH structure[1]. According to Feldman and Sereda the free distance is about 0.5–2.5 nm[1, 12], while Neville[11] mentions a distance of 3 nm. These dimensions are said to be too small to have any influence on the strength and permeability of the hardened cement paste. The water in the gel pores, however, matters with respect to creep and shrinkage. The total volume of gel pores increases with an increasing degree of hydration. In normal conditions the gel pores are filled with water which makes them almost impermeable to fluids and gases[13].

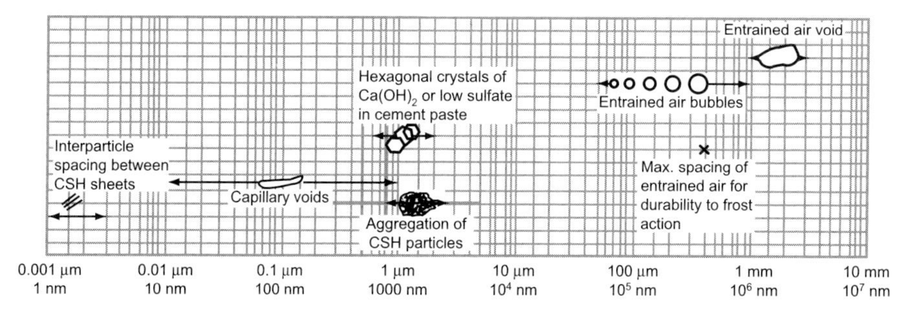

Figure 8.5 *Typical dimensions of the different phases in hardened cement paste*[1].

Capillary pores

During hydration, the space initially taken by water and cement is replaced by hydration products with interlayer water (see below). Capillary voids are created because the solid hydration products occupy less space than the initial volume of water and cement. The capillary voids together with the remaining capillary water are called the capillary pores. Their volume and dimensions depend on the cement content, the w/c ratio and the hydration degree. They have an irregular shape. In well-hydrated cement pastes with a low w/c ratio, the dimensions vary between 10 and 50 nm. For young cement pastes with a high w/c ratio the dimensions can vary in the range of 3–50 μm. A higher w/c ratio increases the amount and the dimensions of the capillary pores. These pores, and their connectivity, strongly influence the permeability of the hardened cement paste. Capillary pores larger than 50 nm, macropores, are said to be detrimental with regard to strength and permeability. The micropores, capillary pores smaller than 50 nm, rather influence the shrinkage and creep due to drying. These voids are the smallest voids which are still optically visible.

Air voids

In contrary to the capillary pores, air voids are mainly spherical. Entrained and entrapped air voids are much larger than the capillary pores and their dimensions in the range 10–200 μm. Entrapped voids can influence the strength and 'permeability' of the material. Entrained voids only influence the strength.

Water in hydrated cement paste

Depending on the ambient conditions (temperature, relative humidity) and the porosity hardened cement paste can hold a certain amount of water within the structure.

Capillary or free water

The capillary or free water is situated in the voids larger than 5 nm (Figure 8.6). This water is available for the ongoing hydration of cement. Removal of the free water causes no change in volume. However, removal of the water in the small capillaries, which is under capillary tension, can induce shrinkage. The capillary water can be determined by drying the material at 105°C.

Adsorbed water

The water molecules are physically adsorbed to the surface of the hydration products (Figure 8.6). The water layer can consist of about six layers of water molecules. In[14] the distance between two adhering surfaces of hydration products is approximately 150–300 nm. Part of this water can disappear when the concrete is dried, causing shrinkage.

Interlayer water

This kind of water is associated with the CSH structure and consists of a monomolecular water film entrapped between the crystallised products (Figure 8.6). This layer will only disappear under strong desiccation, thus causing shrinkage. The

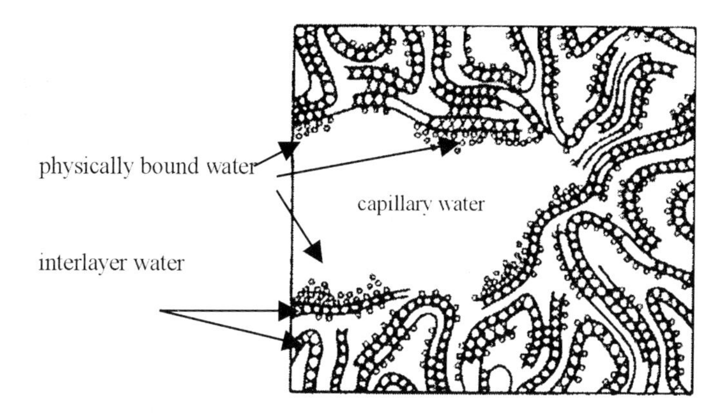

Figure 8.6 *Types of water associated with CSH*[12].

adsorbed water and the interlayer water are located in the gel pores and are referred to as gel water or physically bound water.

Chemically bound water
This water is part of the hydration products and will not disappear at a drying temperature lower than 105°C. From 110°C upwards, the hydration products will decompose and the water will leave the structure. In order to find the total amount of chemically bound water, concrete has to be heated up to a temperature of 1000°C.

Evaporable and nonevaporable water
Depending on whether or not the water in the hardened cement paste is evaporable, two groups can be defined. The evaporable water consists of the capillary water and part of the adsorbed water. The remaining water is considered to be nonevaporable.

8.2.3 Microstructure and pore structure of concrete

The porosity of concrete depends on the hardened cement paste and the transition zone in between the macrocomponents of concrete (aggregates, reinforcement and fibres) and the hardened cement paste. In the interfacial transition zone (ITZ), between the coarse aggregates and the cement matrix, a different structure is noticed compared to the bulk cement paste. The thickness of the ITZ is about 30–50 μm, and can be 30–50% of the total volume of the cement paste in the case of traditional concrete. This zone can be very important for the ultimate properties of concrete. This is due to the higher porosity, the higher amount of CH and the higher vulnerability for cracks in the ITZ. In most cases the porosity of the aggregates is not an important factor.

With respect to freeze and thaw it is important to characterise the air void system of the concrete. The following three types of air voids are the most important ones[15]:

- Entrained air voids: These spherical voids are bigger than the capillary pores and are developed by the addition of an air-entraining admixture during the mixing process. When properly distributed and present in a suitable amount, they contribute to the frost resistance of the concrete.

- Entrapped air voids: These voids are larger than the entrained voids and their shape can be spherical as well as irregular. Their inner surface is smooth, sometimes slightly shiny indicating that the voids were formed by air bubbles or air pockets. As an indication one could separate entrapped and entrained air voids at a dimension of 1 mm. This threshold is only indicative.
- Voids formed by water: Their shape, location and internal surface indicate the original presence of water. They can be caused by bleeding water not being able to rise towards the surface. Their shape is irregular and the inner surface is rough. In most cases they are also larger than the entrapped air voids.

The entrapped air voids and voids formed by water do not contribute to the good properties of concrete, they merely weaken it. The development of these voids is partly dependent on the vibration system.

IUPAC (International Union of Pure and Applied Chemists) has proposed a specific nomenclature for porous materials for the definition of voids: microporous (pore size diameter < 2 nm), mesoporous (2 nm < pore size diameter < 50 nm) or macroporous (pore size diameter > 50 nm). These thresholds have been accepted by several specialists[16]. The division, however, is not based on the physical or chemical properties of the material. Consequently, in the case of hardened cement paste and concrete, alternative classifications have been proposed by different authors. Table 8.2 shows the classification proposed by Mindess and Young which can be found elsewhere[17]. The properties influenced by the voids are also mentioned. Figure 8.5 shows the classification proposed by Mehta and Monteiro[1].

8.3 Calculation of the porosity

8.3.1 Powers' model

In 1948 Powers and Brownyard proposed an empirical model for the distribution of the several phases in the hardening cement paste. The model is mainly based on an

Table 8.2 *Classification of the voids*[17].

Voids	*Diameter*	*Description*	*Properties influenced*
Gel pores	< 0.5 nm 0.5–2.5 nm	'Interlayer' micropores Micropores	Shrinkage and creep Shrinkage and creep
Gel pores and capillary pores	2.5–10 nm	Small capillaries	Shrinkage between 50% and 80% R.H.
Capillary pores	10–50 nm 0.05–10 μm	Medium capillaries Large capillaries	Strength, permeability, Shrinkage at high R.H. Strength, permeability
Air voids	0.005–1.0 mm 0.010 mm – large 0.25–2 mm	Entrained air voids Entrapped air voids Foamed structures	Strength Strength Strength and density

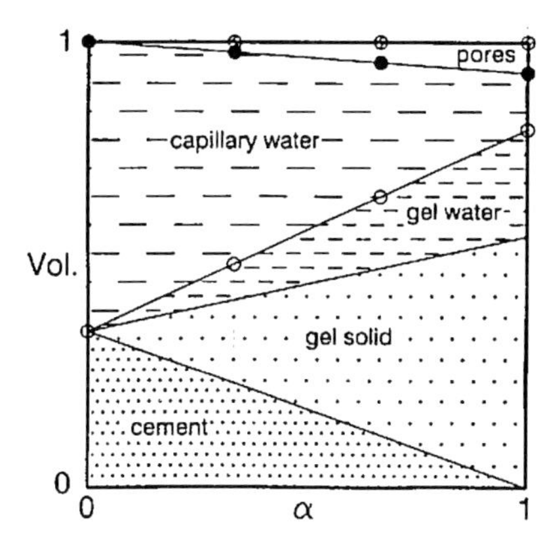

Figure 8.7 *Phase distribution of a hardening cement paste (w/c = 0.6) as a function of the degree of hydration*[20].

extended study of water vapour adsorption isotherms and chemically bound water in hardening cement pastes. In the following years the model was adjusted by Powers. This model is still suitable to determine the volumetric composition of the hardening cement paste. Figure 8.7 gives the volumetric phase distribution of fresh hardening cement paste, with a w/c ratio of 0.6, as a function of the hydration degree α. The assumption is made that there is no external exchange of water and that there is enough water available for the hydration process. No air is assumed to be in the system. In a partially hydrated system ($0 < \alpha \leq 1$) the following phases can be distinguished: nonhydrated cement (V_{nhc}), solid hydration products (V_{shp}), gel water (V_{gw}), capillary or free water (V_{cw}) and the created capillary voids (V_{cv}). In the following the volumes of the different phases and the porosities are expressed as a function of the initial mass of cement C and water W, and the densities of cement ρ_{c} and water ρ_{w}.

8.3.2 Phase distribution in hydrating cement paste

The capillary voids are created because the solid hydration products need less space than the initial cement and water. At complete hydration and when the exact amount of water needed for hydration is added, the volume of the free space is 18.5% of the initial volume of cement. For an intermediate state of the hydration process, a corresponding volume is obtained by taking into account the degree of hydration α (Equation (8.7)). The capillary voids can be either empty or filled with water depending on the amount of mixing water and the external available source of water.

$$V_{\mathrm{cv}} = \frac{0.185\alpha C}{\rho_{\mathrm{c}}} \tag{8.7}$$

The experiments by Powers indicated that approximately 0.23 g water is chemically bound at the hydration of 1 g of cement. The volume of solid hydration products is determined to be the sum of the hydrated cement and chemically bound water minus the volume of capillary voids, and can be calculated by means of Equation (8.8).

$$V_{\text{shp}} = \alpha C\left(\frac{0.815}{\rho_{\text{c}}} + \frac{0.23}{\rho_{\text{w}}}\right) \tag{8.8}$$

The volume of gel water is 28% of the volume of cement gel which consists of the total of solid hydration products and gel water.

$$V_{\text{gw}} = \alpha C\left(\frac{0.317}{\rho_{\text{c}}} + \frac{0.089}{\rho_{\text{w}}}\right) \tag{8.9}$$

The volume of nonhydrated cement can be calculated by means of Equation (8.10).

$$V_{\text{nhc}} = \frac{(1-\alpha)C}{\rho_{\text{c}}} \tag{8.10}$$

The volume of free water can be determined by subtracting the amount of bound water and gel water from the amount of mixing water.

$$V_{\text{cw}} = \frac{W - 0.319\alpha C}{\rho_{\text{w}}} - \frac{0.317\alpha C}{\rho_{\text{c}}} \tag{8.11}$$

The volume of cement gel can be calculated from the combination of Equations (8.8) and (8.9)

$$V_{\text{cement gel}} = \alpha C\left(\frac{1.132}{\rho_{\text{c}}} + \frac{0.319}{\rho_{\text{w}}}\right) \tag{8.12}$$

8.3.3 Capillary pores and gel pores in hydrated cement paste

The capillary pores consist of capillary water and the capillary voids. The volume of capillary pores can be calculated by Equation (8.13).

$$V_{\text{cap pores}} = \frac{W - 0.319\alpha C}{\rho_{\text{w}}} - \frac{0.132\alpha C}{\rho_{\text{c}}} \tag{8.13}$$

The volume of gel pores corresponds to the volume of gel water (Equation (8.9)). It follows that the volume of total porosity $V_{\text{tot pores}}$ can be calculated by means of Equation (8.14).

$$V_{\text{tot pores}} = \alpha C\left(\frac{0.185}{\rho_{\text{c}}} + \frac{\frac{W}{\alpha C} - 0.23}{\rho_{\text{w}}}\right) \tag{8.14}$$

In order to calculate the capillary, gel and total porosity the volumes $V_{\text{cap pores}}$, V_{gw} and $V_{\text{tot pores}}$ have to be divided by the sum of the intial volume of cement and water, V_{tot}.

$$V_{tot} = \frac{W}{\rho_w} + \frac{C}{\rho_c} \tag{8.15}$$

8.3.4 Porosity in concrete

From the previous considerations the porosity of concrete (gel, capillary and total) can be calculated as the relation of the pore volume to the total volume of concrete $V_{concrete}$ (Equations (8.16) and (8.17)). In order to calculate the total volume of concrete, the entrapped air V_{air}, has to be taken into account. The total porosity comprises the capillary pores and the gel pores.

$$V_{concrete} = V_{cement} + V_{water} + V_{coarse\ aggregates} + V_{sand} + V_{fillers} + V_{air} \tag{8.16}$$

$$V_{concrete} = \frac{C}{\rho_c} + \frac{W}{\rho_w} + \frac{G}{\rho_g} + \frac{S}{\rho_s} + \frac{F}{\rho_f} + V_{air} \tag{8.17}$$

Where *C, W, G, S* and *F* are the initial mass of cement, water, coarse aggregates, sand and filler, respectively. The masses are divided by the corresponding densities ρ_c, ρ_w, ρ_g, ρ_s and ρ_f to determine the volumes of the raw materials. It should be noted that the entire pore structure is not accessible from the outside. A distinction has to be made between open and closed porosity. Furthermore, it should be noted that the ITZs are not considered in the calculations according to Equation (8.17).

8.3.5 Interfacial transition zone

The influence of the ITZ on the porosity should also be mentioned. In the case of traditional concrete in this zone there is a higher porosity corresponding to a higher w/c ratio compared to the bulk cement matrix. Often there are also more cracks in the ITZ. Research seems to indicate that a thin layer of hydration products, with a thickness about 1 μm, is formed on the surface of the aggregates. Normally there is no reaction between aggregates and hydration products and the aggregates only act as nucleation sites. In addition to this layer there is a layer with a higher porosity, the transition zone. This zone is created due to the lack of dense packing near the aggregates, and due to the one-sided growth of the hydration products. Rather small cement grains and hollow hydration products, Hadley grains, are noticed in the ITZ. During the mixing procedure the aggregates have a certain water demand and a water film is formed around the aggregates. This also contributes to a more porous structure in the transition zone between cement matrix and aggregates[21]. From quantitative studies it can be assumed that the width of the transition zone is about 30–50 μm. The determining factor for the microstructure of the transition zone is the possibility of the cement grains settling around the aggregates[22, 23]. According to Diamond and Huang[24] the microstructure of the interfacial transition does not differ much from the microstructure of the cement matrix. The enhanced porosity can be attributed to bleeding around the aggregates, which can be avoided by an appropriate mixing of the raw materials. Microbleeding around the aggregates can also be attributed to vibration and compaction. The role of vibration on the ITZ has also been confirmed by Leeman *et al.*[25]. Due to the higher porosity there is more space for large amounts of $Ca(OH)_2$ to be formed[26].

8.4 Hydration of self-compacting concrete

8.4.1 General aspects

In strictly chemical terms hydration is a reaction of an anhydrous compound with water, yielding a new compound, namely a hydrate. In cement chemistry, hydration is understood to be the reaction of a nonhydrated cement or one of its constituents with water, associated with both chemical and physicomechanical changes of the system, in particular with setting and hardening[27].

The progress of the hydration of cement and its kinetics can be considered to be influenced by a variety of factors. In the main they are, in random order:

- the phase composition of the cement
- the fineness of the cement
- the w/c ratio
- the curing temperature

but they also include:

- the presence of foreign ions
- the presence of chemical admixtures
- the presence of additions, i.e. materials interground with cement in larger amounts, such as mineral fillers used to obtain self-compactability

This means that the fillers and chemical admixtures, which are added to the mix composition of SCC to realise the very specific properties, might have an influence on the behaviour and more specifically on the hydration of the hardening concrete.

In the following paragraphs a summary is made of test results of research projects looking into the influence of the addition of different fillers to the mix composition, and in particular:

- limestone powder
- fly ash
- blast furnace slag

As we have seen in Chapter 6, most SCCs contain significant quantities of one or more of these materials.

8.4.2 Limestone powder

Heat of hydration: experimental results

When using limestone powder (also known as ground calcium carbonate, as it mainly consists of calcium carbonate) to realise SCC, it is noticed that the hydration reactions are accelerated and sometimes even altered. This was experimentally verified by means of isothermal conduction calorimetry[28–33]. A few of the findings are illustrated in Figures 8.8 and 8.9[30]. The cement/powder (c/p) ratio as indicated in the legend of these figures is the ratio, where the powder is the sum of the cement and filler content in the paste. The applied cement was a Portland cement with a C_3A content of 8.2%.

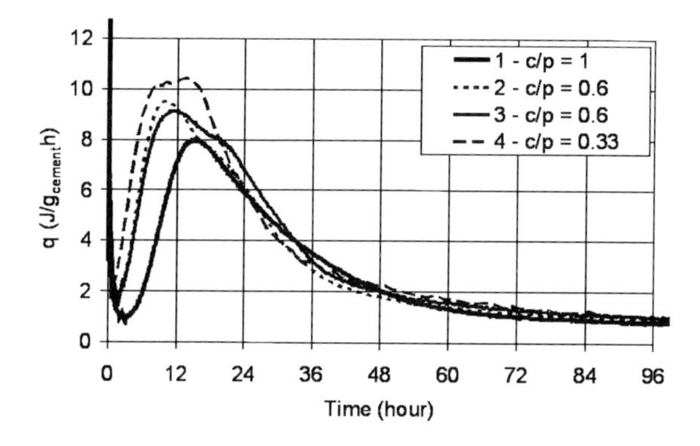

Figure 8.8 *Heat production rate, q, as a function of time during isothermal tests at 20°C* [30].

These figures clearly show that three phenomena might occur:

- the induction period is significantly shortened
- the reactions in the postinduction period are accelerated
- a third peak appears during the heat release which is more prominent than that which often occurs during the hydration of pure cement

These phenomena have been confirmed by means of adiabatic hydration tests on SCC[30]. The shortening of the induction period when limestone filler is added to the cement paste can also be confirmed by Figures 8.10 and 8.11 showing environmental scanning electron microscopy (ESEM) results of C_3S hydration, with and without limestone powder, at the same time after the mixing of the different components.

Based on these test results and literature data on the hydration of blended cements some hypotheses can be formulated which might explain these findings.

Shortening of the induction period and acceleration of the reactions

Looking into the literature on TVC, several theories concerning the reasons for the drastic slowing down of the hydration rate during the induction period have been put forward, for instance[27, 34, 35]:

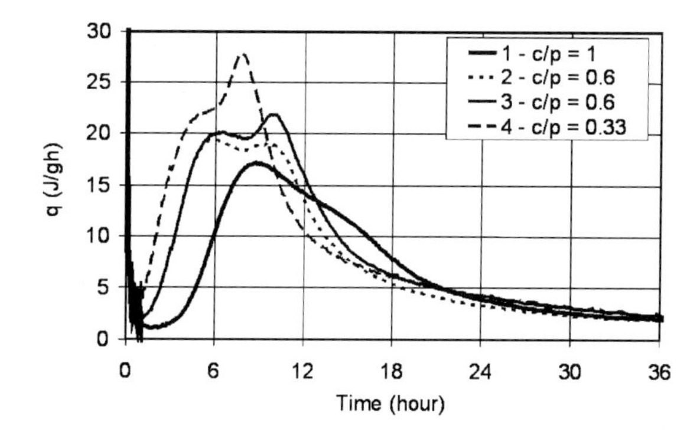

Figure 8.9 *Heat production rate, q, as a function of time during isothermal tests at 35°C*[30].

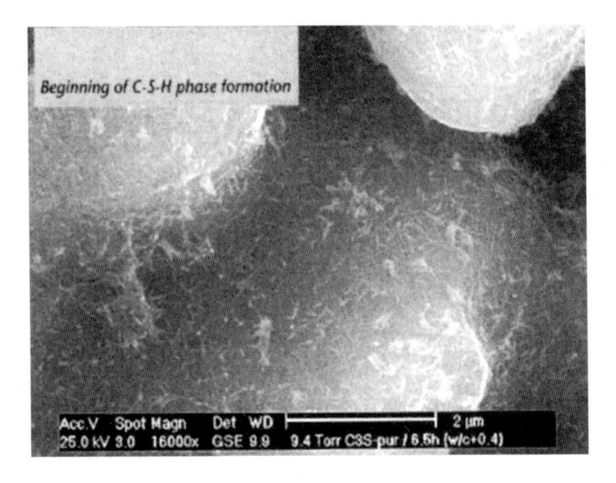

Figure 8.10 *Hydration of C_3S without limestone powder (ESEM). (Reproduced with permission of OMYA, Oftringen, Switzerland.)*

- The impermeable hydrate layer theory, which postulates that the beginning of the induction period is caused by the first-stage hydration products acting as a barrier to further reactions. Towards the end of the induction period the initially formed layer of hydration products undergoes changes that make it more permeable, thus making a renewed, fast hydration possible. There are different theories for the reasons for these changes in the first layer product. The aging of the product is the most widely accepted theory.
- The nucleation of CH (or, in full notation $Ca(OH)_2$) theory, which postulates that the CH formed in the earliest stage of the hydration cannot precipitate in a solid form, even after its concentration in the liquid phase has reached and

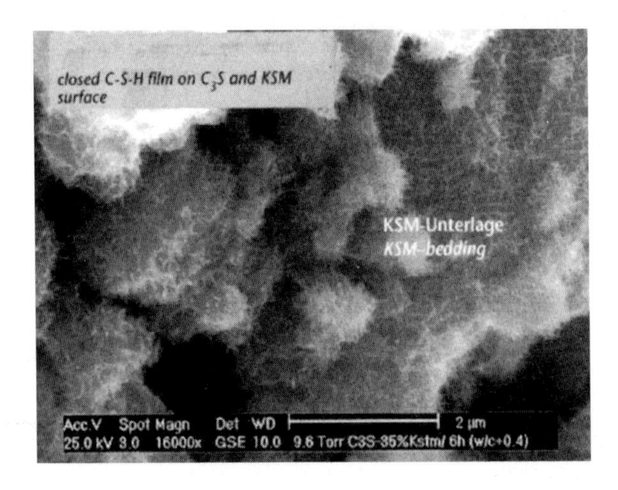

Figure 8.11 *Hydration of C_3S with limestone powder (ESEM). (Reproduced with permission of OMYA, Oftringen, Switzerland.)*

exceeded the saturation level, which slows down the reaction mechanism. It is assumed that this is due to the absorption of silicate ions on the surface of the CH nuclei. Eventually, the concentration of CH becomes high enough to overcome the effect of the silicate ions, and the hydration of C_3S is renewed and intensified.

- The nucleation of CSH theory states that the end of the induction period and the acceleration of the hydration are determined by the nucleation and the growth of 'second-stage' CSH which is different from the 'first-stage' product which was formed initially. The formation rate of the first-stage product is controlled by the CH concentration in the liquid phase and is slowed down as this concentration reaches its saturation level. The 'second-stage' product starts to form after the thermodynamic barrier to its nucleation has been overcome.

Translating this to a cement–filler mix, this discussion can also be extended to SCC. Several researchers have reported results which show a shortening of the induction period and an acceleration of the hydration process during the first hours[30, 33, 36–39]. From results with scanning electron microscopy (SEM), Ramachandran and Zhang[40] postulated that the ending of the induction period can be attributed to the nucleation and growth of hydration products on the surface of $CaCO_3$ molecules and therefore, following the nucleation of CSH theory, the induction period is shortened. This hypothesis was also supported by Kadri *et al.*[41, 42]. Kjellsen and Lagerblad[43] on the contrary support the idea that preferential nucleation of CH on the $CaCO_3$ parts is the reason for the shortening of the induction period and the acceleration of the reactions.

Appearance of the third reaction

The influence of fillers, particularly limestone powder, is often considered to be limited to an alteration of the rate of the reactions. Many authors report an acceleration of the binding and a shortening of the induction period (see previous paragraph) possibly caused by heterogeneous nucleation of the hydration products. This, of course, cannot explain the appearance of an extra peak in the hydration curves of the experimental calorimeter tests. At present two different theories have been formulated to explain this phenomenon. The first theory starts from the hypothesis that limestone powder is inert and therefore not taking part in the reactions chemically. The second theory considers that the limestone powder is not inert and is thus taking an active part in the reactions.

Limestone powder is inert

The reported phenomenon might be related to the hydration of C_3A. Bensted[44] found that the presence of more than 12% of C_3A in cement results in a third peak in the hydration curve during an isothermal hydration test at 20°C. De Schutter[7] established that this peak, probably caused by the transformation of ettringite into monosulfate, could also be found using Portland cement with a lower C_3A content (e.g. 7.5%) during isothermal hydration tests at higher temperatures around 40–50°C.

The fact that limestone powder in SCC could be considered as an activator for this transformation of ettringite into monosulfate with heat release as a consequence can be a possible explanation for the appearance of the third peak, even for temperatures below 40–50°C. This hypothesis might also explain the finding that with a decreasing c/p ratio, the rate of this extra reaction is increasing.

Limestone powder is not inert

In 1956, Farran[45] was among the first to report a study on the nature of the bond between different minerals and cement paste. One of his findings was that the structure of the ITZ between calcite and cement paste is different from the structure of the ITZ between siliceous materials and cement paste, resulting in a better bond strength. Nevertheless, any reaction between C_3S and the limestone could not be observed. Later research showed a reaction between the aluminates in the cement and carbonate rocks resulting in the formation of calcium carboaluminate hydrates. This reaction has also been reported by several investigators suggesting the main reaction product between calcite and the aluminate phases to be $C_3A \cdot CaCO_3 \cdot 11H_2O$[43]. Grandet and Ollivier[46] also confirmed these findings, and they hypothesised from their study that due to the low mobility of the dissolved ions these reactions are limited to the ITZ. These early findings only relate to the use of limestone as coarse aggregate in concrete. For the study of the influence of the powder content on the behaviour of SCC, it is important to extend this discussion to the use of limestone powder in SCC.

Research carried out by Bonavetti *et al.*[47–49] showed that the mixing of Portland cement and limestone powder caused serious changes in the hydration reactions compared to pure cement pastes. First, the formation of ettringite seemed to be accelerated by the presence of the limestone powder, after which the transformation of the ettringite into monosulfate, after the depletion of the gypsum, is delayed or even completely stopped when a large amount of calcium carbonate is present. After three days monocarboaluminate was found in the hardened specimen using X-ray diffraction.

Several researchers[48, 50–53] have reported the presence of AFm phase $C_3A \cdot CaCO_3 \cdot 10\text{–}12H_2O$ as a reaction product of limestone powder with the aluminates in the cement paste. The timing of the appearance of the monocarbonate depends on the concentration of the filler and of the C_3A content of the cement.

In 1968 Seligmann and Greening[54] showed that, in the presence of calcium hydroxide, monosulfate reacts with CO_3^{2-}, which could be handed by finely ground $CaCO_3$, to form ettringite and hemicarbonate. Further reaction gives monocarbonate. The anion exchange within the AFm phase is caused by the fact that monocarbonate is thermodynamically stable at 25°C while the sulfate AFm phase is metastable at this temperature[55, 56]. This transformation reaction, with a heat release as a consequence, might possibly explain the appearance of a third peak in the heat development when cements with moderate to high C_3A content are combined with limestone powder. However, when considering the active role of limestone powder in the formation of carboaluminate, it should be noted that relative to the cement mass, only a minor part of the limestone powder can react with the Portland cement clinker, depending on the Al_2O_3 content[56].

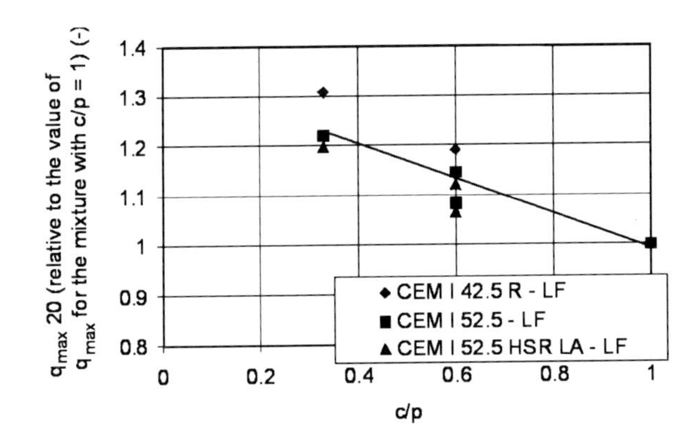

Figure 8.12 q_{max} *at 20°C as a function of c/p ratio*[30].

Heat production rate and accumulated heat

An investigation of the maximum heat production rates during the isothermal tests [29, 30] leads to the finding that this maximum rate also is influenced by the addition of the limestone powder: the more limestone powder is added, the higher the heat production rate peak value q_{max} is. When presenting these values of q_{max} at 20°C in a diagram as a function of c/p, a linear relationship can be found depending on the cement–powder combination. Figure 8.12 graphically combines some results of combinations of cement with limestone powder.

The results of the cumulated heat were combined with the theoretical values of the cumulated heat at complete hydration to give an estimated value of the ultimate degree of hydration. This theoretical approach leads to the finding that the formula of Mill (Equation (8.18)) can still be used as an adequate approximation of the ultimate degree of hydration for SCC (Figure 8.13).

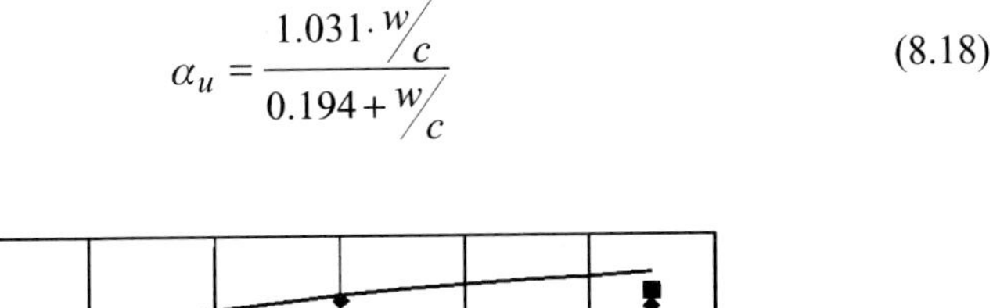

$$\alpha_u = \frac{1.031 \cdot w/c}{0.194 + w/c} \qquad (8.18)$$

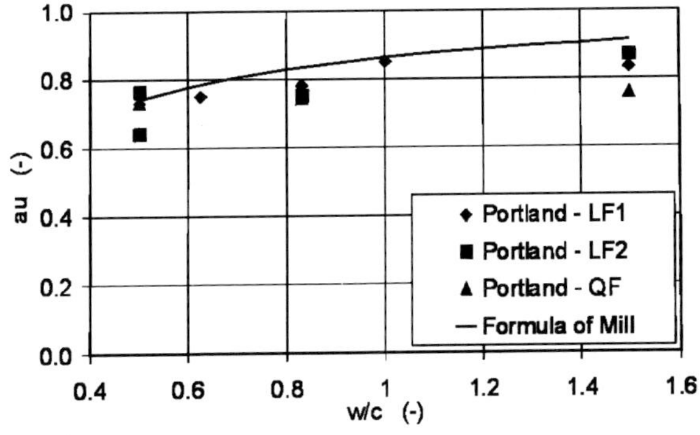

Figure 8.13 *Ultimate degree of hydration as a function of w/c ratio*[30].

Modelling

As a basis for the modelling of the hydration process in SCC in comparison with TVC, Poppe and De Schutter[29, 30, 57] applied the hydration model developed by De Schutter[6, 7]. In this model, the heat production rate of a Portland cement is calculated as follows:

$$q = q_{\max,20} \cdot f(r) \cdot g(\theta) \tag{8.19}$$

$$f(r) = c \cdot \left[\sin(r\pi)\right]^{a} \cdot \exp(-br) \tag{8.20}$$

$$g(\theta) = \exp\left[\frac{E}{R}\left(\frac{1}{293} - \frac{1}{273+\theta}\right)\right] \tag{8.21}$$

Where r is the degree of reaction, θ is the temperature, $q_{\max,20}$ is the maximum heat production rate at 20°C, a, b and c are the model parameters, E is the apparent activation energy and R is the universal gas constant.

When looking into the results of the isothermal tests on the mixes with limestone powder, it can clearly be seen that this hydration mechanism cannot be described as one function as was done for a traditional concrete with Portland cement. The presence of the limestone powder makes a second reaction appear, as has already been discussed. This reaction has to be modelled separately and the superposition principle applied in order to obtain the total cement reaction.

When applying the model as in Equations (8.19)–(8.21) to the first reaction (resulting in the second peak during the hydration process), the parameters are calculated by the least squares method, in the same way as was done for the traditional concretes, seem to be influenced by the addition of the limestone powder and can be described as a function of the c/p ratio (e.g. Figure 8.14 : a_1 as a function of c/p).

The second reaction (resulting in the extra third peak) activated by the presence of the limestone powder in mixes with Portland cement with a considerable C_3A-

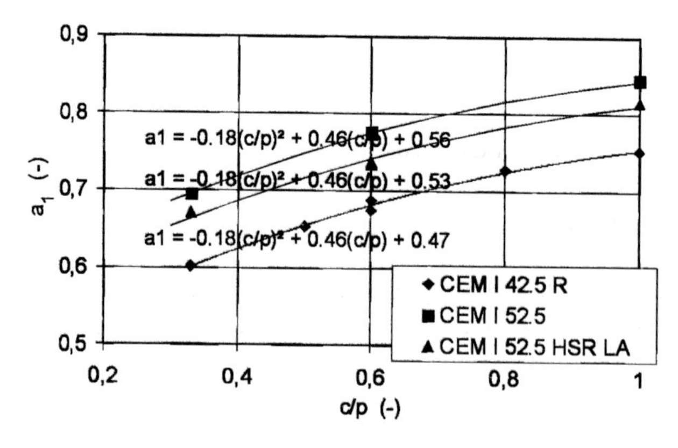

Figure 8.14 *Parameter a_1 as a function of c/p ratio*[30].

content, can be described as a pure sine function. The starting point of this second reaction depends on the temperature during the isothermal hydration test and again on the c/p factor. This leads to the following formulation:

First reaction

$$q_1 = q_{1,\max,20} \cdot f_1(r_1) \cdot g_1(\theta) \quad (8.22)$$

$$f_1(r) = NF_1 \cdot \left[\sin(r_1 \pi)\right]^{a_1} \cdot \exp(-3r_1) \quad (8.23)$$

$$g_1(\theta) = \exp\left[\frac{E_1}{R}\left(\frac{1}{293} - \frac{1}{273+\theta}\right)\right] \quad (8.24)$$

where

r_1 = degree of reaction of the first reaction
θ = temperature
$q_{1,\max,20}$ = maximum heat production rate of first reaction at 20°C (J/hg_{cement})

$$NF_1 = -0.28\,(c/p)^2 + \beta_1\,(c/p) + \beta_0 \quad (8.25)$$

$$a_1 = -0.18\,(c/p)^2 + \gamma_1\,(c/p) + \gamma_0 \quad (8.26)$$

E_1 = apparent activation energy of the first reaction
R = universal gas constant = 0.008 31 kJ/molK
β_1, γ_1 = parameters depending on the limestone powder type (Table 8.3)
β_0, γ_0 = parameters depending on the limestone powder and cement type (Table 8.3)

Second reaction

$$q_2 = q_{2\max,20} \cdot f_2(r_2) \cdot g_2(\theta) \quad (8.27)$$

$$f_2(r_2) = \left[\sin(r_2 \pi)\right]^{a_2} \quad (8.28)$$

$$g_2(\theta) = \exp\left[\frac{E_2}{R}\left(\frac{1}{293} - \frac{1}{273+\theta}\right)\right] \quad (8.29)$$

where

r_2 = degree of reaction of the second reaction
θ = temperature
$q_{2,\max,20}$ = maximum heat production rate of second reaction at 20°C (J/hg_{cement})
a_2 = 0.71
E_2 = apparent activation energy of the second reaction
R = universal gas constant = 0.008 31 kJ/molK

Complete reaction

$$q = q_1 + q_2 \quad (8.30)$$

where

$q_2 = 0$ as long as:

$r_1 < \delta_1 \cdot c/p - 0.0032\,\theta + \delta_0$ or $r_1 > \varepsilon_1 \cdot c/p - 0.0039\theta + \varepsilon_0$

δ_0, δ_1, ε_0 and ε_1 are parameters depending on the filler type (Table 8.3).

As can be seen from the model, the second reaction does not start immediately after the addition of water. Further research is needed to obtain a more detailed view on the nature of the second reaction, and how and at which moment during hydration it is activated.

8.4.3 Fly ash

Hydration process

Fly ash is a by-product from the burning of pulverized coal in the furnaces of power stations. The fly ash is filtered electrostatically from the smoke. The chemical composition of fly ash is very similar to that of natural pozzolans. The chemical and physical properties of fly ash are strongly dependent on the kind of coal from which it is derived and from the type of furnace. It is a fine powder that mainly consists of glassy hollow spherical particles. An important difference between cement and fly ash is the much lower content of CaO in fly ash, which means that fly ash has pozzolanic instead of hydraulic properties. A pozzolan mainly consists of SiO_2, or of SiO_2 and Al_2O_3 which in the presence of water will react with $Ca(OH)_2$ (or CH in short notation) to form compounds which have cementitious properties. This principle is demonstrated in the following reactions:

Portland cement: $C_3S + H_2O \rightarrow CSH + CH$ (rapid reaction)

Pozzolan: pozzolan $+ CH + H_2O \rightarrow CSH$ (slow reaction)

The alkalinity of the pore fluid does not change much because of the chemical binding of the CH.

Table 8.3 *Parameters for different cement–limestone powder combinations.*

	CEM I 42.5 R		*CEM I 52.5*	*CEM I 52.5 HSR LA*
	LF1	*LF2*	*LF1*	*LF1*
β_1	0.69	0.57	0.69	0.69
β_0	2.30	2.42	2.43	2.39
γ_1	0.46	0.38	0.46	0.46
γ_0	0.47	0.55	0.56	0.53
δ_1	0.28	0.36	–	–
δ_0	0.18	0.13	–	–
ε_1	0.28	0.36	–	–
ε_0	0.61	0.46	–	–

LF1 = limestone powder with an average particle diameter of 5 μm
LF2 = limestone powder with an average particle diameter of 10 μm

The cement clinker and the fly ash follow different reaction processes and react at different rates. It is generally agreed that the pozzolanic reaction becomes apparent 3–14 days after mixing with water, meaning, as soon as 70–80% of the alite contained in ordinary Portland cement has reacted. Some researchers even report starting times of 28 days after mixing[58]. The rate of the pozzolanic reaction depends on the properties of the fly ash and of the mix, as well as on temperatures. The incubation period of the pozzolanic reaction is explained by the strong dependence of the solubility of the glassy part of the fly ash and on the alkalinity of the pore solution. The pozzolanic reaction will not start until the pH has reached a required value[59].

However, it should be recognised that the effect of the fly ash in the fly ash – cement binder will not only be the pozzolanic activity, but also its ability to promote the hydration of the cement which means that the two phenomena will interfere with each other[59–61]. The presence of the fly ash affects many aspects of the hydration, i.e.:

- the kinetics of reaction
- the formation of portlandite
- the composition of the hydrates

Fly ashes tend to lengthen the dormant period. The rate of heat evolution varies with the type of fly ash used, it can be either lower[62–64] or higher[65]. Researchers have pointed out that both hydration processes tend to accelerate each other[60, 61]. The height of the second peak, when referred to the content of the Portland cement of the blend, can be either higher or lower[63] than that of the parent Portland cements. Because of this interference of both processes and the dependence of the interference on the type of fly ash, it is extremely difficult to distinguish the contributions of cement and fly ash in the overall hydration reaction. Several researchers have tried to separate the two influences each using their own measuring techniques, but it still remains difficult to assess from the measured results when the start of the pozzolanic reaction occurs and when the hydration of the cement clinker reaches an ultimate value. Two techniques often used are isothermal calorimetry and thermal analysis (thermogravimetric analysis (TGA)/derivative thermogravimetric analysis (DTA)).

Several research projects report isothermal calorimetry and thermal analysis to evaluate the hydration of fly ash – cement binder systems[60–62, 64, 66, 67]. A few of the findings are given in Figures 8.15 and 8.16.

These figures show the rate of heat development q and the cumulative heat generation Q, both normalised by the weight of the cement of each binder. By normalizing the rate of heat development by the total weight of cement, all measured heat is considered to be derived from the hydration of cement and all water is considered to be available for this reaction. The active effect of the fly ash in this case is limited to its promoting role in the hydration of the cement clinker. These results show that both at 20°C and at 35°C, with increasing fly ash content:

(1) The acceleration period is more retarded, probably caused by a dispersion effect due to the larger water availability for the cement hydration.
(2) The second peak is decreasing.

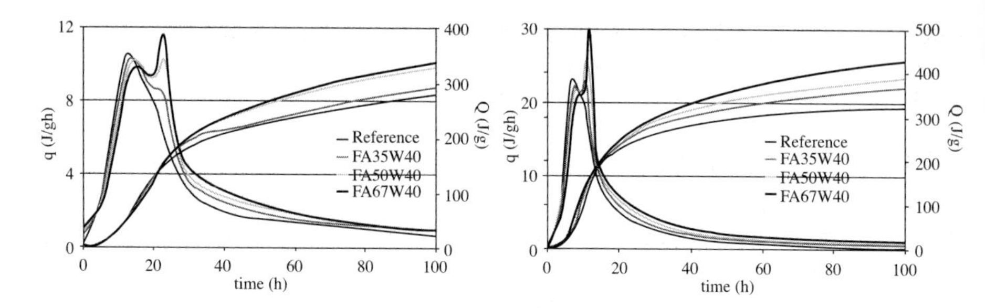

Figure 8.15 *Rate of heat production q as a function of time during the isothermal tests at 20°C (left) and 35°C (right), for combinations of Portland cement and fly ash. The percentage of fly ash is given in the legend, after the letters "FA". W40 means a w/c ratio of 0.40.*

(3) The third peak, which appears during the hydration of the cement–fly ash binder, is increasing. Comparison with thermogravimetric tests on pastes at the same age suggests that this peak is not caused by the pozzolanic reaction of the fly ash, but by the influence of the fly ash on the hydration of cement[44, 68].

Modelling

In the same way as for the modelling of cement–limestone combinations, the model of De Schutter[6, 7] (Equations (8.19)–(8.21)) was used as a basis for the prediction of the heat production rate of the cement–fly ash binders[66].

The heat production rate of a cement–fly ash paste, similar to the limestone powder–cement combinations, can be divided into two reactions: a reaction very similar to the reaction of Portland cement (P-reaction) and a second reaction which corresponds with the third hydration peak (F-reaction). Baert *et al.*[66] have presented the suggested model as follows:

$$q = q_P + q_F = \sum_{i=P,F} q_i \tag{8.31}$$

with

$$q_i = q_{i,\max 20} \cdot f_i(r_i) \cdot g_i(\theta) \tag{8.32}$$

$$f_i(r_i) = \mathrm{NF}_i \cdot \left[\sin(r_i \pi)\right]^{a_i} \tag{8.33}$$

$$g_i(\theta) = \exp\left[\frac{E_i}{R}\left(\frac{1}{293} - \frac{1}{273+\theta}\right)\right] \tag{8.34}$$

with i = P or F

The normalised curves of the F-reaction have the shape of a pure sine function. The model constants $\mathrm{NF_F}$ and b_F are therefore kept zero. The modelling of different mix compositions shows that with increasing fly ash content of the binder, the model constants $\mathrm{NF_P}$ and a_P decrease, while the model constant of the second reaction a_F increases. The average activation energy E_P of the first reaction of all mixes has a value of 38.4 kJ/mol. This corresponds with values often found in literature[5–7]. The average activation energy E_F of the second reaction of all mixes is 46.4 kJ/mol. More details about this model can be found in[66].

8.4.4 Blast furnace slag

Hydration process

Blast furnace slag is a by-product of steel production. It is formed by the reaction of limestone with materials rich in SiO_2 and Al_2O_3 associated with the ore or present in ash from the coke[69]. Granulated blast furnace slag, produced by rapid cooling to below 800°C, is a glassy material, which is latently hydraulic. After grinding, it is mixed with Portland cement or added separately to the concrete. The chemical composition of blast furnace slag is similar to that of Portland cement. However, compared with Portland cement, the contents of SiO_2, Al_2O_3 and MgO in blast furnace slag increase and the content of CaO decreases. According to Lea[70], the composition of blast furnace slag varies over a wide range depending on the nature of the ore, the composition of the limestone flux, the coke consumption and the kind of iron being made.

Taylor[69] has reported that in blast furnace slag 90–100% of the principal elements occur as amorphous oxide networks similar to silicate glass, this was confirmed by Williams[71]. Through SEM and backscattered electron microscopy (BEM) images, Williams found that the blast furnace slag was almost entirely amorphous glass material with a dense centre of calcium, aluminium, silicon and magnesium without any crystalline structure. X-ray diffraction also demonstrated the high glass content with no significant crystalline structure[71].

The implementation of slag as a filler material in SCC is not as widespread as that of limestone powder or fly ash. According to some resources, it is quite common in China, although not much information is available. In Europe, blast furnace slag is more frequently used as a component within blended cements (e.g. cement of type CEM III).

The hydration of Portland cement–slag mixes is more complex than that of Portland cement, since both constituents react with the water[72]. The hydration of slag cement is composed of two reactions: a Portland reaction (P-reaction) and a slag reaction (S-reaction), as illustrated in Figure 8.16[73]. The slag reaction is mainly activated by the lime made available during hydration of the Portland cement fraction (and by sulfates and alkalis)[6, 7, 73]. Blast furnace slag reacts with the CH produced by the cement, thereby introducing some microstructural and chemical advantages, e.g. improved strength and durability[74]. The blast furnace slag grains become surrounded by rims of hydration products. Blast furnace slag has a composition with nearly enough Ca and Si to sustain the formation of CSH gel. Most of the cement grains are totally hydrated in the blended cement pastes incorporating ground granulated blast furnace slag. This can probably be attributed to enhanced cement hydration. It is well-known that slags react more slowly with water and that the slag reaction is more sensitive to heat than the Portland clinker[75, 76].

Modelling

Figure 8.16 also makes it clear that modelling is not possible using just one function. The twofold character of the hydration (P- and S-reactions) is approached and used as a basis

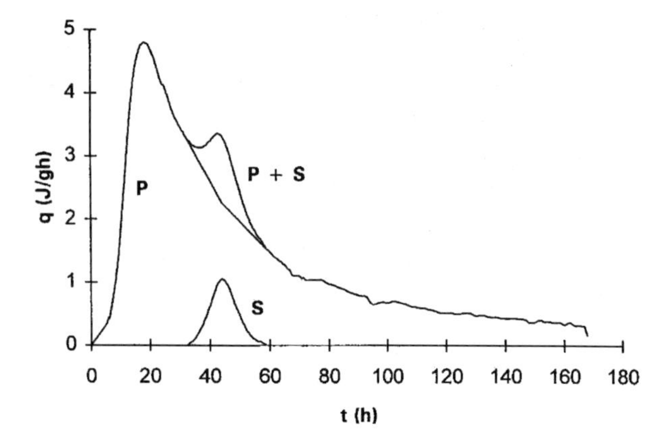

Figure 8.16 *Superposition of Portland reaction (P) and slag reaction (S)*[73].

to develop a model for the hydration of slag cements. Taking this into consideration a general hydration model for slag cement was presented by De Schutter[6, 7], based on the superposition of the heat production of the P- and S-reactions:

$$q = q_P + q_S \tag{8.35}$$

with

$$q_P = q_{P,\max 20} \cdot f_P(r_P) \cdot g_P(\theta) \tag{8.36}$$

$$f_P(r_P) = \text{NF}_P \cdot \left[\sin(r_P \pi)\right]^{a_P} \cdot \exp(-b_P \cdot r_P) \tag{8.37}$$

$$g_P(\theta) = \exp\left[\frac{E_P}{R}\left(\frac{1}{293} - \frac{1}{273+\theta}\right)\right] \tag{8.38}$$

and

$$q_S = q_{S,\max 20} \cdot f_S(r_S) \cdot g_S(\theta) \tag{8.39}$$

$$f_S(r_S) = \left[\sin(r_S \pi)\right]^{a_S} \tag{8.40}$$

$$g_S(\theta) = \exp\left[\frac{E_S}{R}\left(\frac{1}{293} - \frac{1}{273+\theta}\right)\right] \tag{8.41}$$

In these equations $q_{P,\max20}$ and $q_{S,\max20}$ are the maximum heat production rate of the P, respectively, S reaction at 20°C. E_P and E_S are the apparent activation energy of the the P- and S-reaction, and a_P, b_P, NF_P and a_S are parameters. For more details on this hydration model reference is made to De Schutter[6, 7, 73].

8.5 Microstructure of self-compacting concrete

8.5.1 General aspects

Influencing factors

The microstructure of concrete is dependent on time, humidity and temperature. The hydration process changes with time. The humidity of the environment influences the amount of water available for the hydration process. An increasing temperature changes not only the hydration rate but also the degree of polymerisation of the CSH. When hydrating cement paste is located in an environment of 0–50°C, no significant changes of the stoechiometry of the products occur, only the hydration rate increases. This means that there is no change of volume proportions and capillary pores. However, the pore size distribution can be changed by the formation of more coarse capillary pores and fewer small capillary pores at higher temperatures[77].

Changes in the concrete composition also have their influence on the microstructure and pore structure and the use of a different cement type or class can change the situation. Also the addition of mineral fillers, fly ash or admixtures and deviations of the traditional volume proportions of the ingredients of concrete all have an influence. The factors related to the production of SCC are discussed below.

Type of cement

As well as typical different chemical composition, each type of cement also has a different fineness. Both aspects influence the hydration process and the ultimate hydration degree[78]. The main different cement types defined in the European Code EN 197-1 (2004) are Portland cement (I), Portland composite cement (II), blast furnace slag cement (III) and composite cement (V). Due to differences in fineness several strength classes are available for every type of cement.

As well as Portland clinker and a small amount of calcium sulfate, blast furnace slag cement also consists of alkaline blast furnace slag. The mean proportions of the most important oxides in slag are 38% CaO, 37% SiO_2, 15% Al_2O_3 and 10% MgO. The following components are mainly present: CS ($CaO.SiO_2$), C_2S ($2CaO.SiO_2$), C_2AS ($2CaO.Al_2O_3.SiO_2$) and C_2MS_2 ($2CaO.MgO.2SiO_2$). Blast furnace slag has latent hydraulic properties. The hardening of blast furnace slag in contact with water is induced by the presence of CH. The hydration process of blast furnace slag cement is slower compared to a Portland cement with the same fineness. In Europe, based on the amount of blast furnace slag, three types of blast furnace slag cement are defined: CEM III/A (36–65%), CEM III/B (66–80%) and CEM III/C (81–95%).

Additions

Mineral additions, fly ash and blast furnace slag can be used in SCC. Mineral additions are defined as fine distributed material, mainly smaller than 63 μm, obtained by the mechanical preparation of stony material of artificial or natural nature[79]. These additions can improve the grain size distribution and the physical properties of concrete (workability, water retention). They can be inert, hydraulic, latent hydraulic or pozzolanic[80]. In SCC the fillers can represent an important part of the total volume. Their main purpose is to improve the workability of the fresh concrete and to attain an improved grain size distribution in order to obtain a better packing[81].

According to theoretical models the packing density should be higher and the water demand should be lower when a mix is made of two grain-sized materials with different grain-size distributions. This is due to the filling ability of the smaller particles in the interstitial voids of the coarser particles. In[82] the effect of the addition of mineral fillers on the packing density and the water demand has been investigated. It was concluded that besides the fineness the morphology of the particles, like shape and texture, also has to be considered. By adding natural carbonates with round particles a lower water demand and a higher packing density has been achieved. The higher the fineness, the higher the water demand. The use of angular quartzite grains resulted in a higher water demand (up to four times) and a lower packing density.

When part of the cement is replaced by inert mineral additions, the percolation will not be improved. When an additional amount of additions is used at a constant cement content, however, there is an improvement because of the lower proportion of water compared to the total amount of solid material. When reactive mineral additions are used for the partial replacement of cement, some changes in the percolation properties can occur due to extra reactions[83].

Admixtures

The first chemical reactions cause the appearance of positively charged ettringite particles. These particles can start an electrostatic interaction with the negatively charged cement particles and other hydration products. Furthermore, nonhydrated cement particles tend to attract each other. This creates a flocculated structure with a diminished flowability. Because of the coagulation, large amounts of mixing water can be entrapped, which has some consequences for the w/c ratio. In the case of severe flocculations, not only the shape and the dimensions of the pores change but also different hydration products can be formed[1, 84].

Plasticisers and superplasticisers are used to give the same workability at lower water content or for a higher workability at the same water content. Tensio-active material is adsorbed by cement particles, yielding a negative charge. This causes the cement particles to repel each other, preventing coagulation. As a consequence, fewer cement particles remain unhydrated. Furthermore, a water film is formed around the cement particles with a lubricating effect. Due to the more homogeneous distribution of the cement particles a higher initial strength and sometimes also a higher ultimate strength is achieved.

The influence of a sulfonated melamine-formaldehyde condensate superplasticiser has been investigated by means of mercury intrusion porosimetry[85]. When superplasticiser is added the maximum amount of intruded mercury decreases. The threshold diameter decreases and the amount of pores smaller than 100 nm increases. Since the amount of large pores is smaller, there is an increase in the small pores. This use of a superplasticiser provides a refinement of the pore structure. This has also been confirmed by Sakia *et al.*[86]. Especially in the case of polycarboxylate types of superplasticiser, the pore volume of hardened paste represented by pores smaller than 100 nm increases significantly, while the pore volume represented by larger pores decreases significantly.

Volume fractions of the components

Compared to TVC, the volume fractions of the components in SCC are significantly different, as has been discussed in Chapter 6. The fraction of coarse aggregate is much decreased making the fraction of the cement matrix even more important than it was. The structure of the cement matrix itself is also changed due to the application of additions and admixtures.

8.5.2 Self-compacting concrete without an addition

The influence of the compaction process on the ITZ between aggregates and cement matrix has been investigated[25]. TVC and SCC with a w/c ratio of 0.39 are compared (Table 8.4). In none of the mixes are additions used to establish the self-compacting properties. By means of optical microscopy an increased porosity could be observed in the ITZ for TVC as well as for SCC. The ITZ was 50 μm thick in the case of SCC, and 70 μm thick in the case of TVC. By means of electron microscopy in SCC the appearance of CH was noticed in the proximity of the aggregate grains, but the layer was thinner compared to TVC. There is a decrease of the gradient of the pore volume at a distance to the aggregates of 10 μm which is related to the deposition of calcium hydroxide at the edge of the aggregates. The reason for the higher porosity in the ITZ when the concrete has been vibrated is thought to be the accumulation of pore fluid due to vibration during the casting of the concrete. Considerable differences in the porosity of the lower, lateral and upper ITZ can occur. This anisotropic distribution of pores in the ITZ may vary from concrete to concrete.

8.5.3 Powder-type self-compacting concrete

Limestone powder

Addition of limestone powder can have different consequences for the hydration process. Several researchers have concluded that there is an increased hydration rate of the cement clinkers at young age (especially of C_3S)[48, 87], an improved packing density[48], a higher development of CH due to new nucleation sites[48, 51] and the formation of calcium carboaluminate hydrate ($3CaO.Al_2O_3.3CaCO_3.32H_2O$) due to the reaction between $CaCO_3$ of the limestone and C_3A of the Portland clinker[48, 51, 88]. In a later stage there can be a transformation of calcium carboaluminate hydrate into a more stable monocarboaluminate hydrate. It has been found that the addition of limestone powder induced an extra hydration peak[30]. The possibility was also considered that the mineral additions might form new nucleation sites. Some of these aspects have already been discussed in Section 8.4.2.

When fine fillers are used, for example with a mass of 20% below 2 μm, there is a high probability that the fillers do not distribute well over the complete cement matrix and will flocculate at certain spots. This can cause more porous zones in the concrete. This problem can be avoided by optimising the mixing procedure so that the particles are homogeneously distributed over the concrete mass[88].

Trägårdh[26] and Rougeau[89] have investigated limestone powder based SCC. The microstructure of SCC containing limestone powder has been investigated by means of electron microscopy and optical microscopy[26]. The same type of cement and aggregates were used for both TVC and SCC. The amount of cement is 401 kg/m^3 for SCC (Table 8.4) and 430 kg/m^3 for TVC. The water contents are determined in order to obtain a w/c ratio of 0.40 for both mixes. In the case of SCC a superplasticiser and limestone powder were added, namely 170 kg/m^3. Air entrainment was used in both mixes. It was concluded that the porosity is less in SCC than in TVC, in the bulk cement matrix as well as in the ITZ. The stable paste and the absence of vibration energy probably reduce the risk of bleeding inducing a lower porosity. The chemical composition of the ITZ is found to be different for SCC and TVC. In SCC, first there is a lower sulfate content in the phases; second the layer of CH, preferably formed in the ITZ, is thinner. The CH layer was about 3–5 μm thick in the SCC, while it was about 5–10 μm thick in the TVC.

The Hadley grains are evenly distributed over the complete cement matrix in SCC, in contrast to the preferable appearance in the ITZ in TVC. This may indicate that the effect of microbleeding, which may produce a local increase of the ratio at the interfacial zone, was much less in the SCC. The amount of large, above 20 μm, nonhydrated cement grains is higher in the case of SCC. This indicates that the limestone powder affected the mode of hydration in SCC. Study by light microscopy revealed that any defects in the air-pore structure due to handling and processing of concrete, such as clustering of air pores at aggregate walls and formation of air pockets, were not observed in SCC, but were commonly observed in conventional concretes. Although both SCC and TVC were produced with the same w/c of 0.40, light microscopy study showed that the SCC had a lower porosity than a reference concrete with a w/c of 0.40, while the TVC had a higher porosity.

Reinhardt *et al.*[90] have given an overview of the pore structure of SCC. Okazawa [mentioned in[90]] concluded by means of mercury intrusion porosimetry (MIP) that the pore volume of several SCC mix was not dependent on the type of superplasticiser. At the same time it was found that the microstructure of SCC is denser than that of TVC.

Table 8.4 *Mix proportions applied in some selected references (kg/m^3 concrete).*

Reference	*Type of cement*	*Mix*	*C (kg/m^3)*	*Type of filler*	*F (kg/m^3)*	*W (kg/m^3)*
Leemann[25]	CEM I 42.5	SCC	465	–	0	181
	CEM I 42.5	CVC 1	465	–	0	181
Rougeau[89]	CEM I 52.5 R		352	limestone powder	101	172
	CEM I 52.5 R		350	limestone powder	100	165
Trägårdh[26]	ASTM type 5		401	limestone powder	170	161
Liu[39]	CEM I 52.5 N	TC	350	–	0	165
	CEM I 52.5 N	SCC01	400	limestone powder	200	165
	CEM I 52.5 N	SCC02	400	limestone powder	300	192
	CEM I 52.5 N	HPC	400	–	0	132

Liu[39] has compared two mixes of SCC to one mix of HPC and one mix of TVC (called TC, as noticed in Table 8.4). In order to obtain the self-compacting properties of the SCC mixes limestone filler and polycarboxylic ether as a superplasticiser have been used. Among those mixes, TC and SCC02 have the same w/c ratio of 0.48, while HPC, representing a high performance concrete, has a lower w/c ratio of 0.33. Both SCC mixes have a similar w/p ratio. In order to study the microstructure of the SCC, several tests have been performed on the corresponding cement pastes. From these tests it was concluded that: both the cumulative heat release and the rate of heat release are higher in SCC pastes containing limestone powder than in the TC and HPC pastes. The development of the CH phases in the SCC pastes is different from that in the HPC and TC pastes. The CH content in SCC decreases to a certain age and then increases. Results from BEM image analysis and MIP shows that the pore structure, including the total pore volume, pore size distribution and critical pore diameter, in the SCC pastes is very similar to that in the HPC. The fact that limestone powder does not participate in the chemical reaction can be confirmed from both thermal analysis and BSE image analysis. On the other hand, limestone powder acts as accelerator during cement hydration, at early age.

Although the experiments show that the limestone powder does not react chemically during hydration, its influence on the microstructure or other properties is still not clear. By introducing the computational analysis of concrete, the effect of limestone powder can be studied more clearly. Based on the simulation, the influence of limestone powder could be classified into two periods along the hydration process. At the initial hydration stage, as limestone powders fill the space between the hydrating cement particles, the microstructure of samples with added limestone powder is denser than those without limestone powder. The effective contact area, which is linked with the strength of the material, in the SCC at the initial stage is higher for higher limestone powder content samples at the same w/c ratio. This indicates that the samples with limestone powder should have good properties at initial age. As the hydration proceeds, the expanding cement particles come to connect and form a network, while at the same time, limestone powder particles do not expand during the entire process. During a later hydration stage the contact area for samples without the addition of limestone becomes larger, while the porosity reaches comparable values[39].

As the limestone powder acts as inert filler, its addition could bring a denser microstructure at the initial hydration stage. The porosity and microstructural development are better compared with samples without limestone powder but with the same w/c ratio. On the other hand, as the hydration proceeds, the limestone powder has a decreasing influence on the microstructure because the cement particles expand to form a network. Furthermore, at the later hydration stage, the limestone powder even obstructs somewhat the formation of a denser microstructure[39].

Ye *et al.*[91] studied similar mixes, at cement paste level, to those mentioned in Table 8.4. The paste mixes are called TCP (corresponding to TC), SCCP1 (corresponding to SCC1), SCCP2 (corresponding to SCC2) and HSCP (corresponding to HSC). By thermal analysis the amount of each individual phase can be determined. The results of the thermal decomposition of the four cement pastes are shown in Figure 8.17. When comparing the mass loss in the different mixes, two main differences can be observed. First, a much higher mass loss is found in TCP compared to SCCP and HPCP at 100–200°C. This is due to the higher water content in the traditional concrete. Secondly, a larger mass loss is observed only on the samples of SCCP1 and SCCP2 at 750°C. The mass losses of SCCP1 and SCCP2 are due to the escape of CO_2 from the cement pastes. The amount of CO_2 escaping from the cement paste can be calculated exactly from the thermogravimetric analysis (TGA) tests and is compared with theoretical calculations. Comparing the TGA analysis and the theoretical calculations, the mass loss measured with TGA is slightly larger than the mass loss from the theoretical calculations. If one takes into account that a small part of the mass losses is due to the decomposition of the calcium silicate hydrates in the cement paste, the main part of the mass loss at this temperature is due to decarbonation of the limestone. According to the mass balance law, it can be found that almost no limestone powder participated in the chemical reaction during cement hydration. The limestone powder only acts as inert filler in the SCCP.

It could be argued that if carboaluminate were to be present due to some chemical activity of the limestone powder, it would most probably also decompose during the TGA test. This would question to some extent Ye's conclusion about the inert character of the limestone powder, as concluded from TGA tests. However, Ye *et al.* also further illustrates their conclusion by means of BSE images[91]. One of the BSE images from each mix at the ages of 28 days is shown in Figure 8.18. It can be seen that even very small limestone particles still exist in the pastes even after 28 days hydration. Porous gaps can be found in samples SCCP1 and SCCP2. These large pores are mainly around the large limestone particles. Quantitative calculation from BSE analysis has been used to investigate the phase distribution of the four pastes at different curing ages. The development of CSH in the SCCP mixes is similar in spite of their different w/c ratios. The HPCP shows a higher CSH content due to the low w/c ratio. However, the development of CH is very different if SCCP is compared with HPCP and TCP. The CH content in SCCP decreases between days 3 and 7 and slightly increases afterwards. In HPCP and TCP, the CH content increases for 7 days and decreases afterwards. In the SCCP samples, considering the accuracy obtained

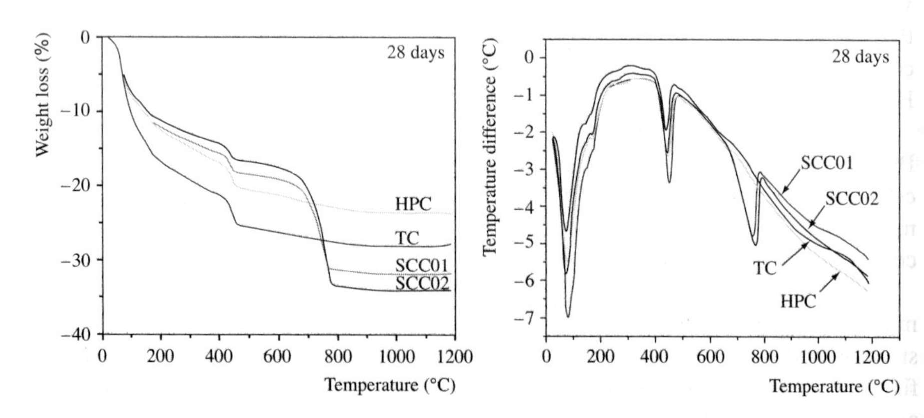

Figure 8.17 *Thermal decomposition of pastes by TGA and DTG* [91].

TCP 28 days

SCCP1 28 days

SCCP2 28 days

HPCP 28 days

Figure 8.18 *BSE images at 28 days*[91].

with the image analysis, the total amount of limestone powder is almost unchanging during 28 days of hydration. From image analysis it is found that the interface between the limestone and hydrates is somewhat porous, even on samples after 28 days of hydration. However, within the HPCP samples, the interface around the unhydrated cement is much denser.

In addition to the BSE observation tests on MIP have also been performed[91]. The comparison of pore size distributions and differential curves of pore size distribution of four mixes at 28 days are presented in Figure 8.19. As expected, the total pore volume in traditional concrete is much higher than in SCCP and HPCP. However, there is only a slight difference of the total pore volume in SCCP and HPCP. This is consistent with BSE image analysis. The critical pore diameters of the SCCP and HPCP samples are almost the same as can be seen from Figure 8.19.

The microstructure of hardened cement paste and concrete has been studied by Boel using several techniques[92]. An overview of the mixes of cement paste and concrete are given in Tables 8.5 and 8.6. The proportions of materials used for making the cement paste correspond to the amounts of constituents in 1 m^3 of concrete. The following conclusions were drawn.

From MIP on hardened cement pastes it was concluded that the self-compacting mixes have a denser structure compared to the traditional mixes. The more dense structure and decreased porosity can be explained by the physical presence of mineral fillers. Furthermore, the degree of hydration and the w/c ratio in combination with the amount of water strongly influence the pore structure. The hydration products fill up more and more space when the degree of hydration increases. The dimensions of the

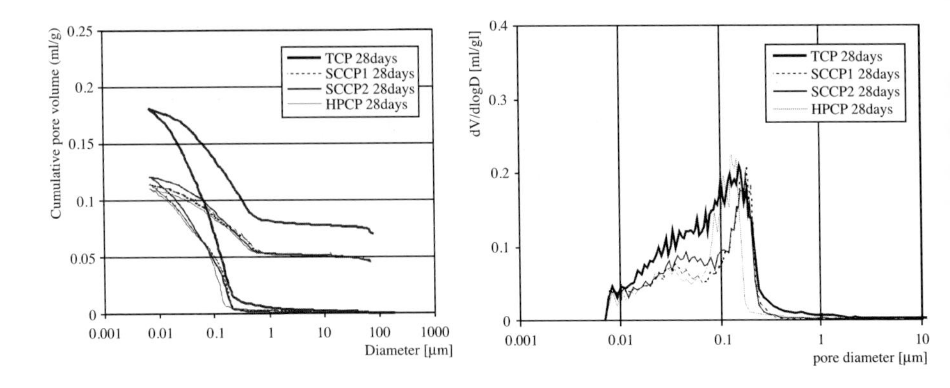

Figure 8.19 *Comparison of pore size distributions at 28 days*[91].

pores decrease inducing a lower connectivity. A lower w/c ratio gives a lower capillary porosity and connectivity. By means of the porosity measured by MIP, it was illustrated that the development of the pore structure of hardened self-compacting cement paste with limestone powder as a filler follows Powers' model[18, 19] (Figure 8.20). As such it is possible to calculate the total capillary porosity from the composition of the cement paste or concrete. The volume of the limestone powder is assumed to be constant in time. Possible reactions with limestone powder were not taken into account. The good agreement between the theoretically calculated and experimentally determined capillary porosity indicates that the limestone powder is almost inert and does not participate in the hydration process. This does not exclude the function as a catalyst accelerating the hydration reactions as mentioned by Poppe[30], and as discussed in Section 8.4.2.

Several parameters have been discussed including: the total intruded volume, critical diameter, specific surface, retention ratio and mean pore diameter[92, 93]. The intrusion and extrusion curves and the pore size distribution are also discussed. The difference between the critical pore size of traditional and self-compacting paste can

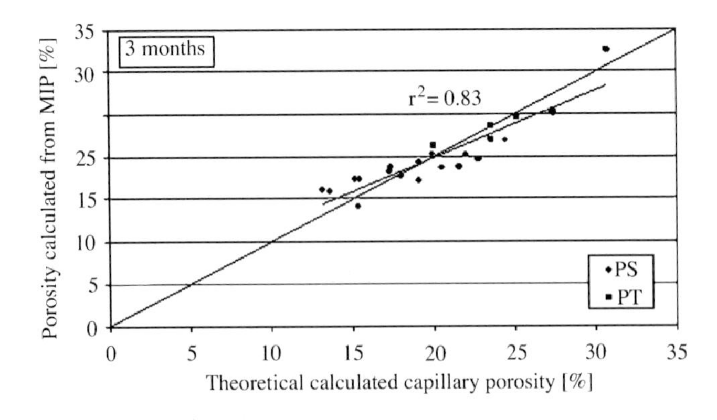

Figure 8.20 *Porosity from MIP versus the theoretical capillary porosity at 3 months (PS = paste from SCC, PT = paste from TVC).*

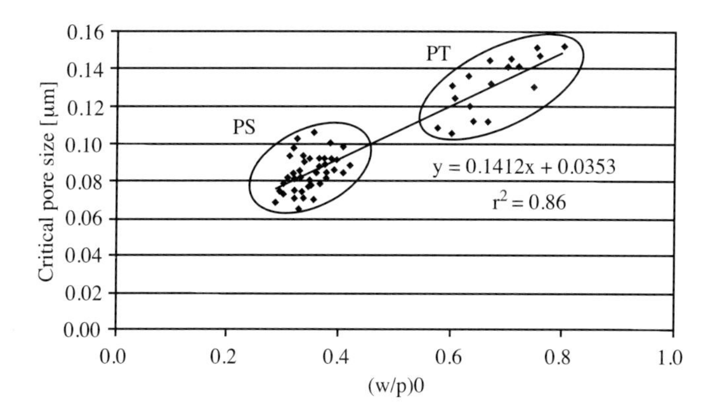

Figure 8.21 *Critical pore size as a function of the w/p ratio (PS = paste from SCC, PT = paste from TVC*[92]*).*

be explained by considering the critical pore size as a function of the w/p ratio (Figure 8.21).

When the pore size distribution curves of the traditional cement paste mixes PT1, PT2 and PT3 are compared with the self-compacting cement paste mixes PS1, PS2 and PS3 (Figure 8.22), the following differences can be seen. When there is almost no intrusion yet for the PS mixes, there is already a clearly visible intrusion for the PT mixes. This indicates a less dense and more permeable microstructure of the PT mixes. For the PT mixes, the largest peak, indicating the critical pore size, is shifted towards larger pore diameters compared to the PS mixes. This means that for traditional mixes the microstructure percolates at a higher pore diameter. For PS mixes at pore diameters smaller than the critical pore size there is only a small increase in intruded volume compared to the PT mixes. This means that for self-compacting pastes, once the microstructure is percolated there are few pores left to be filled. There are still a lot of pores to be filled in the traditional mixes. This also indicates a more dense structure of the self-compacting pastes.

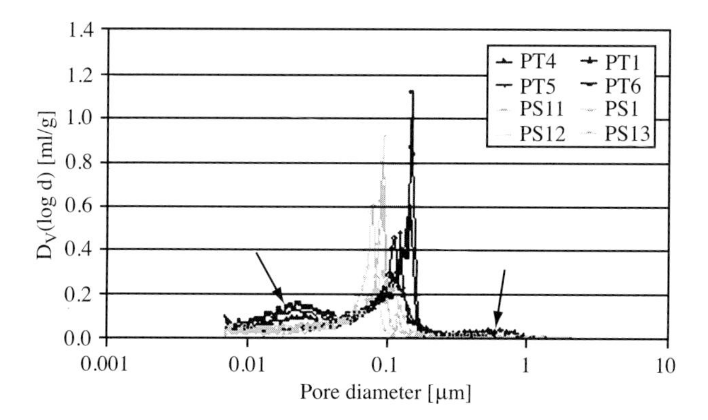

Figure 8.22 *Pore size distribution of traditional and self-compacting mixes*[92].

Analysis by means of optical light microscopy on SCC1 indicates that the adhesion between the aggregates and the cement matrix is good. No defects or plastic cracks were noticed. The macroporosity can be attributed to the small air voids. Some of the aggregates are rather porous and allow the cement paste to intrude into the aggregate. Most of the cement matrix is very dense. Typically under plane polarised light and crossed nicols, a cement matrix looks 'dark'. When carbonation has occurred the cement matrix is brownish due to the presence of $CaCO_3$. When a thin slice of TVC is compared to SCC, one can see that the cement matrix of the SCC looks 'brownish' (Figure 8.23). Not knowing the actual composition of the concrete, the conclusion could be that the cement matrix is carbonated. The limestone powder used in these mixes consists of 95% $CaCO_3$. Due to the important amount of limestone powder there is already a significant amount of $CaCO_3$ present in the noncarbonated concrete. It was illustrated by TGA that the amount of CH developed during hydration of an equal amount of Portland cement should be more or less the same for SCC as for TVC. The paste content of a SCC is higher compared to TVC which means that the amount of CH per unit volume of paste/cement matrix is lower. The space available for CH to deposit is smaller and more distributed over the paste volume. The CH can be found in smaller concentrations and more spread over the cement matrix.

The ITZ between aggregates and cement matrix was studied by means of BSE images. The extent and structure of the ITZ was depending on the type of concrete (SCC or TVC) and on the initial amount of water. Compared to TVC, SCC has a denser matrix and the ITZ differs less from the bulk cement matrix. A higher w/c ratio increases the extent and the porosity of the ITZ. The concentrations of the nonhydrated cement grains were within the expectations according the concrete composition.

By means of tomography and automated air void analysis it was possible to study the air void system and the appearance of microcracks. From these results it was concluded that the parameters could not be linked directly to the concrete composition. On the other hand there was a clear influence of the consistence of the fresh concrete on the air content and the air void distribution. As the consistence of the fresh concrete increases, the air voids have a greater chance to escape from the

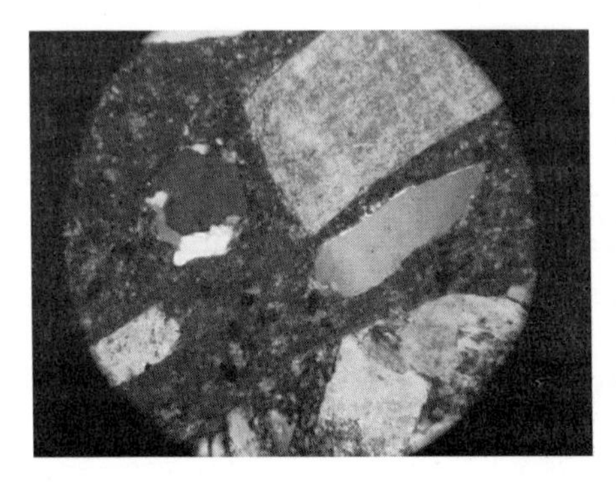

Figure 8.23 *Plane polarised light and crossed nicols, magnification ×200, TC1 (left) and SCC1 (right).*

concrete, and a lower air content is observed. In contrast to the information in the literature, there was no important increase in air content, and no enhanced air void system due to the application of a superplasticiser. Sometimes, in the case of SCC, isolated large air voids are noticed. The results obtained from air void analysis do fit with what was found with microscopy and tomography.

The same mixes as mentioned in Table 8.6 have been studied by Audenaert[94]. In this research the gel porosity, capillary porosity and total porosity of SCC and TVC were calculated by means of Powers' model[18, 19]. It was concluded that the vacuum saturation with preceding drying at 40°C and 105°C is a suitable test method to estimate the capillary porosity and the sum of the capillary porosity and the gel porosity. The drying implies that the samples are dried until constant mass. When this is not the case a correction has to be made taking into account the amount of water left in the pore structure.

The ITZ has an important influence on the transport properties of concrete[95]. The question is whether this zone is still as important for SCC as for TVC. The importance of the transition zones depends on the density of the intervening cement matrix. When the matrix has a dense structure, it is difficult for the potential aggressive agents to move from one transition zone to another. In SCC, the paste volume is also larger which reduces the possibility of meeting transition zones. This interpretation leads to the idea that the transport properties and durability aspects of SCC can be linked very well to the capillary porosity of the cement matrix, without taking into account the ITZs. This conclusion has already been confirmed[94, 96]. It was found that the relationship with the capillary porosity is better in the case of SCC than of TVC. This can be explained by the improved microstructure of SCC, as mentioned before. The transport mechanisms and durability are considered in Chapter 10.

Fly ash

When a certain amount of cement is replaced by the same amount of fly ash, the development of the hydration heat and the strength slows down. However, the strength development between 28 and 90 days is more pronounced. The reaction products fill the coarse capillary pores and thus increase the strength of the cement matrix as well as the resistance against chemical attack and intrusion. Due to the presence of fly ash the cement particles tend to flocculate less. The fly ash particles are hydrophilic which induces a lubricating effect.

A mix of SCC with fly ash has been studied (SCC6)[92, 94]. The overall conclusion from this research was that the presence of fly ash improved the microstructure. A more dense structure was developed. It has to be taken into account that the development of the microstructure is slower when fly ash is added.

Unreacted fly ash contributes to the microstructural behaviour of cement stone as it appears to be acting as microaggregates in the paste. Zhang[97] has investigated paste mixes with 0.3 water to cementitious material ratio and with 58% mass replacement of the Portland cement by fly ash. He concluded that high volume fly ash (HVFA) paste may be considered as a composite material microscopically with fly ash particles as reactive microaggregates embedded in a matrix of hydration and reaction products[97]. Crack propagation generally deviates around fly ash particles, since fly ash particles are much harder, stronger, and of higher modulus of elasticity

Table 8.5 *Mix proportions of cement paste (kg/m^3 concrete).*

	PS1	*PS2*	*PS3*	*PS4*	*PS5*	*PS6*	*PS7*	*PS8*	*PS9*	*PS10*	*PS11*
CEM I 52.5 N (kg/m^3)	360	0	0	360	300	400	450	0	0	0	360
CEM III/A 42.5 LA	0	360	0	0	0	0	0	300	400	450	0
CEM I 52.5 HSR (kg/m^3)	0	0	360	0	0	0	0	0	0	0	0
Limestone powder P2	240	240	240	0	300	200	150	300	200	150	240
Quartzite powder	0	0	0	240	0	0	0	0	0	0	0
Water	165	165	165	165	165	165	165	165	165	165	144
Superplasticiser (l/m^3)	2.8	2.3	2.5	4.1	2.3	3.1	3.4	1 .8	2.5	2.7	3.2
Water/cement (-)	0.46	0.46	0.46	0.46	0.55	0.41	0.37	0.55	0.41	0.37	0.40
Cement/powder (-)	0.60	0.60	0.60	0.60	0.50	0.67	0.75	0.50	0.67	0.75	0.60
$F_{ccub1\ 50,28}$ (MPa)	77.6	69.2	74.5	101.1	62.8	79.3	80.0	62.3	78.1	74.4	87.3

	PS12	*PS13*	*PS14*	*PS15*	*PS16*	*PT1*	*PT2*	*PT3*	*PT4*	*PT5*	*PT6*
CEM I 52.5 N (kg/m^3)	360	360	0	0	0	360	0	0	360	360	360
CEM III/A 42.5 LA	0	0	360	360	360	0	360	0	0	0	0
CEM I 52.5 HSR (kg/m^3)	0	0	0	0	0	0	0	360	0	0	0
Limestone powder P2	240	240	240	240	240	0	0	0	0	0	0
Quartzite powder	0	0	0	0	0	0	0	0	0	0	0
Water	180	198	144	180	198	165	165	165	144	180	198
Superplasticiser (l/m^3)	2.6	2.3	2.7	2.1	1.8	0	0	0	0	0	0
Water/cement (-)	0.50	0.55	0.40	0.50	0.55	0.46	0.46	0.46	0.40	0.50	0.55
Cement/powder (-)	0.6	0.6	0.6	0.6	0.6	1	1	1	1	1	1
$F_{ccub1\ 50,28}$ (MPa)	70.2	73.3	79.3	68.4	60.1	57.3	54.6	54.8	67.0	49.3	48.3

Table 8.6 *Mix design.*

	CEM I 42.5 R (kg/m³)	*CEM I 52.5 (kg/m³)*	*CEM III/A 42.5 LA (kg/m³)*	*CEM I 52.5 HSR (kg/m³)*	*Limestone powder P2 (kg/m³)*	*Water (kg/m³)*	*Sand 0/5 (kg/m³)*	*River gravel 4/14 (kg/m³)*	*Calcareous rubble 2/14 (kg/m³)*	*Glenium 51 (l/m³)*	*w/c (-)*	*c/p (-)*	*Compressive strength (MPa)*
SCC1	360				240	165	853	698		2.3	0.46	0.60	57.3
SCC2		360			240	165	853	698		2.5	0.46	0.60	68.0
SCC3			360		240	165	853	698		2.3	0.46	0.60	66.1
SCC4				360	240	165	853	698		2.2	0.46	0.60	70.1
SCC5	360				240 *	165	853	698		2.8	0.46	0.60	56.9
SCC6	360				240 **	165	853	698		2.8	0.46	0.60	66.2
SCC7	300				300	165	853	698		2.2	0.55	0.50	46.5
SCC8	400				200	165	853	698		2.9	0.41	0.67	64.2
SCC9	450				150	165	853	698		3.0	0.37	0.75	68.7
SCC10	300				200	137	923	755		3.4	0.46	0.60	60.1
SCC11	400				300	192	782	640		2.6	0.48	0.57	55.9
SCC12	450				350	220	712	583		2.7	0.49	0.56	50.9
SCC13	360				240	144	865	707		3.6	0.40	0.60	68.7
SCC14	360				240	198	835	683		1.8	0.55	0.60	46.6
SCC15	360				240	216	825	675		2.0	0.60	0.60	40.3
SCC16	360				240	165	816		734	1.8	0.46	0.60	74.7
TC1	360					165	640	1225			0.46	1.00	48.6
TC2			360			165	640	1225			0.46	1.00	49.7
TC3				360		165	640	1225			0.46	1.00	50.2
TC4	400					165	626	1200			0.41	1.00	53.7

* Limestone powder S instead of limestone powder P2. ** fly ash instead of limestone powder P2

than the matrix of hydration and reaction products because of their glassy nature[97]. Wang *et al.*[98] further studied the microaggregate effect of fly ash by micromechanics, the hypothesis of the centre particle and pore size distribution. Particles of unreacted fly ash may be considered to be centre particles that can influence the behaviour of the medium[98]. According to Wan *et al.*[98], fly ash cannot refine the pore structure at early age, while fly ash may refine the pore structure relatively at a later age: the hydrated product of the pozzolanic reaction fills in the pores, which not only reduces the porosity, but also changes larger pores into smaller ones.

The refining role was reported earlier by Takemoto and Uchikawa *et al.*[99]. They noted a remarkable decrease in the volume of capillary space with the curing time, and meanwhile an increase of the gel pores in the early stage[99]. Between days 91 and 180 the gel pores become smaller. Along with filling the capillary space by hydration products formed in the early and middle age, fine gel pores are newly formed in this filled capillary space. These gel pore spaces begin to decrease in the later age, when the amount of filling of gel pore with fine crystal, such as CSH, exceeds the volume of gel pore that is formed. Therefor the ratio of the amounts of gel pore and total porosity, increases rapidly till 91 days, then decreases after that. Though the total porosity will decrease with curing time, it will still show a higher value than that of Portland cement at 180 days[99]. By 180 days, significant densification takes place, although residual, typically spherical fly ash particles remain in large quantities in the system[97]. Compared with the condition after 28 days, the microstructure of the paste becomes more unified and the reaction products are more evenly distributed in the system in general. The porous zone around the fly ash particles appears to have densified substantially, although some regions of higher porosity persist[97]. The capillary pores are markedly discontinuous in pastes made with fly ash. The main reason is the lower Ca/Si ratio of the CSH[69]. Uchikawa has reported a value of 1.01 for a 4-year-old paste with 40% replacement of the cement by fly ash[100]. The dominant effect, however, responsible for the higher porosities of pastes made from cements with fly ash is the relatively low degree of reaction of the the fly ash[69]. The effect increases with the percentage replacement and is marked at early ages[69].

Studies on the pore structures of pastes of cements with fly ash have presented difficulties, which arise from the discontinuity of the pores. Mercury intrusion appears to alter the structure with fly ash even more than with plain Portland cement[69]. The pastes of composite cements appear to contain relatively large, but discontinious pores, separated by walls that are broken by the mercury[69]. Pore size distribution in pastes with 40% replacement of the Portland cement by fly ash and a water to cementitious material ratio of 0.3 shifts to small radius with curing time and the peak appears at about 10 Å radius at 91 days[99]. At early ages, the distribution of the pore entry sizes in fly ash cement pastes and the porosity recorded by the method at maximum pressure is higher because of the slow pozzolanic reaction. At later ages the distribution of pore entry sizes is finer than in comparable Portland cement pastes, and the curves do not tend to flatten out at high intrusion pressures, which suggest that even at the highest attainable pressures, the mercury does not enter all the pore space[69]. Uchikawa[100] studied pastes of Portland cement and composite cements by N_2 adsorption. The pore size distribution of both peaked at 2 nm, but partial replacement of the cement by fly ash decreased the height of the peak[99]. Xu *et al.* have investigated the morphology of cement–fly ash pastes and mortars by means of secondary electron imaging[101, 102]. In cement pastes with 60% replacement of the cement by fly ash and a water-to-cementitious material ratio of 0.465 the paste structure became denser from 90 days onwards[101]. From that age, the space between the hydration products is less than 0.5 μm indicating a much less porous structure. In mortar, the equal weight replacement of cement by fly ash lowers the total initial volumetric porosity of the mortar and increases the neat w/c ratio[102]. The consumption of calcium hydroxide by fly ash at the aggregate–paste interface decreases the flaws in the mortar and makes the matrix more homogeneous[102].

Blast furnace slag

When blast furnace slag is present a reaction similar to the pozzolanic reaction occurs. Portland cement with a certain amount of fly ash or blast furnace slag has a higher ultimate strength compared to plain cement. This is due to the pore refinement and the increase of CSH[1].

XRD and SEM studies have shown that the principal hydration products are essentially similar to those given by pure Portland cement. XRD and SEM images show that there are three distinct layers formed on the surface of blast furnace slag[71]. The interior of the blast furnace slag is rich in magnesium oxide. The inner layer is a hydrotalcite-like phase (an aluminium-magnesium-hydroxycarbonate), which contains mostly magnesium and aluminium. The outer edge has an increase in calcium oxide and a relatively lower amount of silicon and aluminium. Outside the blast furnace slag boundary, in the outer layer, there is a paste, which contains relatively high amounts of CaO, and similar SiO_2 and a decrease in aluminium and silicon.

In blended cements, there are no abrupt changes in the general pattern of the microstructure of hardened pastes as the composition of the blend changes. In mature pastes, inner and outer products, CSH, form the dominant component of the microstructure with a decreasing Ca/Si ratio as the blast furnace slag content increases. The texture of the outer product CSH appears to become finer as the blast furnace slag content increases and the magnesium content of the blast furnace slag is retained within inner product where thin magnesium-rich precipitate particles develop. The CH content decreases as the slag content increases and becomes small for slag/Portland ratios in excess of 3:1.

The fineness of the clinker influences the microstructure. Finely ground blended cements hydrate quickly and show compact structures[103]. Separate grinding of the blast furnace slag and the clinker can lead to a high degree of fineness.

On ageing, all blended cement pastes show a decrease in overall porosity and pore size, due to the infilling of interstices by hydration products[104]. It has been reported that blending blast furnace slag results in 10–25% reduction in total porosity and 35–60% reduction in fine porosity. With increasing slag content the pores become much smaller after 28 days curing[105]. Smolczyk[106] has reported that hydrated blended cement pastes contain more gel pores and fewer capillary pores than Portland cement pastes.

The differences in pore size distribution in blended cement pastes and Portland cement pastes are functions of differences between the hydration processes in these pastes. Bakker[107] explained that in addition to the hydration formation around the BFS and clinker particles there are additional (and identical or related) hydrate precipitations in the 'gaps' between adjacent particles. The permeability of blended cement pastes is lower than that of Portland cement pastes and decreases as the hydration progresses[108]. Permeability coefficients are reduced by 60% after 180 days curing.

8.6 Conclusions

Extensive studies over many years of both normal and high-performance TVCs have shown the importance of hydration and microstructure in controlling important material characteristics such as mechanical properties, the transport of material through the concrete and durability. Similar studies by several researchers have been carried out on SCC, often comparing the hydration processes and microstructure to those of TVC. Hydration has been studied by isothermal and adiabatic tests, and microstructure by MIP, optical and electronic microscopy, tomography and TGA.

The overall conclusions from these studies are that the combination of lower w/p ratios, additions and superplasticisers necessary to give satisfactory fresh properties of SCC lead to a denser structure, higher strength and decreased porosity of the hardened concrete in comparison to TVC. The total porosity of both the bulk paste and the paste/aggregate ITZ are significantly reduced, and there are fewer defects and less crack formation. Thus the benefits of SCC extend to its hardened as well as its fresh properties.

References

[1] Mehta P.M. and Monteiro P.J.M. *Concrete Microstructure, Properties and Materials*, 2nd edn., University of California, CA, USA, 2001.

[2] Hewlett P.C. (Ed.) *Lea's Chemistry of Cement and Concrete*, Arnold, London, UK, 1998.

[3] Stutzman P.E. National Institute of Standards and Technology, Gaithersburg, MD, USA, http://ciks.cbt.nist.gov/.

[4] Dela B. Eigenstresses in hardening concrete. In: *Hydration and Hardening of Cement Paste*, DTU Department of Structural Engineering and Materials, Series R, No. 64, 2000, pp 9–19.

[5] van Breugel K. Simulation of hydration and formation of structure in cement-based materials. Doctoral thesis, Technical University Delft, the Netherlands, 1991.

[6] De Schutter G. General hydration model for Portland cement and blast furnace slag cement. *Cement and Concrete Research*, 1995, 25(3) 593–604.

[7] De Schutter G. Fundamental and practical study of thermal stresses in hardening massive concrete elements (in Dutch). Doctoral thesis, Ghent University, Ghent, Belgium, 1996.

[8] Locher F.W. and Richarts (1976) mentioned in: Dela B. Eigenstresses in hardening concrete. In: *Hydration and Hardening of Cement Paste*, DTU Department of Structural Engineering and Materials, Series R, No. 64, pp 9–19.

[9] Garboczi E.J. and Bentz D.P. The microstructure of Portland cement-based materials: computer simulation and percolation theory. *Materials Research Society Symposium Proceedings* 1998, 529, 89–99.

[10] Mailvaganam N.P., Grattan-Bellow P.E. and Pernica G. Deterioration of concrete: symptoms, causes and investigation. Institute for Research in Construction, Ottawa, Ontario, Canada, 2000.

[11] Neville A. *Properties of Concrete*, 4th edn, Longman, UK, 1995.

[12] Feldman R.F. and Sereda P.J. A new model for hydrated Portland cement and its practical implications. *Engineering Journal (Canada)*, 1970, 53(8/9) 53–59.
[13] Romberg H. Cement paste porosity and concrete porosity (in German). *Beton-Information*, 1978, 18(5) 50–55.
[14] Hansen T.C. Physical structure of hardened cement paste – A classical approach. *Matériaux et constructions*, 1986, 19(114) 423–435.
[15] US FHA, Petrographic methods of examining hardened concrete: a petrographic manual. U.S. Department of Transportation, Federal Highway Administration, 1997, www.tfhrc.gov/pavement/pccp/petro/chaptr06.htm.
[16] Schaefer D.W. Engineered porous materials. *MRS Bulletin* 1994, 19(4) 14–17.
[17] St John D., Poole A.W. and Sims I. *Concrete Petrography, A Handbook of Investigative Techniques*, Arnold, London, UK, 1998.
[18] Powers T. and Brownyard L. Studies of the physical properties of hardened Portland cement paste (nine parts). *Journal of the American Concrete Institute*, 1946–1947, 43.
[19] Powers T. The non-evaporable water content of hardened Portland cement paste: its significance for concrete research and its method of determination. *ASTM Bulletin*, 1949, 158, 68–76.
[20] Jensen O.M. and Hansen P.F. Water-entrained cement-based materials I. Principles and theoretical background. *Cement and Concrete Research* 2001, 31(4) 647–654.
[21] Nokken M.R. and Hooton R.D. Discontinuous capillary porosity in concrete – does it exist? In: *Proceedings of the International RILEM Symposium on Advances in Concrete through Science and Engineering*, Evanston, IL, USA, March 2004.
[22] Scrivener K.L. The microstructure of concrete. In: Skalny J.P. (Ed.) *Materials Science of Concrete*, American Ceramic Society, Westerville, OH, USA, 1989, Vol. 1, pp 127–161.
[23] Elsharief A., Cohen M.D. and Olek J. Influence of aggregate size, water cement ratio and age on the microstructure of the interfacial transition zone. *Cement and Concrete Research*, 2003, 33(11) 1837–1849.
[24] Diamond S. and Huang J. The ITZ in concrete – A different view based on image analysis and SEM observations. *Cement and Concrete Composition*, 2001, 23(2) 179–188.
[25] Leemann A., Münch B., Gasser L. and Holzer L. Influence of compaction on the interfacial transition zone and the permeability of concrete. *Cement and Concrete Research*, 2006, 36, 1425–1433.
[26] Trägårdh J. Microstructural features and related properties of self-compacting concrete. In: *Proceedings of the First International RILEM Symposium on Self-compacting Concrete*, Stockholm, Sweden, 1999, pp 175–186.
[27] Odler I. Hydration, setting and hardening of Portland cement. In: Hewlett P.C. (Ed.) *Lea's Chemistry of Cement and Concrete*, 4th edn., Arnold, London, UK, 1998.
[28] Poppe A. and De Schutter G. Heat of hydration of self-compacting concrete. In: Kovler K. and Bentur A. (Eds.) *International RILEM Conference on Early Age Cracking in Cementitious Systems*, Haifa, 2001, pp 71–78.
[29] Poppe A. and De Schutter G. Cement hydration in the presence of high filler contents. *Cement and Concrete Research*, 2005, 35, 2290–2299.
[30] Poppe A. Influence of fillers on hydration and properties of self-compacting concrete (in Dutch). Doctoral thesis, Ghent University, Ghent, Belgium, 2004.
[31] Xiong X. and Van Breugel K. Effect of limestone powder and temperature on cement hydration processes. In: *Proceedings of ICPCM – A New Era of Building*, Cairo, Egypt, 2003, pp 231–240.
[32] Xiong X. and Van Breugel K. Hydration processes of cements blended with limestone powder: experimental study and numerical simulation. In: *Proceedings of the 11th*

International Congress on the Chemistry of Cement, Durban, South Africa, 2003, pp 1983–1992.

[33] Ye G., Liu X., De Schutter G., Poppe A.M. and Taerwe L. Influence of limestone powder used as filler in SCC on hydration and microstructure of cement pastes. *Cement and Concrete Composites*, 2007, 29(2) 94–102.

[34] Scrivener K. Cement chemistry: Importance of aluminates. PhD – advanced course on advances in cement-based materials, Technical University, Delft, the Netherlands, 2003.

[35] Jennings H. Introduction to the chemistry of hydration: a materials scientist's view. PhD – advanced course on advances in cement-based materials, Technical University, Delft, the Netherlands, 2003.

[36] Billberg P. Influence of filler characteristics on SCC rheology and early hydration. In: Ozowa K. and Ouchi M. (Eds.) *Proceedings of the Second International RILEM Symposium on Self-compacting Concrete*, Tokyo, Japan, 2001, pp 285–294.

[37] Kratochvil A., Urban J., Stryk J. and Hela R. Fine filler and its impact to a cement composite life cycle. In: Bilek V. and Kersner Z. (Eds.) *Proceedings of the Symposium on Non-traditional Cement and Concrete*, Brno, Czech Republic, 2002, pp 260–267.

[38] Tsivilis S., Chaniotakis E., Kakali G. and Batis G. An analysis of the properties of Portland limestone cements and concrete. *Cement and Concrete Composites* 2002, 24, 371–378.

[39] Liu X. Microstructural investigation of self-compacting concrete and high-performance concrete during hydration and after exposure to high temperatures. Doctoral thesis, Ghent University, Belgium, and Tongji University, China, 2006.

[40] Ramachandran V.S. and Zhang C.M. Influence of CaCO3 on hydration and microstructural characteristics of tricalcium silicate. *Il Cemento* 1986, 3, 129–152.

[41] Kadri E., Aggoun S. and Duval R. Influence of grading and diameter size of admixture on the mechanical properties of cement mortars. In: Bilek V. and Kersner Z. (Eds.) *Proceedings of the Symposium on Non-traditional Cement and Concrete*, Brno, Czech Republic, 2002, pp 306–313.

[42] Kadri E. and Duval R. Effect of ultrafine particles on heat of hydration of cement mortars. *ACI Materials Journal* 2002, 99(2) 138–142.

[43] Kjellsen K. and Lagerblad B. Influence of natural minerals in the filler fraction on hydration and properties of mortars. CBI Report 3:95, 1995.

[44] Bensted J. Some applications of conduction calorimetry to cement hydration. *Advances in Cement Research* 1987, 1(1) 35–44.

[45] Farran J. Mineralogical contributon to the study of adherence between hydration products and aggregates (in French). *Revue des Matériaux de Construction* 1956, 490, 155–172.

[46] Grandet J. and Ollivier J.P. Study of the formation of monocarboaluminate in Portland cement paste containing calcarious aggregates (in French). *Cement and Concrete Research* 1980, 10(6) 759–770.

[47] Bonavetti V., Donza H., Rahhal V. and Irassar E. Influence of initial curing on the properties of concrete containing limestone blended cement. *Cement and Concrete Research* 2000, 30, 703–708.

[48] Bonavetti V., Rahhal V. and Irassar E. Studies on the carboaluminate formation in limestone filler blended cements. *Cement and Concrete Research* 2001, 31(6) 853–862.

[49] Bonavetti V., Donza H., Menendez G., Cabrera O. and Irassar E. Limestone filler cement in low w/c concrete: A rational use of energy. *Cement and Concrete Research* 2003, 33, 865–871.

[50] Sawicz Z. and Heng S. Durability of concrete with addition of limestone powder. *Magazine of Concrete Research* 1996, 48, 131–137.
[51] Sharma R. and Pandey S. Influence of mineral additives on the hydration characteristics of ordinary Portland cement. *Cement and Concrete Research* 1999, 29, 1525–1529.
[52] Vuk T., Tinta V., Gabrovsek R. and Kaucic V. The effects of limestone addition, clinker type and fineness on properties of Portland cement. *Cement and Concrete Research* 2001, 31, 135–139.
[53] Petersson O. Limestone powder as filler in SCC – frost resistance, compressive strength and chloride diffusivity. In: *Proceedings of the First International North American Conference on the Design and Use of Self-consolidating Concrete*, 2002, 391–396.
[54] Famy C. and Taylor H. Ettringite in hydration of Portland cement concrete and its occurrence in mature concretes. *ACI Materials Journal* 2001, 98(4) 350–356.
[55] Matschei T., Lothenbach B. and Glasser F.P. The role of calcium carbonate in cement hydration, (submitted for publication).
[56] Matschei T., Lothenbach B. and Glasser F.P. The Afm-phase in Portland cement. *Cement and Concrete Research* 2007, 37, 118–130.
[57] Poppe A. and De Schutter G. Analytical hydration model for filler rich binders in self-compacting concrete. *Journal of Advanced Concrete Technology* 2006, 4(2) 259–266.
[58] Hanehara S., Tomosawa F., Kobayakawa M. and Hwang K. Effects of water/powder ratio, mixing ratio of fly ash, and curing temperature on pozzolanic reaction of fly ash in cement paste. *Cement and Concrete Research* 2004, 31, 31–39.
[59] Massaza F. Pozzolana and pozzolanic cements. In: Hewlett P.C. (Ed.) *Lea's Chemistry of Cement and Concrete*, 4th edn., Arnold, London, 1998.
[60] Wang A., Zhang C. and Sun W. Fly ash effects II. The active effect of fly ash. *Cement and Concrete Research* 2004, 34, 2057–2060.
[61] Pane I. and Hansen W. Investigation of blended cement hydration by isothermal calorimetry and thermal analysis. *Cement and Concrete Research* 2005, 35, 1155–1164.
[62] Langan B.W., Weng K. and Ward M.A. Effect of silica fume and fly ash on heat of hydration of Portland cement. *Cement and Concrete Research* 2002, 32, 1045–1051.
[63] Bai J. and Wild S. Investigation of the temperature change and heat evolution of mortar incorporating PFA and metakaolin. *Cement and Concrete Composites* 2002, 24, 201–209.
[64] Baert G., Van Driessche I., Hoste S., De Schutter G. and De Belie N. Interaction between the pozzolanic reaction of fly ash and the hydration of cement. *Congress ICCC2007*, Montreal, Canada, July 2007.
[65] Sakai E., Miyahara S., Ohsawa S., Lee S.H. and Daimon M. Hydration of fly ash cement. *Cement and Concrete Research* 2005, 35, 1155–1164.
[66] Baert G., De Schutter G. and De Belie N. Modelling of isothermal hydration heat of cement pastes with fly ash. *Fifth International Essen Workshop – Transport in Concrete*, Essen, Germany, 2007.
[67] Sanchez de Rojas M. and Frias M. The pozzolanic activity of different materials, its influence on the hydration heat in mortars. *Cement and Concrete Research* 1996, 26(2) 203–213.
[68] Schindler A.K. and Folliard K.J. Influence on supplementary cementing materials on the heat of hydration of concrete. In: *Proceedings of Advances in Cement and Concrete IX Conference*, CO, USA, August 2003.
[69] Taylor H.F.W. *Cement Chemistry*, Academic Press, London, UK, 1990.

[70] Lea F.M. *The Chemistry of Cement and Concrete,* 3rd edn., Chemical Publishing Co., Inc., New York, 1970.
[71] Williams J. Characterization of blast furnace slag and blast furnace slag reactions for use in blended slag-cement systems. Department of Chemical Engineering, USA, ChE499, 1999.
[72] Moranville-Regourd M. Cements made from blast-furnace slag. In: Hewlett P.C. (Ed.) *Lea's Chemistry of Cement and Concrete*, 4th edn., Arnold, London, UK, 1998.
[73] De Schutter G. Hydration and temperature development of concrete made with blast-furnace slag cement. *Cement and Concrete Research* 1999, 29, 143–149.
[74] Escalente-Garcia J.I. and Sharp J.H. The microstructure and mechanical properties of blended cements hydrated at various temperatures. *Cement and Concrete Research* 2001, 31, 695–702.
[75] Roy D.M. and Idorn G.M. Hydration, structure and properties of blast furnace slag cements, mortars and concrete. *ACI Journal* 1982, 444–457.
[76] Wu X., Roy D.M. and Langton C.A. Early stage hydration of slag cement. *Cement and Concrete Research* 1983, 13, 277–286.
[77] Igarashi S.-I., Watanabe A. and Kawamura M. Effects of curing conditions on the evolution of coarse capillary pores in cement pastes. In: *Proceedings of the International RILEM Symposium on Concrete Science and Engineering*, 2004, pp 105–116.
[78] Mather B. and Hime W.G. Amount of water required for complete hydration of Portland cement. *Concrete International*, 2002, 24(6) 56–58.
[79] KOMO, National guideline for the KOMO product certificate for stone powder applied as filter in concrete and mortel (in Dutch), 2002, (BRL 1804 d.d. 2002-11-04).
[80] Jones M.R., Zheng L. and Newlands M.D. (2003), Estimation of the filler content required to minimise voids ratio in concrete. *Magazine of Concrete Research* 2003, 55(2) 193–202.
[81] Lukas W. and Lottner S. Alternatives for cement: new binders for the concrete industry (in German). *BFT* 2002, 7, 24–28.
[82] Bigas J.P. and Gallias J.L. Effect of fine mineral additions on granular packing of cement mixtures. *Magazine of Concrete Research* 2002, 54(3) 155–164.
[83] Bentz D.P. and Garboczi E.J. Percolation of phases in a three-dimensional cement paste microstructural model. *Cement and Concrete Research* 1991, 21(2) 325–344.
[84] Walraven J.C. and Pelova G.I. Workability of concrete (in Dutch). *Cement*, 1998, 9.
[85] Khatib J.M. and Mangat P.S. Influence of superplasticizer and curing on porosity and pore structure of cement paste. *Cement and Concrete Composites*, 1999, 21, 431–437.
[86] Sakai E., Kasuga T., Sugiyama T., Asaga K. and Daimon M. Influence of superplasticizers on the hydration of cement and the pore structure of hardened paste. *Cement and Concrete Research* 2006, 36, 2049–2053.
[87] Asaga K. and Kuga H. Effect of various particle size of calcium carbonate powders on the hydration of Portland cement. *Semento Konkurito Ronbunshu* 1997, 51, 20–25.
[88] Petersson Ö. Limestone powder as filler in self-compacting concrete – frost resistance and compressive strength. In: *Proceedings of the Second International RILEM Symposium on Self-compacting Concrete*, Tokyo, Japan, 2001, 277–284.
[89] Rougeau P., Maillard J.L. and Mary-Dippe C. Comparative study on properties of self-compacting and high performance concrete used in precast construction. In: *Proceedings of the First International RILEM Symposium on Self-compacting Concrete*, Stockholm, Sweden, 1999, pp 251–261.
[90] Reinhardt H.-W. *et al. Sachstandsbericht Selbstverdichtender Beton*, Deutscher Ausschuss für Stahlbeton, Vol. 516, Beuth, Berlin, 2001.

[91] Ye G., Liu X., De Schutter G., Taerwe L. and Vandevelde P. (2006), Phase distribution and microstructural changes of *SCC* at elevated temperature. *Cement and Concrete Research* 2007, 37, 978–987.
[92] Boel V. Microstructure of self-compacting concrete in relation with gas permeability and durability aspects (in Dutch). Doctoral thesis, Ghent University, Ghent, Belgium, 2006.
[93] Boel V., Audenaert K. and De Schutter G. Characterization of the pore structure of hardened self-compacting concrete. *Twelfth International Congress on the Chemistry of Cement*, July 2007, Montreal, Canada.
[94] Audenaert K. Transport mechanisms in self-compacting concrete in relation to carbonation and chloride penetration (in Dutch). Doctoral thesis, Ghent University, Ghent, Belgium, 2006.
[95] Scrivener K.L. and Nemati K. The percolation of pore space in the cement paste/aggregate interfacial zone of concrete. *Cement and Concrete Research* 1996, 26(1) 35–40.
[96] Boel V., Audenaert K. and De Schutter G. Modelling of gas permeability in self-compacting concrete. In: *Fifth International Essen Workshop*, Essen, Germany, June 2007.
[97] Zhang M.H. Microstructure, crack propagation and mechanical properties of cement pastes containing high volumes of fly ashes. *Cement and Concrete Research* 1995, 25(6) 1165–1178.
[98] Wang A., Zhang C. and Sun W. Fly ash effects III. The microaggregate effect of fly ash. *Cement and Concrete Research* 2004, 34, 2061–2066.
[99] Takemoto K. and Uchikawa H. Hydration of pozzolanic cement. In: *Seventh International Congress on the Chemistry of Cement*, Paris, 1980, IV-2, pp 1–29.
[100] Uchikawa H. Effect of blending components on hydration and structure formation. In: *Eight International Congress on the Chemistry of Cement*, 1986, 1, pp 249.
[101] Xu A. and Sarkar S.L. Microstructural study of gypsum activated fly ash hydration in cement paste. *Cement and Concrete Research* 1991, 21, 1137–1147.
[102] Xu A., Sarkar S.L. and Nilsson L.O. Effect of fly ash on the microstructure of cement mortar. *Materials and Structures* 1993, 26, 414–424.
[103] Satarin V.I. Slag Portland cement. In: *Sixth ICCC*, Moscow, USSR, 1974.
[104] Pandey S.P. and Sharma R.L. The influence of mineral additives on the strength and porosity of OPC mortar. *Cement and Concrete Research* 2000, 30(1) 19–23.
[105] Pigeon M. and Regourd M. Freezing and thawing durability of three cements with various granulated blast furnace slag contents. In: *Proceedings CANMET/ACI Conference*, Montebello, Canada. ACI Publication SP-79, vol. 2, pp 979–998.
[106] Smolczyk H.G. Slag structure and identification of slag. In: *Seventh ICCC*, Paris, France, 1980, vol. 1, pp 1–17.
[107] Bakker R. On the cause of increased resistance of concrete made from blast furnace cement to the alkali-silica reaction and to sulphate corrosion. Ph.D. thesis, University of Aachen, Germany, 1980.
[108] Hootton R.D. Permeability and pore structure of cement pastes containing fly ash, slag and silica fume. In: Frohnsdorf G. (Ed.) *Blended Cements*, ASTM Special Publication 897, USA, 1986, pp 128–143.

Chapter 9
Engineering properties

9.1 Introduction

Experimental evidence available indicates that most basic relationships between the most important properties of hardened SCC such as strength and its composition (mix design), production (charging sequence and mixing process), placing and curing remain the same as for TVC. The microstructure of hardened SCC may differ from that of a TVC, as shown in Chapter 8, but the relationships between the characteristics of the microstructure and performance of hardened concrete remain the same, regardless of it being compacted by vibration or not. Such relationships have been explained in the basic texts and reference books on concrete[1] and are beyond the scope of this book.

However, as discussed in Chapter 6, there is an inescapable difference in practical composition between SCC mixes and TVC ones, which guarantees the self-compaction when the mix is fresh. This difference is inevitably, and to a varying degree, reflected in performance of the hardened SCC. The most common and usually the most significant difference lies in the proportion of powder (cement plus additions) which is significantly higher in SCC and, conversely, the proportion of coarse aggregate which tends to be lower in SCC. However, the engineering properties of hardened SCC arise from the basic characteristics and properties of its microstructure as with any kind of concrete.

The high content of fine particles affects all of the microstructure of the concrete, as has been discussed in Chapter 8. The differences in the ITZ between the hardened binder (paste) and the particles of aggregate, which are considered to be of critical importance in development of basic properties of hardened concrete, are particularly important. The ITZ of practical, ordinary SCC mixes with higher powder contents shows a densified, stronger ITZ when compared with that of ordinary TVC[2]. This, in turn, explains the good performance of hardened SCC mixes with regard to strength and durability. This chapter will review the effects of producing a SCC on its engineering properties. Durability is of equal importance to strength and will be dealt with in Chapter 10.

It appears to be useful, and there are frequent demands, to compare the strength of hardened SCC with that of a TVC, and many attempts at doing this have been reported[3–5]. However, few of the researchers, more in early stages of SCC development but some even very recently, fully appreciate the difficulty of making direct comparisons. This is due to the inescapable fact that none of the TVCs are self-compacting (see Chapters 2 and 4).

In order to modify a given mix design, which produces a specific level of strength when hardened after an adequate compaction (see also Chapter 11), its composition must be altered. It is possible for TVCs of very high workability (consistence) to be made self-compacting simply by an extra dosage of one or more admixtures. Such an approach would maximise the true comparability of the TVC and SCC, but only for the traditional concrete mixes of very high workability (e.g. slump > 150 mm).

A more realistic comparison would be with ordinary vibrated concrete mixes with a compressive strength in the range 30–40 MPa and with a slump of 50–100 mm. However, such comparisons become increasingly unreliable because even if the cement contents and water/cement ratios were kept constant, additives have to be introduced or an overall grading of the aggregate changed and more admixtures added to obtain self-compaction.

It is very tempting to extrapolate from results of experiments on selected (e.g. 'typical-ordinary', 'high-performance' etc.) self-compacting and vibrated concretes and propose generalised conclusions, suggesting their applicability to any SCC mix. This is particularly the case when the experiments have been rigorously carried out and evaluated. However, unless the results are linked to a general, underlying condition (e.g. a widespread feature of the microstructure), the conclusions will apply only to the SCC (and vibrated concrete) mixes examined, no matter how well supported by evidence. Practically all of the observations presented in this chapter are therefore only examples of what differences, if any, in engineering properties, can occur when SCC and TVC mixes of selected compositions (and production methods) are used.

9.2 Compressive strength and modulus of elasticity

Basic relationships between composition of concrete, such as water/cement ratio and compressive strength apply, in principle, equally to SCC as to TVC. It is essential to appreciate that SCCs of any compressive strength, from very low to very high, can be produced. Self-compaction does not mean that the concrete will be automatically 'high-performance', having a very high compressive strength when hardened.

Further complications regarding the strength of hardened SCC and its comparison with vibrated concrete have arisen in countries where existing standard methods for assessment of compressive strength (of all concrete) are legally binding for its determination. Such standard methods had been developed for TVC and invariably insist on 'full' compaction being achieved. This would not exclude SCC, but there is often an integral clause attached, prescribing the manner and extent of the compaction effort to be applied to the freshly made test specimens. It may require a cube mould to be filled with concrete in layers, each thoroughly vibrated (usually by an external vibration from a vibrating table). Applied to fresh SCC, such a process may lead to severe settlement/segregation and produce test results unrelated to the observed performance of the same mix in practice. If the specimens were not vibrated, the standard procedure would not be followed and the test result would fail to obtain the required legal standing.

A separate, independent assessment of properties of a proposed hardened SCC mix by an independent testing authority became necessary. This is a costly and lengthy process, hindering the use of SCC in the countries affected, namely in Germany. The situation improved once a relevant formal national guidance for SCC was developed (see Chapter 11) and was given a seminormative legal status. It has been easier in countries, where only the state of full/adequate compaction of the test specimen was normatively prescribed and no vibration had to be applied to an SCC test specimen in order for the test result to be accepted. There was usually no precise, quantitative characterisation of features identifying the specimen as being 'fully' compacted when fresh. This explained why it was often enforced indirectly by the added prescription of a compacting effort.

A typical SCC contains a high proportion of additions (fines), which 'densify' its microstructure when hardened (see Chapter 8). The densification by the additions, which may be also chemically 'active', tends to increase the compressive strength. An increase in characteristic compressive strength of up to approximately 10% over that indicated by traditional, established, relationships with w/c ratio for vibrated concrete at 28 days was not uncommon[6–8]. For this reason, it is sometimes more difficult to produce a 'low-strength' (<25 MPa) than a high strength (>70 MPa) SCC.

The rate of gain of compressive strength of SCC with age is largely similar to that of a TVC containing the same proportion of cement and having the same water/cement ratio[6, 7], except when limestone-type additions are used. An extensive survey of test data reported over more than a decade by Domone[9] has shown that a significant increase in compressive strength was observed across a wide range of limestone powder contents for all ages up to 28 days.

It has been widely assumed that with a change from rounded to coarse aggregate, while all other parameters of a concrete mix were kept constant, the mix with the coarse aggregate would achieve higher compressive strength. The survey by Domone[9] showed that the difference due to the shape of the coarse aggregate has narrowed for SCC, when compared with TVC (Figure 9.1).

Further analysis of data on the compressive strength of hardened SCC[9] revealed that a significant difference existed between the cube-to-cylinder strength ratio assumed in current structural codes for TVC and the ratio obtained from test results on SCC. The cube/cylinder strength ratio for SCC in the strength range of 20–40 MPa

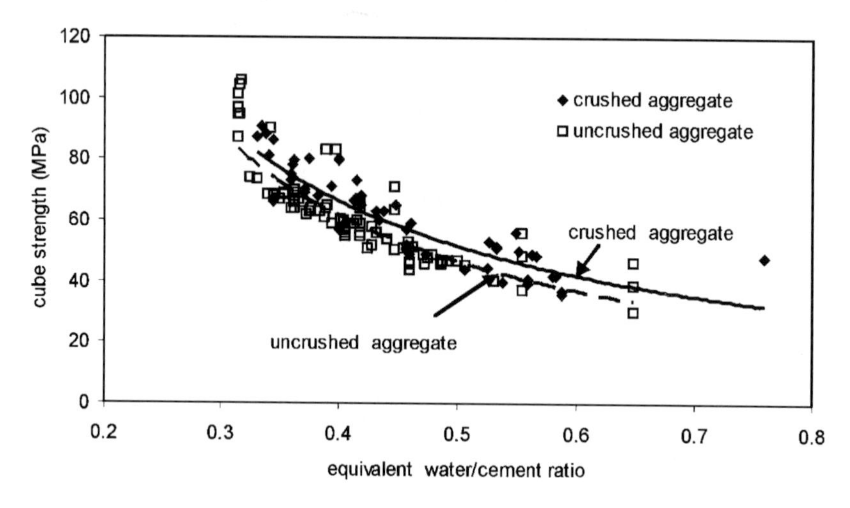

Figure 9.1 *Cube compressive strength versus equivalent w/c ratio*[9].

was 0.80, about the same as for vibrated concrete. However, it increased with increase of strength, reaching 1.0 for SCCs at about 90 MPa; a value much greater than the generally assumed ratio of approximately 0.85 for TVC.

The only general conclusion which can be drawn, is that by using the same content of cement and water as for traditional vibrated mix, but producing a SCC (Chapters 2 and 4) the strength of the concrete when hardened is likely to be the same or greater than that of the vibrated one.

9.2.1 In situ compressive strength, near-surface strength and surface hardness

The amount of information about these properties of hardened SCC is very limited, compared with that for hardened vibrated concrete. Most of the data available[10] is from a program of tests on columns and beams made of SCC and vibrated concrete of similar strength classes (C35 and C60). Cores were extracted for in situ strength, pre-embedded inserts (LOK test[11–12]) were pulled-off and rebound values (Digi-Schmidt hammer) were obtained. The principal aim was to examine the uniformity of these properties. The columns were cast using skip and crane (Figure 9.2), with a maximum free-fall of 3 m when the formwork was empty. The columns contained

Figure 9.2 *Full-scale structural elements (3 m tall columns) cast from SCC and TVC using crane and skip.*

reinforcement of medium and high density (Figure 9.3) through which the fresh mix had to pass. The beams were cast using chutes, directly from truck-mixers. The vibrated concrete was compacted using standard practice, by external vibrators (columns) and poker vibrators (beams). Additional standard cube specimens were made from batches used for casting of the structural elements and tested for compressive strength at different ages, after both standard curing and air curing alongside the full-scale elements.

Results indicated[11] that:

- In situ strength achieved in the columns for all the SCC and reference vibrated mixes was in the range of 80–100% of the 28-day strength for standard cube specimens made for the same batch (Figure 9.4). The percentage was higher for the C60 grade than the C35 grade mixes. The SCC mix containing limestone powder achieved a significantly higher ratio of in situ/cube strength than the corresponding vibrated mix, probably due to the densifying and other effects of this additive.
- The in situ strength of the concrete increased from top to bottom of the columns, the difference being greater for the lower strength (C35) mixes. The differences were significantly smaller when SCC instead of TVC was used.
- The surface hardness of SCC also varied from top to bottom of the columns. It was equal or, sometimes, higher than that of the TVC of similar grade of compressive strength, when assessed using a Schmidt hammer.
- Near-surface hardness varied in a pattern similar to that of the strength of the cores; it was higher towards the bottom of the columns.

It is important to note, that variations in strength with depth of concrete placed are also dependent on segregation resistance of the concrete used. SCC mixes with the high segregation resistance (both static and dynamic – see Chapters 4 and 5) will have the lowest variation. This assumes that the segregation resistance will be adequate for the placing/casting system adopted.

Figure 9.3 *Reinforcement for the columns ('medium' and 'dense').*

9.2.2 Effect of curing conditions on compressive strength

A limited amount of evidence is available regarding any differences between effects of curing on compressive strength of hardened TVC and SCC. Experiments on identical full-scale structural elements made of two strength classes of both SCC and vibrated concrete[7, 10] confirmed that in both cases the compressive strength of air-cured specimens is lower than that of the corresponding water-cured specimens, as expected. However, the extent of the reduction of strength when cured only in air (up to 90 days) was less in the case of the SCC. The least amount of reduction was observed for SCC mixes with limestone powder as an additive.
For example, air curing reduced the strengths at 28 days and 90 days ages to:

- 85% and 79% for SCC
- 71% and 65% for TVC

It is important to appreciate that such conclusions cannot be generalised for all types of SCC, where the type and proportion of additives (fines) influences the nature and development of microstructure, which, in turn, affects the level of compressive strength and its development.

A change of the additive, from limestone powder to ground granulated slag (GGBS) has increased the reduction of compressive strength due to inadequate curing, a result that was not unexpected as the continued presence of an adequate quantity of water in the mix is required for the reaction between cement hydrates and ground slag to continue.

The lesser sensitivity to curing conditions of many SCC mixes is related to a reduction in the free movement of moisture in a fresh and early-age mix, inherent to 'ordinary', 'cohesive' SCC mixes. This has been confirmed by Khayat *et al.*[13], who detected varying effects of VMAs. This may be exploited by the development of SCCs, which may be to a large extent 'self-curing', or require less protection during curing. Until more evidence is to hand and reliable advice has been formulated, it is advisable to also apply the existing current good practice for curing TVC to fresh SCC.

9.2.3 Modulus of elasticity and toughness

Considering hardened concrete as a composite with coarse aggregate being the principal component contributing to the modulus of elasticity, a reduction in the content of coarse aggregate should lower the value of the modulus. Several experiments[8, 10] and a survey[9] have been reported indicating that the modulus of elasticity of most SCC mixes tested was lower than that of TVC of the same compressive strength. Klug and Holschemacher[8] reported that the scatter of the values for different SCCs was smaller, causing all the results to remain within the band acceptable for design using CEB-FIB model code 90[14]. The common relationship between the modulus E and the square root of characteristic compressive strength f_c could still be used. A more recent survey[9] showed that the difference in the modulus of elasticity was greater for lower compressive strengths (Figure 9.4).

Too few results for tests for fracture process in SCC are available to draw conclusions regarding the fracture toughness of SCC in comparison with that of

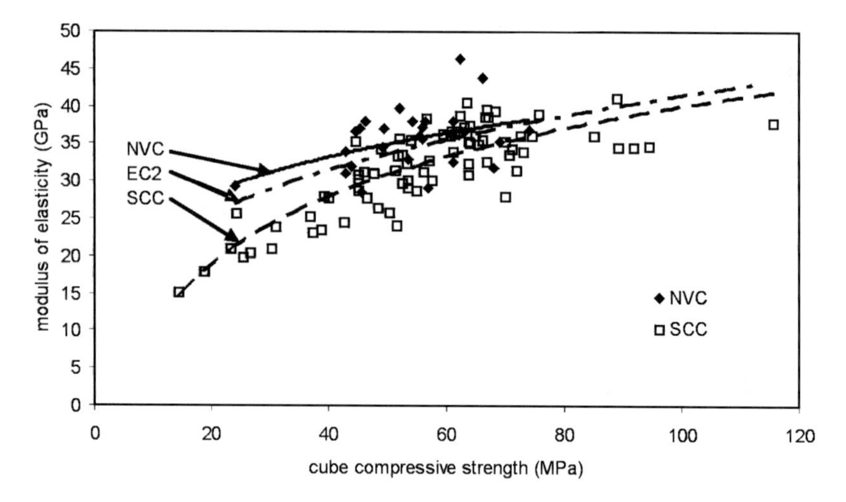

Figure 9.4 *Elastic modulus versus compressive strength. From the survey by Domone*[9].

vibrated concrete[9]. As with TVC, toughness can be increased by adding suitable fibres into the mix. Reduction of workability (consistence) during production of original fibre reinforced concrete had initially been considered too great to enable them to be produced as fresh self-compacting mixes. However, such doubts were dispelled by results of the first European research project on SCC[6] which demonstrated that concrete mixes could be made self-compacting with a substantial content of steel fibres. Steel fibres are known for their capacity to significantly improve the toughness of hardened concrete and provide useful pseudoductility. The SCC mix design can be adjusted sufficiently to counteract the reduction of filling ability (flow) and produce genuine fresh SCC mix, albeit with its passing ability limited by geometry of formwork and reinforcement in relation to the type of fibre, quantity added and method of production used. The toughness of steel fibre-reinforced SCC was included in the development of guidelines for structural design using steel fibre reinforced concrete established by Vandewalle *et al.*, published by RILEM TC 162-TDF[15, 16].

9.3 Tensile and shear strength

In contrast to our knowledge on compressive strength, there is relatively little information available on the effects of change in SCC composition on the tensile and shear strengths.

All data available regarding the tensile strength of SCC are expressed as the indirect tensile strength. This is usually obtained from results of splitting tensile tests, mostly on cylindrical specimens. Comparisons between SCC and TVC are based on the ratio of (indirect) tensile strength to compressive strength. Data available indicate, that for the concretes tested, hardened SCC tended to have a tensile strength the same or better than that of TVC. Substantial differences were encountered, with values being up to 40% higher than for vibrated concrete[6–8,10,11,17–20]. The data available were analysed by Domone[9] who concluded that the ratio remained within the range expected for vibrated concrete, tending to be within its upper half (Figure 9.5)

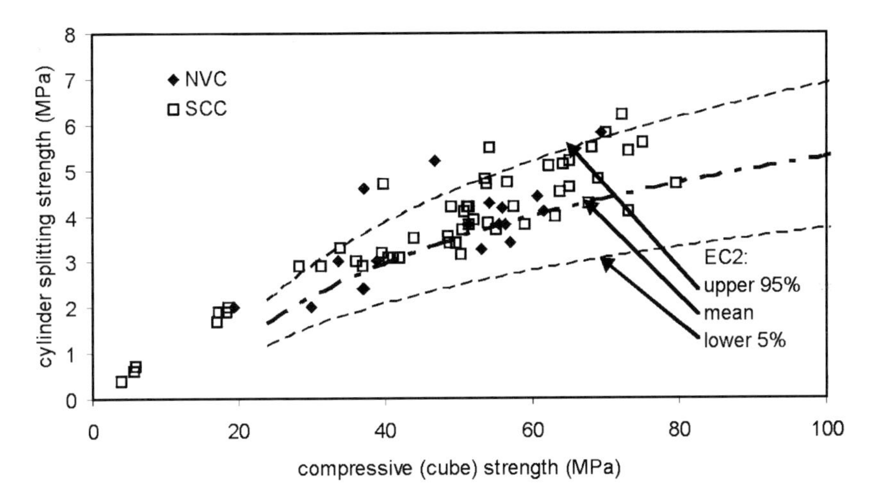

Figure 9.5 *Ratio of the indirect tensile strength (from cylinder splitting) to compressive strength (from*[9]*).*

The shear strength of hardened SCC appears to have received little direct attention. It is conceivable, that the reduced content of coarse aggregate in a 'typical' SCC may reduce the contribution of the 'aggregate-interlock' to shear strength of plain concrete but there is no evidence currently available to clarify this matter. Indirectly, evidence from results of load tests to destruction of typical full-scale structural elements (beams, columns)[10] designed according to the current concrete Eurocode did not show signs of tensile or shear strength below that expected from TVC of the same strength.

It is very likely that any potential reduction in the shear strength of hardened SCC due to fewer particles of coarse aggregate interlocking across the shear fracture surface may in reality be compensated for, or even outweighed by, the gain in strength, coming from denser and therefore stronger hardened binder and mortar.

The overall structural performance of elements made of SCC invariably exceeded that expected and in some instances the higher tensile strength appeared to reduce the amount of cracking during the fracture process[10]. Klug and Holschemacher[8] have also reported compliance in structural behaviour of a wide range of hardened SCCs with that expected from current Eurocodes.

9.4 Creep

The same factors, which govern the rate of development and magnitude of creep of hardened vibrated concrete apply to SCC. The amount of water in the mix, the cement content and the nature/properties of aggregate remain the most important factors.

Initial concerns, based on a perception of the high workability of fresh SCC being achieved by very high water contents, have been shown to be erroneous. The cement content and water content of an ordinary SCC tend to be very similar to those for the TVC, the difference relevant to creep being in the content of coarse aggregate. Testing of SCC mixes both in a research laboratory[21–23] and in trials on mixes for significant practical projects (e.g. for the construction of a prestressed bridge)[24, 25] support the expected conclusions[21]:

- creep decreases with decreasing water/cement ratio or with an increase in cement/powder ratio while the water content remains constant
- the fineness of the additions has no significant effect on creep
- the type of cement affects creep

The magnitude of creep was shown to be the same as for vibrated concrete, which allows the results from tests on SCC to be used in relevant design codes.

Estimates of creep values from functions in current design codes for concrete are applicable to SCCs with the normal range of cement content and moderate-low water/cement ratios. However, SCC mixes can have high cement contents, particularly when for reasons of expediency (in the production process) the increased total powder content is not achieved by suitable additions but by using more cement. The ACI code was then shown to be appropriate while the CEB-FIB code tended to underestimate the deformations[21]. The current structural design procedures for concrete bridges were found to be applicable for the same structures made of SCC, however, more data is required for a more reliable comparison to be made.

Creep can be inadvertently negatively affected if a SCC is produced with an increase in cement content instead of suitable additions.

9.5 Shrinkage

Most predictions of drying shrinkage of TVC focus on the role of the water/cement ratio. A typical SCC may have the same water/cement ratio but very different contents of additions compared with vibrated concrete. The relationship between total shrinkage and composition therefore becomes more complex for SCC. This also explains why the results from tests for the shrinkage of SCC mixes have produced both higher and lower values – apparently conflicting results.

As with comparisons of other characteristics and properties between SCC and vibrated concrete, it is essential to appreciate that direct comparisons are not practical; it is possible to have vibrated and self-compacting concretes of the same shrinkage at the same age in similar conditions, but which have different compositions.

The fundamental relationships have been summarised by Poppe and De Schutter[21]:

- Increase in the water/cement ratio alone increases the pore content and permits higher shrinkage (a prime factor in vibrated concrete with a low content of additions).
- Increase in the water/cement ratio in SCC is less effective, as there is a substantial amount of fine particles (cement and additions), which also require to be 'coated' by a layer of water.

The value of the total shrinkage is therefore dependent not only on the water/cement ratio but also on the water/powder ratio. It can be both greater and smaller that that of vibrated concrete and subject to the same secondary influences.

Attempts have been made to verify the validity of predictive relationships for shrinkage from different codes for concrete design and construction. According to Poppe and De Schutter[21], the CEB-FIB model code 90 underestimates the values (after approximately 50 days), while the ACI model tends to overestimate it. It can be concluded, that shrinkage of SCC is of a magnitude entirely comparable with that of TVC. As with TVC, SCC can be produced with extreme values of high or low shrinkage.

9.6 Bond with reinforcement

9.6.1 Ordinary steel reinforcement

The bond between the standard reinforcement and concrete is generally linked to the compressive strength and toughness (resistance to cracking/splitting) of the concrete. This is expected to apply for any type of concrete, but the very high filling ability (fluidity) of fresh SCC raised concerns regarding bonding at the very beginning of the introduction of SCC into Europe. There was a particular concern regarding horizontal bars, where a potentially significant weak layer of locally high water/cement ratio could form below them in the absence of mechanical compaction. A secondary concern arose from previous findings[26] that the bond between deformed reinforcement and concrete was lower when the concrete contained significantly less coarse aggregate. However, the change to SCC introduced potentially positive effects regarding bonding; a typical SCC has a denser microstructure and its cohesion when fresh reduces or eliminates bleeding, which could reduce the strength of the concrete around reinforcing bars.

A significant investigation of the bond was therefore included in the original European SCC project[6, 10] where the performances of selected 'typical' TVCs and SCCs were compared. Evaluation of large numbers of standard pullout tests[10] initially indicated that there was a distinctly higher bond between the ordinary deformed reinforcement and SCC. However, when adjusted for the higher strength of the SCC, the bond between SCC and reinforcement became equal or slightly higher than that measured on specimens from TVC. The same differential was observed regarding the position of the bars and microstructural investigations were carried out[2, 27–29]. Results revealed that the interfacial zone below the bars was indeed weaker than the one above the bars, although it was in both cases stronger and the difference between top and bottom was less in the case of SCC. Bonds with reinforcement have been investigated as part of other research projects[30–32]; however, the use of SCC did not show any adverse results.

Bond strength is often expressed in terms of the tensile strength of the concrete or the square root of compressive strength f'_c and this expression is also useful for estimates of bond between reinforcement and hardened SCC. Bond values associated with TVC of the same strength category as that of the SCC can be therefore be safely used in structural design.

The results of the first European project on SCC[6, 10] demonstrated that useful amounts of steel fibres could be incorporated in concrete, which is self-compacting when fresh. This has opened possibilities for faster and less expensive production of structural elements in which steel fibres replaced traditional shear reinforcement.

Bonding between such fibre-reinforced SCC and the reinforcement was therefore investigated. There were concerns that the alignment of the fibres in the direction of flow during casting, and their potential settlement and the anisotropy of the material would adversely affect the strength of the bond. Extensive research into the fracture mechanism of structural concrete elements made of SCC with steel fibres by Schumacher[33] and others[34, 35] confirmed that the alignment/orientation of the fibres had to be considered. Fibre reinforcement overall made the bond failure more likely to lead to a pullout rather than to a splitting failure of the surrounding concrete. Steel fibres, even in small quantities appeared to improve the confinement capacity of the concrete around reinforcing bars. A general increase in pullout load bearing capacity was observed when the SCC contained steel fibres. Investigations focusing on the bonding when SCC contains fibres other than steel ones have not been reported.

9.6.2 Prestressing steel

Bond related parameters such as, 'transmission length' and anchorage bond are critical for the development and maintenance of prestress, namely in pretensioned rather than posttensioned concrete. Different types and arrangement of steel bars, tendons or strands are used in current practice. Arbelaez *et al.*[17] and Khayat *et al.*[36] carried out comparative assessments which concluded that for the self-compacting concretes examined:

- no significant difference were found when compared to bonding with vibrated concretes of similar strength
- anchorage lengths were sometimes shorter with SCC, while transmission lengths remained the same
- top-bar effect was less significant in SCC and depended on the viscosity modifying admixture used

There were no fundamental adverse effects of SCC on the bonds in prestressed concrete. Overall, the SCC tended to produce a slightly better performance than TVC.

9.7 Conclusions

Results from an overall survey of extensive data available on the engineering properties of hardened SCC[9] confirmed conclusions, which were drawn from individual sections of this chapter.

The very wide range of materials and mixes already used for SCC produced a significant scatter of data, but clear trends have been obtained between cylinder and cube compressive strength, between tensile and compressive strengths and between elastic modulus and compressive strength, which differed from those for TVC.

It also confirmed that limestone powder, a common addition to SCC mixes, made a substantial contribution to the rate of gain of strength, particularly at early ages. This conclusion provides additional evidence that exploration of environmentally more acceptable SCCs, which may achieve the required practical performance levels with cement contents lower than in TVC may be well justified.

The strength of the bond between SCC and reinforcing and prestressing steel was found to be similar to or higher than that when TVC was used. The variation of the in situ properties in structural elements cast with SCC was similar or lower, when compared to TVC. The performance of structural elements made of SCC was largely as predicted by the measured material properties; no hidden problems were detected. Overall, the conclusions provide full confidence in hardened SCC reaching the required engineering properties and following a predictable structural behaviour.

References

[1] Neville A.M. *Properties of Concrete*, 4th edn., Longman, London, UK, 1995.

[2] Zhu W., Sonebi M. and Bartos P.J.M. Bond and interfacial properties of reinforcement in self-compacting concrete. *Materials and Structures* 2004, 37(8-9) 442–448.

[3] Petersson O. and Skarendahl Å. (Eds.) *Self-compacting Concrete, Proceedings of the First International Symposium*, Stockholm 1999, RILEM Publications, Cachan, France, 2000.

[4] Shah S.R. *et al.* (Eds.) SCC 2005, *Proceedings of the Second North-American Conference on Self-consolidating Concrete and Fourth International RILEM Symposium on Self-compacting Concrete*, Chicago, IL, USA, October–November 2005.

[5] Yu Z., Shi C., Khayat H. and Xie Y. *SCC 2005 China*. RILEM Publications, Bagneux, France, 2005.

[6] Grauers M. *et al.* Rational production and improved working environment through using self-compacting concrete. EC Brite-EuRam Contract No. BRPR-CT96-0366, 1997–2000.

[7] Gibbs J.C. and Zhu W. Strength of hardened self-compacting concrete. In: *Proceedings of the First International RILEM Symposium on Self-compacting Concrete*, Stockholm, Sweden, RILEM Publications, Cachan, France, 1999, pp 199–209.

[8] Klug Y. and Holschemacher K. Comparison of the hardened properties of self-compacting and normal vibrated concrete. In: Wallevik O. and Nielsson I. (Eds.) *Self-compacting Concrete, Proceedings of the Third International Symposium*, RILEM Publications, Cachan, France, 2003, pp 663–671.

[9] Domone P.L. A review of mechanical properties of hardened self-compacting. *Cement and Concrete Composites*, 2007, 29(1) 1–12.

[10] Bartos P.J.M. *et al.* Task 4 in Rational production and improved working environment through using self-compacting concrete. EC Brite-EuRam Contract No. BRPR-CT96-0366, 1997–2000.

[11] Zhu W., Bartos P.J.M. and Gibbs J.C. Uniformity of in-situ properties of self-compacting concrete. *Cement and Concrete Composites* 2001, 23(1) 57–64.

[12] prEN 12504: Testing concrete in structures: Determination of pullout force 2000.

[13] Khayat K.H., Petrov N., Attiogbe E.K. and See H.T. Uniformity of bond strength of prestressing strands in conventional flowable and self-consolidating concrete mixtures. In: Wallevik O. and Nielsson I. (Eds.) *Self-compacting Concrete, Proceedings of the Third International Symposium*, RILEM Publications, Cachan, France, 2003, pp 703–712.

[14] Comite Euro-International du Beton: *CEB-FIP model code 1990: Design code*, Thomas Telford, London, UK, 1993.

[15] Vandewalle L. *et al.* Test and design methods for steel fibre reinforced concrete: Part 1. *Materials and Structures* 2000, 33, 3–5.

[16] Vandewalle L. *et al.* Test and design methods for steel fibre reinforced concrete: Part 2. *Materials and Structures*, 2002, 35, 178–262.

[17] Arbelaez J.C.A., Rigueira V.J.W., Marti Vargas J.R., Serna R.P. and Pinto Barbossa M. Bond characteristics of prestressed strands in self-compacting concrete. In:

Wallevik O. and Nielsson I. (Eds.) *Self-compacting Concrete, Proceedings of the Third International Symposium*, RILEM Publications, Cachan, France, 2003, pp 684–691.

[18] Sonebi M., Tamimi A.K. and Bartos P.J.M. Performance and cracking behaviour of reinforced beams cast with self-consolidating concrete. *ACI Materials Journal*, 2003, 100(6), 492–500.

[19] Zhu W.Z. and Gibbs J.C. Use of different limestone and chalk powders in self-compacting concrete. *Cement and Concrete Research*, 2005, 35(8) 1457–1460.

[20] Brouwers H.J.H. and Radix H.J. Self-compacting concrete: theoretical and experimental study. *Cement and Concrete Research*, 2005, 35(11) 2116–2136.

[21] Poppe A.M. and De Schutter G. Creep and shrinkage of self-compacting concrete. In: Yu Z., Shi C., Khayad H. and Xie Y. (Eds.) *SCC 2005 China*, RILEM Publications, Bagneux, France, 2005, pp 329–336.

[22] Pons G., Proust E. and Assie S. Creep and shrinkage of self-compacting concrete: A different behaviour compared with vibrated concrete? In: Wallevik O. and Nielsson I. (Eds.) *Self-compacting Concrete, Proceedings of the Third International Symposium*, RILEM Publications, Cachan, France, 2003, pp 645–654.

[23] Vieira M. and Bettencourt A. Deformability of hardened SCC. In: Wallevik O. and Nielsson I. (Eds.) *Self-compacting Concrete, Proceedings of the Third International Symposium*, RILEM Publication, Cachan, France, 2003, pp 637–644.

[24] Vitek J.L. Long-term deformations of self-compacting concrete. In: Wallevik O. and Nielsson I. (Eds.) *Self-compacting Concrete, Proceedings of the Third International Symposium*, RILEM Publications, Cachan, France, 2003, pp 663–671.

[25] Bartos P.J.M. Self-compacting concrete in bridge construction – guide for design and construction. Concrete Bridge Development Group, Technical Guide 7, Camberley, UK, 2005.

[26] Martin H. Bond performance of ribbed bars: Influence of concrete composition and consistency. In: Bartos P. (Ed.) *Bond in Concrete*, Applied Science Publishers, London, UK, 1982, pp 289–299.

[27] Bartos P.J.M. and Sonebi M. Hardened SCC and its bond with reinforcement. In: *Proceedings of the First International Symposium on Self-compacting Concrete*, Stockholm, 1999, RILEM Publications, Cachan, France, pp 275–279.

[28] Zhu W. and Bartos P.J.M. Micromechanical properties of the Interfacial Bond in Self-compacting Concrete. In: Balasz G., Bartos P.J.M., Borsonyi A. and Cairns J. (Eds.) *Bond in Concrete: from Research to Standards*, Budapest University Press, 2002, pp 387–394.

[29] Sonebi M., Gibbs J.C. and Zhu W. Bond of reinforcement in self-compacting concrete. *Concrete*, 2001, 35(7) 26–28.

[30] Daoud A., Lorrain M. and Laborderie C. Anchorage and cracking behaviour of self-compacting concrete. In: Wallevik O. and Nielsson I. (Eds.) *Self-compacting Concrete, Proceedings of the Third International Symposium*, RILEM Publications, Cachan, France, 2003, pp 692–701.

[31] Chan Y.W., Chen Y.S. and Liu Y.S. Development of bond strength of reinforcement steel in self-compacting concrete. *ACI Materials Journal*, 2003, 100(4), pp 490–498.

[32] Konig G., Holschemacher K., Dehn F. and Weisse D. Bond of reinforcement in self-compacting concrete under monotonic and cyclic loading. In: Wallevik O. and Nielsson I. (Eds.) *Self-compacting Concrete, Proceedings of the Third International Symposium*, RILEM Publications, Cachan, France, 2003, pp 939–94.

[33] Schumacher P. *Rotation capacity of self-compacting steel fibre reinforced concrete*. Delft University Press, 2006.

[34] Groth P. Fibre-reinforced concrete: Fracture mechanics applied on self-compacting Concrete and energetically modified binders. Ph.D. thesis, University of Lulea, Sweden, 2000.

[35] Grunewald S. *Performance based design of self-compacting fibre reinforced concrete.* Delft University Press, 2004.

[36] Khayat K.H., Petrov N., Attiogbe E.K. and See H.T. Uniformity of bond strength of prestressing strands in conventional flowable and self-consolidating concrete mixtures. In: Wallevik O. and Nielsson I. (Eds.) *Self-compacting Concrete, Proceedings of the Third International Symposium*, RILEM Publications, Cachan, France, 2003, pp 703–712.

Chapter 10
Durability

10.1 Introduction

A properly designed concrete in agreement with the environment in which it has to be used, is an extremely durable material. Inadequate design or production, however, may lead to deterioration of the material. To design a concrete for a specified minimum service life, it is necessary to understand the processes that cause deterioration, including the rates at which these will occur under the conditions to which concrete will be subjected[1]. These processes are strongly dependent on the ease with which potentially aggressive fluids enter the material, expressed by the transport properties of the material. For this reason we will discuss both transport properties and deterioration processes. The transport properties include gas transport (permeability, diffusion), liquid transport (permeability, capillary absorption) and ion transport. The deterioration processes that we will discuss are: carbonation, chloride penetration, freeze–thaw attack, sulfate attack, alkali–aggregate reaction (AAR/ASR), chemical attack and fire resistance.

The transport properties and deterioration processes are to a large extent dependent on, among other things, the composition of the material and therefore the type of concrete (e.g. self-compacting). This chapter summarises experimental results on both the transport properties and durability aspects of SCC.

10.2 Transport properties

Movements of gases, liquids and ions through concrete occur due to various combinations of differentials in air pressure, water pressure, humidity, concentration or temperature. Depending on the driving force of the process and the nature of the transported matter, the transport processes for deleterious substances through concrete may be diffusion, absorption and permeation[2].

10.2.1 Gas transport

Permeability

Theoretical background

Gas permeability is a measure of gas flow through a porous material caused by a pressure head. This property depends to a large extent on the open porosity prevailing in the material, which is also related to the degree of saturation. The gas pressure acting in the pores is not sufficient to move the water, so the pores will remain blocked and are unable to let the gas pass. For this reason it is important to take into account the moisture content of the concrete. Other parameters such as the pore connectivity, tortuosity and isotropy of the pore network also have an influence.

Different test methods are commonly used to determine the gas permeability both in situ test methods and in the laboratory. A distinction is also made as to whether the gas permeability is defined by means of a permeability index, or is calculated on the basis of the laws of fluid dynamics. The determination of the permeability index can be established by inducing an initial standard pressure and measuring the loss of pressure in a given time. The disadvantage of this method is the dependence of the result on the apparatus, the type of gas and the initial pressure. This makes the comparison of different methods difficult[3–5]. Test methods which are used in situ are 'Figg and Kasai', 'Paulmann', 'the German method (GGT)', 'Parrott', 'Reinhardt–Mijnsbergen', 'Schönlin', 'Torrent' and 'Surface Air Flow (SAF)'[3, 6, 7].

The most common laboratory test method consists of admitting gas under pressure to one side of the specimen and measuring the flow either at the inlet or at the outlet. Depending on the apparatus a constant pressure gradient may or may not be maintained. One of the methods commonly used is the RILEM–Cembureau method[8]. The apparatus maintains a constant pressure gradient. An important issue is the preparation of the test specimens. Several procedures have been proposed in recent years[8–10].

Gas permeability is mostly determined by means of the apparent gas permeability k_a which can be calculated by means of Equation (10.1). This formula is based on the Hagen–Poiseuille relationship for laminar flow of a compressible fluid through a porous body with small capillaries under steady-state conditions[11].

$$k_a = \frac{4.04Q}{A}\frac{LP_2}{(P_1^2 - P_2^2)}.10^{-16} \quad (\mathrm{m}^2) \tag{10.1}$$

with Q the volume flow rate of the fluid measured during the test with a bubble flow meter (ml/s), L the thickness of the specimen in the direction of the flow [m], A the cross-sectional area of the specimen (m^2), P_1 and P_2 are respectively, the inlet and outlet pressure of oxygen (bar). This formula is valid for tests performed at 20°C for which the viscosity of oxygen is 2.02×10^{-5} Nsm^{-2}.

Experimental results for self-compacting concrete

Zhu *et al.*[12] and Zhu and Bartos[13] compared a range of different SCC mixes to TVC mixes of the same strength grade (C40 and C60). The gas permeability was investigated by means of the Cembureau method[8]. The SCC mixes listed in Table 10.1 were designed to either contain limestone powder, PFA or a viscosity agent. Portland cement has been used for all the mixes. For C40 concrete, the SCC mixes have a significantly lower coefficient of gas permeability than the corresponding TVC mixes. For the SCC mixes containing PFA and limestone powder, the coefficient is even only 30–40% of the corresponding TVC mixture. The same trend can be noticed for the C60 mixes. When there is no additional powder added and only a viscosity agent has been used, a considerably higher coefficient of gas permeability is achieved. According to the authors the lower gas permeability can be attributed to the improved ITZ and the denser microstructure of SCC.

Rougeau *et al.*[14] have presented results of two SCC mixes (w/c ratios of 0.49 and 0.47) and one high performance concrete (HPC). After 7 days of drying at 80°C the

Table 10.1 *Overview of the SCC compositions of whose gas permeability has been determined.*

References	*Type of cement*	*C(kg/m³)*	*Type of filler*	*F(kg/m³)*	*W (kg/m³)*
12, 13	CEM I 42.5	285	Limestone	265	180
	CEM I 42.5	320	Limestone	230	167
	CEM I 42.5	335	PFA	145	195
	CEM I 42.5	410	PFA	100	177
	CEM I 42.5	360	–	–	210
	CEM I 42.5	475	–	–	196
14	CEM I 52.5 R	352	Limestone	101	172
	CEM I 52.5 R	350	Limestone	100	165
15	CEM II/A-LL 32.5 R	315	Limestone	150	240
	CEM II/A-LL 32.5 R	315	Limestone	150	200
	CEM I 52.5 N	350	Limestone	140	200
	CEM I 52.5 N	450	Limestone	70	190
16	CEM I 42.5	340	PFA	150	196
	CEM I 42.5	290	Limestone	260	180
	CEM I 42.5	365	–	–	212

gas permeability of the SCC mixes was higher than of the HPC. However, after drying at 105°C the gas permeability of the mix with a w/c ratio of 0.47 was closer to the permeability of the HPC. It was mentioned that the porosity of the SCC mixes was higher compared to HPC.

Investigations based on completely dry samples of SCC and TVC of strength grades C20, C40 and C60 have been reported by Assié *et al.*[15, 17]. The intrinsic gas permeability is found to decrease with increasing age. Compared to TVC mixes, the SCC mixes exhibit lower coefficients of gas permeability. This is probably due to a finer pore network of SCC.

Boel[18] investigated the gas permeability of 16 self-compacting mixes and 4 traditional mixes (see Table 8.6 in Chapter 8). In this research the influencing factors were: type of cement, type of filler, amount of powder (cement and filler), amount of water, type of coarse aggregate and w/c ratio. The gas permeability was measured at several saturation degrees. The results indicate that traditional concrete has a much higher gas permeability than SCC. It was also found that the gas permeability of completely dry specimens can be calculated as a function of the capillary porosity raised to the power of 2.4. Other factors influencing the gas permeability are the inlet pressure and the morphology of the pore network. At the same porosity a material with finer pores has lower gas permeability. More information about the microstructure of SCC can be found in Chapter 8.

Equation (10.2), based on the capillary porosity, has been proposed in order to estimate the apparent gas permeability $k_{a(s=0)}$. The capillary porosity is calculated based on the theory of[19], as mentioned in Chapter 8.

$$k_{a(s=0)} = c.a.\varepsilon_c^{\ 2.4}.k_{a,ref} \quad (m^2) \tag{10.2}$$

With $k_{a,ref}$ a reference value equal to 10^{-16} m^2, ε_c the capillary porosity (-), the factor a takes into account the inlet pressure and the factor c takes into account the difference in microstructure between SCC and TVC (-). The difference in microstructure can be seen as a function of the critical pore size. In Figure 10.1, at an inlet pressure of 3 bars, the estimated apparent gas permeability is plotted versus the calculated apparent gas permeability. A good correlation is apparent.

Using Van Genuchten's model, it has been proved possible to calculate the gas permeability of a partly saturated concrete. The influence of the saturation degree at a given inlet pressure can be investigated by means of the introduction of the relative gas permeability $k_{r(S,P)}$ (Equation (10.3))[20]. For this purpose the assumption is made that the moisture in the sample is homogeneously distributed. The drying procedure was carried out so this assumption was met.

$$k_r(S,P) = \frac{k_{(S=x)}}{k_{(S=0)}} \quad (-) \tag{10.3}$$

Where $k_{(S=x)}$ and $k_{(S=0)}$, respectively, the apparent gas permeability at a certain inlet pressure at a certain saturation degree ($S = x$) and at the completely dry state ($S = 0$) of the sample. Van Genuchten, mainly active in the field of soil physics, proposed the following equation to relate the relative gas permeability to the saturation degree[20].

$$k_r(S) = \sqrt{(1-S)}.(1-S^b)^{2/b} \tag{10.4}$$

Where b is a model constant and S is the saturation degree. In Figure 10.2, for several inlet pressures, the relative gas permeability of SCC1 (see Table 8.6 in Chapter 8) is

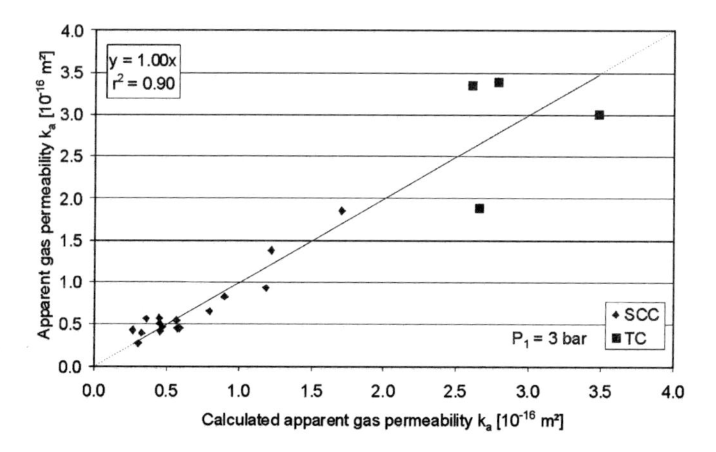

Figure 10.1 *Apparent gas permeability versus the calculated apparent gas permeability*[18].

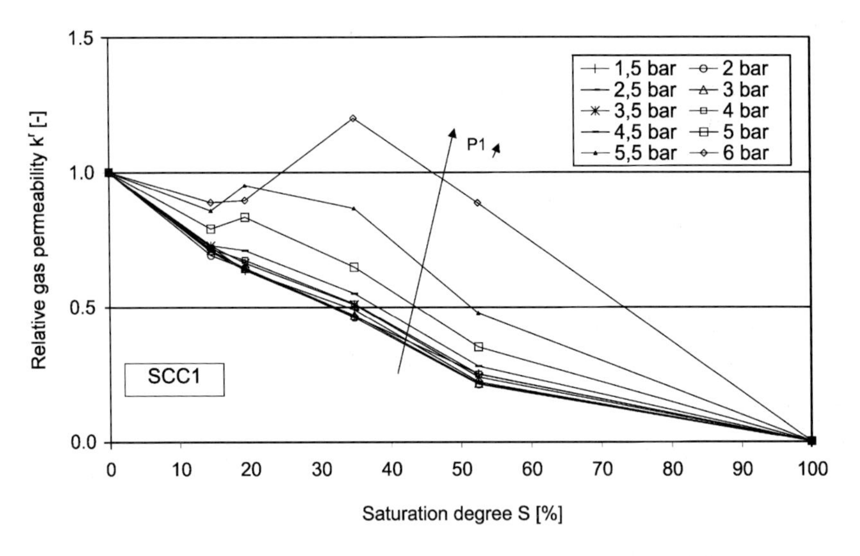

Figure 10.2 *Relative gas permeability plotted versus the saturation degree*[18].

plotted versus the saturation degree. It can be seen that at low pressures the curves approximately coincide with each other. The same was concluded for traditional concrete 1 (TC1) (Table 8.6). At higher inlet pressures, above approximately 5 bars, the curves no longer overlap.

Figure 10.3 has been used to investigate whether or not Van Genuchten's equation can be used to write the relative gas permeability as a function of the saturation degree. As can be seen from Figure 10.3, this works quite well. This has been found not only for SCC1 and TC1, but for all investigated mixes. The values of b are

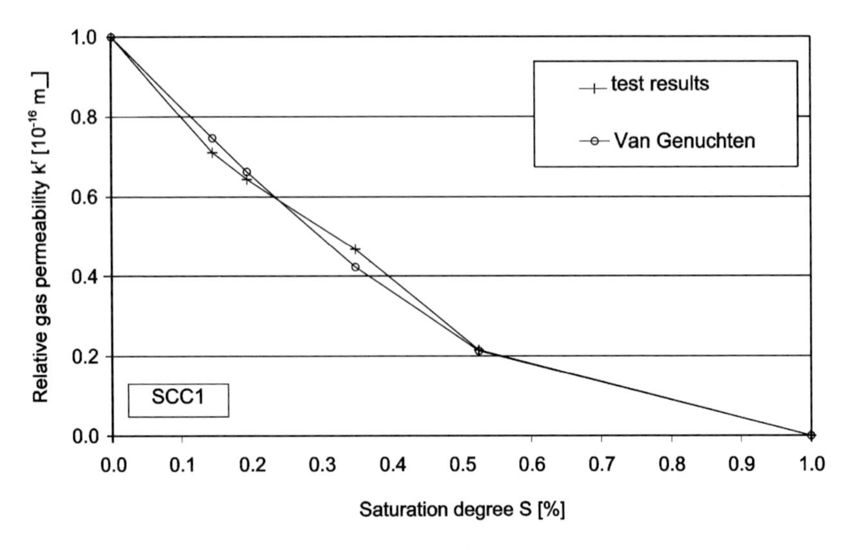

Figure 10.3 *Testing of the Van Genuchten model*[18].

typically of the order of 1 and tend to have slightly higher values for SCC than for TVC. From these results it follows that it is possible to determine the gas permeability at a certain saturation degree once the gas permeability at the dry state, $k_{(s=0)}$, is known, according to Equation (10.5).

$$k_{(s=x)} = k_{(s=0)} \cdot \sqrt{(1-s)}.(1-s^{b})^{2/b} \quad (\mathrm{m}^2) \tag{10.5}$$

A permeability cell following the Cembureau method has been used to obtain all the results mentioned above. Sonebi and Ibrahim[16] studied the gas permeability using the Autoclam method. The tests were conducted on the surface of concrete slabs which were preconditioned at 40°C for 14 days. An air permeability index is obtained, and the results reflect the same trends as mentioned above.

Conclusions

The results mainly indicate that, when fillers are used, the gas permeability of SCC is significantly lower than of the corresponding TVC. The reason can be found in a denser microstructure of the SCC[18]. The gas permeability is one of the parameters giving an indication of the resistance of concrete against several kinds of deterioration. From the parameters studied it follows that the gas permeability is mainly influenced by the relative volume of cement matrix, the amount of water, the w/c ratio and the w/p ratio. These parameters also influence the capillary porosity and the pore structure of the cement matrix and the concrete. Based on the capillary porosity and the critical pore size it is possible to estimate the apparent gas permeability of both SCC and TVC. The influence of the saturation degree on the gas permeability can be taken into account by applying Van Genuchten's proposed model.

Diffusion

Theoretical background

If a different concentration of a gas is present at the concrete surface and in the concrete, or a concentration gradient exists through the concrete element, diffusion of the gas through the concrete will take place in order to obtain equilibrium. In concrete practice, two different cases are occurring continuously: the diffusion of water vapour through concrete and the diffusion of CO_2 molecules into the concrete.

The diffusion of water vapour into or out of the concrete leads to changes in the humidity balance in the pores of the concrete. These changes influence the velocity of some deterioration processes, e.g. a high humidity in the concrete leads to slower carbonation or to higher chloride ingress. The diffusion of CO_2 molecules leads to carbonation of the concrete, inducing corrosion of the reinforcement if air and water are present. Diffusion of a gas in concrete could be described by Fick's law, describing the proportionality between the amount of molecules passing through a certain surface area, J (kg/(m^{2}s)), and the concentration gradient:

$$J = -D\frac{dc}{dx} \tag{10.6}$$

where D is the diffusion coefficient (m^2/s), c is the concentration of the gas (kg/m^3) and x is the distance in the flow direction (m). This equation can be written as:

$$J = D\frac{c_1 - c_2}{x} \qquad (10.7)$$

where c_1 is the concentration of the gas at the concrete surface and c_2 is the concentration at depth x in the concrete. In the case of water vapour diffusion, the concentration can be calculated from the different partial pressures or from the relative humidity. In the case of carbonation, the concentration in the air is known: approximately 0.03 vol.%, increasing to 0.3% in cities and exceptionally to 1% in industrial zones. The concentration inside the concrete equals 0 in the uncarbonated zone.

Experimental results for self-compacting concrete

The determination of the diffusion coefficient of gases through SCC is very limited. Many research programmes are investigating carbonation behaviour, unfortunately without determining the diffusion coefficient itself. Mostly, the carbonation depth is obtained experimentally and a carbonation coefficient is derived (see Section 10.3.1).

However, one research project which has determined the diffusion coefficient of CO_2, has been carried out at Chalmers Technical University and reported in[21]. In this research one SCC, with limestone powder, and one TVC with the same w/c ratio are compared. The diffusion coefficient to CO_2 was determined: the SCC had a lower diffusion coefficient than the TVC. This result is explained by the denser microstructure of the SCC. The microstructure was investigated with SEM-EDS, image analysing and light microscope techniques, which indicated that the porosity of the ITZ and of the bulk paste is reduced.

Research on the water vapour diffusion of SCC has been carried out by Reinhardt and Jooss[22] and by Audenaert[23]. Reinhardt and Jooss determined the water vapour diffusion coefficient at different temperatures between 20°C and 80°C and with two different concentration gradients: in the ranges 0–50% and 50–93%. Several types of concrete were tested, including two SCC in two different strength classes (C30/37 and C55/67). The diffusion coefficient of SCC was similar to the values for TVC in the same strength class.

Audenaert[23] investigated the water vapour diffusion, among other transport properties, of 16 self-compacting mixes and 4 traditional mixes (see Table 8.6 in Chapter 8). The following influencing factors were considered: type of cement, type of addition, amount of powder (cement and addition), amount of water, type of coarse aggregate, w/c ratio. The water vapour diffusion was measured at 20°C with a concentration gradient in the range 93–60%. In the literature[24], the gas diffusion is a function of the capillary porosity to the power 1.8. The capillary porosity is calculated based on the theory of Powers and Brownyard[19], as explained in Chapter 8. The diffusion coefficient as a function of the capillary porosity is given in Figure 10.4, and the equation describing this relation is:

$$D = 2100\, \varphi_{cap}^{1.8}\, 10^{-8} \qquad (\mathrm{m}^2/\mathrm{s}) \qquad (10.8)$$

Conclusions

The diffusion of a gas through concrete mostly occurs in two cases: the diffusion of water vapour and the diffusion of CO_2. The first one influences the pore humidity, the latter the carbonation process.

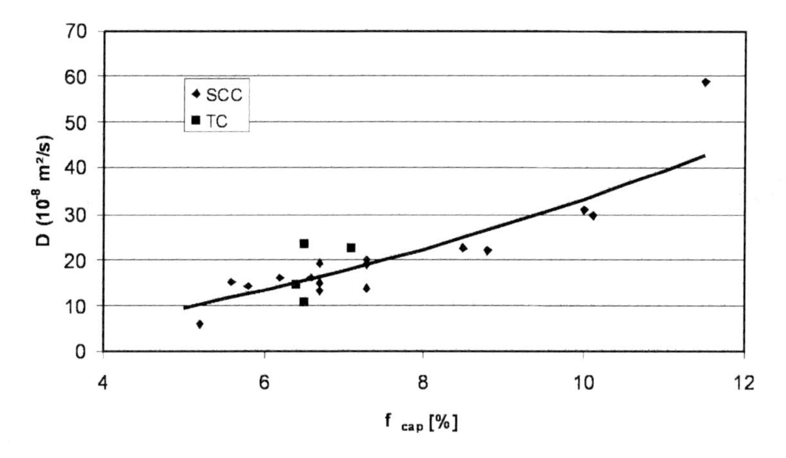

Figure 10.4 *Relationship between capillary porosity and water vapour diffusion coefficient for SCC and TVC*[23].

The diffusion coefficient is determined by the microstructure of the concrete. The denser the pore structure, the smaller the diffusion coefficient. As the ITZ of SCC is more dense, the diffusion of gases is slower in SCC, although not all research projects described in the literature came to the same conclusion.

10.2.2 Liquid transport

Permeability

Theoretical background

In the presence of a pressure gradient through the concrete, for steady-state saturated flow of liquid the flow will be laminar for small pressure differences, and Darcy's law can be applied. In the case of one-dimensional flow of a Newtonian fluid, this can be written as:

$$Q = AK\frac{h}{L} \tag{10.9}$$

Where Q the flow (m^3/s), A the surface (m^2), K the permeability coefficient (m/s), Δh the difference in total potential head (m) and L the thickness of the concrete (m).

In concrete it takes a long time to reach the steady-state regime for small pressure gradients. As well as the long test duration the main disadvantage of this type of test is that part of the unhydrated cement will hydrate during the test, which will cause a decrease in the water permeability coefficient with time.

For this reason, nonsteady-state tests are often used, in which the test specimen is split after a certain test duration and the penetration depth determined. The depth is shown by the discolouration of the broken surface. The main advantage of this type of test is the short duration of the test, but the main disadvantage is that the water permeability cannot be determined because the specimens are not saturated before the start of the test. This implies that part of the penetrated water is absorbed by capillary forces.

Experimental results for self-compacting concrete

Most of the research programs described in the literature (see Table 10.2) determined the water penetration depth instead of the water permeability, as described above. This water penetration depth is determined after exposure to a certain water pressure during a certain period and finally splitting of the test specimen. This type of test[25–27] is carried out, following, respectively, the German (DIN 1048[28]), Chinese (GBJ 82-85[29]) and Dutch (NEN EN 12390-8[30]) standards. In these research projects, respectively 4, 1 and 3 SCC compositions were tested. No traditional concrete was made in order to make a comparison. The conclusion was that the SCC was 'watertight'.

Reinhardt and Jooss[22] have determined the water penetration depth for two SCC mixes and four traditional mixes of different strength classes. The conclusion is that the water penetration depth of SCC is equal to the water penetration depth of TVC in the same strength class.

Audenaert[23] determined the water permeability under a pressure height of approximately 500 mm of 16 self-compacting mixes and 4 traditional mixes (see Table 8.6 in Chapter 8). The investigated parameters are: type of cement, type of filler, amount of powder (cement and filler), amount of water, type of coarse aggregate, w/c ratio. Figure 10.5 shows the water permeability of the SCC and TVC mixes as a function of the capillary porosity, which is calculated based on the theory of Powers and Brownyard[19] explained in Chapter 8.

Conclusions

The water permeability of concrete is a property that can be determined with different test methods. The methods differ strongly with regard to the applied water pressure gradient and the preparation, and all have a significant influence on the resulting permeability coefficient, making it very difficult to compare the results. Most

Table 10.2 *Overview of the SCC compositions of whose water permeability has been determined.*

References	*Type of cement*	*C (kg/m³)*	*Type of filler*	*F(kg/m³)*	*W (kg/m³)*
25	CEM II/A-L 32.5	450	–	–	170
	CEM II/A-L 32.5	450	–	–	183
	CEM I 42.5	340	Fly ash	105	185
	CEM I 42.5	330	Fly ash	105	173
26	OPC 525R	303	–	–	181
22	CEM II/A-L 32.5	320	Fly ash	180	144
	CEM II/A-L 32.5	370	Fly ash	170	166
27	CEM III/B 42.5 N LH/HS	310	Limestone powder	189	170
		315	Limestone powder	164	173
		320	Limestone powder	153	174

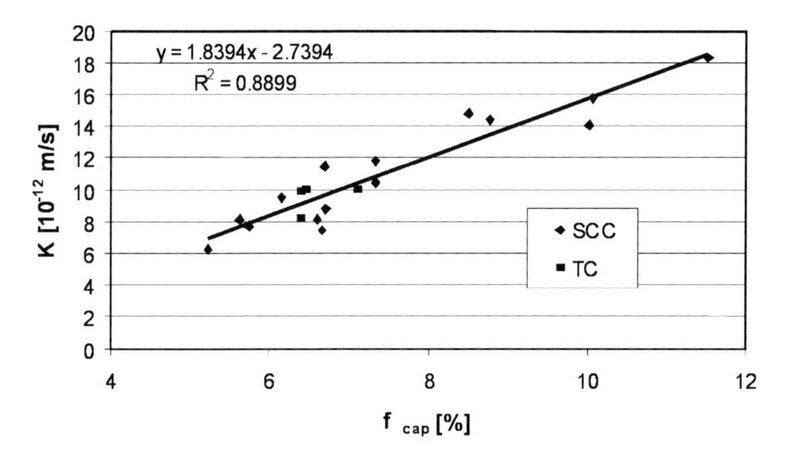

Figure 10.5 *Relationship between capillary porosity and water permeability coefficient*[23].

research described in the literature focuses on the comparison of SCC mixes with traditional concrete. The conclusion is that the water permeability of SCC is comparable to that of TVC of the same strength class.

Capillary absorption

Theoretical background

If a nonsaturated material is in contact with a fluid, flow of the fluid is created by capillary absorption. This flow is determined by the properties of the fluid, such as viscosity, density, surface tension etc. and by the properties of the material, such as the pore structure (tortuosity, pore size distribution, connections between pores etc.) and the surface energy.

Two possibilities exist:

- The material is in contact with the fluid for a long period and a permanent regime is reached. For example, this is the case if a concrete structure is permanently in contact with water on one side and on the other side, the water can evaporate.
- The material is not permanently in contact with water and a nonpermanent regime is present. This is the case, for example, for surfaces exposed to rain.

The absorbed water often contains other substances, e.g. chlorides. In this way, capillarity plays a very important role in the penetration of damaging constituents into concrete.

Equation (10.9) can also be written as a function of the flux q (m/s), total potential h (m) and stream length x (m):

$$q = \frac{Q}{A} = K\frac{\partial h}{\partial x} \tag{10.10}$$

for an unsaturated medium, this formula can be written as:

$$q = \frac{Q}{A} = K(\theta)\frac{\partial h}{\partial x} \tag{10.11}$$

In this formula, K is no longer a constant but a function of the volume fraction of liquid θ (-). This equation can be combined with the one-dimensional continuity equation:

$$\frac{\partial \theta}{\partial t} = \frac{\partial q}{\partial x} \tag{10.12}$$

leading to:

$$\frac{\partial \theta}{\partial t} = \frac{\partial}{\partial x}\left[K(\theta)\frac{\partial h}{\partial x}\right] \tag{10.13}$$

In some cases, the influence of gravity can be neglected. This is the case for example for capillary absorption into a vertical surface, or for construction elements being too small to reach an equilibrium state, e.g. dimensions smaller than 2 m in the direction of suction etc. For these applications, Equation (10.13) can be simplified and solved as written in[31], in which the amount of absorbed fluid as a function of time t is:

$$w = St^{1/2} \tag{10.14}$$

where w is the increase in mass per surface unit (kg/m^2) and S the absorption coefficient in kg/m^{2}s$^{1/2}$. If S is written in mm/s$^{1/2}$, w is a mean penetration depth of the wet front (m). Equation (10.14) is often described and is frequently applied in the literature.

Experimental results for self-compacting concrete

Many test methods are described in the literature, some are test methods to be used in situ[3, 32, 33]. The main disadvantages of these test methods are that the flow patterns are complicated and difficult to calculate and that the moisture content is unknown and difficult to determine.

Other test methods can only be used in laboratories. The main differences can be found in the preparation of the test specimens (coating, temperature and duration of drying etc.), the direction of the flow (downward, upward, horizontal flow), the duration of the water uptake and the quantity that is measured (mass increase or penetration depth).

Mostly, the experimental programmes described in the literature focus on the difference in capillary absorption behaviour between SCC and TVC. Some are reporting a smaller water uptake[12–14, 34, 35] and some describe an equal behaviour [15, 36–38]. Audenaert[23] investigated the water uptake by capillary absorption of 16 self-compacting mixes and 4 traditional mixes (see Table 8.6 in Chapter 8). The test was carried out for one week. The test results, giving the mass increase in function of $t^{1/2}$, clearly demonstrate a bilinear behaviour. The first steep part is found to be determined by the porosity of the outer zone of the concrete (15 mm), where more mortar is present and less coarse aggregate due to the wall effect. A slower uptake is noticed when the water front penetrated more than 15 mm into the concrete. The inner zone of the concrete has a lower porosity than the outer zone due to the presence of the coarse aggregate leading to a smaller water uptake. It is concluded that the outer zone of TVC is more susceptible to drying out than the outer zone of SCC. This leads to a larger water uptake in the first hours and a larger total water uptake for TVC.

Conclusions

Due to the denser pore structure of SCC, the water uptake due to capillary absorption is found to be larger or equal in TVC than in SCC in laboratory testing. Based on these results, it is expected that SCC will also perform well with respect to capillary absorption in in situ measurements.

10.2.3 Ion transport

Theoretical background

The degradation of concrete structures often involves the coupled transport of fluids (water, oxygen etc.) and ions (chlorides, sulfates etc.) through the pore structure. While the phenomena of gas and liquid transfer appear to be reasonably understood, many papers have emphasised the complexity of ionic transport mechanisms. For instance, it has been clearly established that the penetration of chloride ions through the concrete pore structure is affected by a wide range of parameters such as the transfer of liquid water, the movement of all other ions in solution and the local ionic concentration of the electrolytic solution. It is well known that the mechanisms of diffusion are also influenced by the physical and chemical interactions of ions with the solid phases of the material.

Ions in any cement system can be found in two different forms. They can be free in solution or bound to the solid matrix. In most cases, the movement of the free ions through the system occurs strictly in the liquid phase and the ionic flow in the solid phases can be neglected.

Ions cannot exist alone, the presence of other ions is necessary to keep the solution electrically neutral. In case of chloride diffusion, it has been found that cations, e.g. Na+, K+,... have a lower diffusion rate than Cl– ions[39–41]. Once chloride ions move forward, a counter electrical field between the chloride ions and the surrounding slower cations will be formed. This counter electrical field tends to draw back the chloride ions[41, 42]. The combination of the driving force (chemical potential) and the drawback force (counter electrical field), is called ‘electrochemical potential’. The actual ionic diffusion is in fact under the gradient of electrochemical potential $\nabla\mu$

$$\nabla\mu = -RT\left(\frac{\partial \ln\gamma}{\partial x} + \frac{\partial \ln c}{\partial x}\right) - zF\frac{\partial\phi}{\partial x} \tag{10.15}$$

where $\nabla\mu$ is the gradient of electrochemical potential (J/mol), R is the gas constant (8.314 J/molK), T is the temperature (K), γ is the activity coefficient (-), c is the molar concentration, x is the distance (m), z is the valence of the ions (-), F is the Faraday constant (9.648 10^4 J/V mol) and ϕ is the counter electrical potential (V).

The diffusion process can be expressed by the Nernst–Plank equation:

$$= -\frac{D}{RT}c\nabla\mu = -D\frac{\partial c}{\partial x}\left(1 + \frac{\partial \ln\gamma}{\partial \ln c}\right) - cD\frac{zF}{RT}\frac{\partial\phi}{\partial x} \tag{10.16}$$

where J is the flux of ions (kg/m^{2}s) and D is the diffusion coefficient (m^2/s).

The terms $\frac{\partial \ln \gamma}{\partial \ln c}$ and $\frac{\partial \phi}{\partial x}$ are often neglected in the literature. Thus Equation (10.17) becomes a general form of Fick's first law:

$$J = -D\frac{\partial c}{\partial x} \tag{10.17}$$

Experimental results for SCC

The experimental results of ion transport in SCC, discussed in the literature, describe the transport of chloride ions through concrete. These experimental results are presented in Section 10.3.2, dealing with chloride penetration in concrete as a deterioration process.

As far as is known, no results of transport of ions in SCC other than chloride ions have yet been published.

Conclusions

Ion transport in concrete is, in comparison to gas and liquid transport, a very complex transport mechanism that is not perfectly understood. This transport is affected by a wide range of parameters. The equations describing this transport are given and simplified. No test results are given in this section since only chloride penetration in SCC has been described so far. This topic is discussed in Section 10.3.2, dealing with chloride penetration as a deterioration process.

10.3 Deterioration processes

The deterioration processes can have a physical or a chemical origin. This distinction is merely artificial since physical and chemical causes are often combined. In most cases the mobility of fluids (water, carbon dioxide, oxygen) is involved. This is why, in the first part of this chapter, the transport properties have been discussed.

The physical causes of concrete deterioration can be divided into:

- surface weathering and mass loss due to abrasion, erosion and cavitation
- crack formation due to normal changes of temperature and humidity, crystallisation pressures of salts in the pores, structural loading and exposure to extreme temperatures by frost and fire[43]

In the cement matrix there is a high pH (12.5–13.5) due to large amounts of OH^- ions in the pore solution. When concrete is in contact with an acid environment, the balance is disturbed. A decrease in alkalinity can cause a destabilisation of the hydration products and as such each environment with a pH lower than 12.5 can be considered aggressive. The rate of attack depends on the pH of the aggressive agent and the transport properties of the concrete. The harmful effects due to acid attack are revealed by an increase in porosity and penetrability, a decrease in strength, crack formation and mass loss. The chemical causes can be classified into three categories: hydrolysis of phases of the cement matrix due to the presence of soft water, cation exchange between aggressive fluids and the hardened cement paste, and reactions which lead to the formation of expansive reaction products.

10.3.1 Carbonation

Theoretical background

Carbonation is the chemical reaction of the hydrated cement paste with CO_2 molecules, diffusing from the air into the concrete.

$$Ca(OH)_2 + CO_2 + H_2O \rightarrow CaCO_3 + 2H_2O \tag{10.18}$$

Carbonates are formed and OH^- ions are consumed by this chemical reaction. This leads to a drop in the pH value of the pore water. Once this value decreases below 9, the steel becomes depassivated, and can corrode in the presence of water and oxygen.

The CO_2 molecules diffuse into the concrete through the pores, where they react with the hydrated cement paste. Fick's first law (Equation (10.19)) is applicable, stating that the velocity of the molecules diffusing through a unit section, J (mol/m^{2}s), is proportional to the concentration gradient dc/dx. In this, c is the concentration of CO_2 (mol/m^3) and x the distance from the concrete surface (m). In case of one-dimensional diffusion, this law is:

$$J = -D\frac{dc}{dx} \tag{10.19}$$

Where D is the diffusion coefficient (m^2/s). In most cases, the carbonation depth x is determined with a colour indicator such as phenolphthalein, which when sprayed onto a fresh concrete surface will be coloured pink where the concrete is not carbonated and remains colourless in the carbonated zone. This is due to the change in pH of the pore water in the carbonated zone from approximately 13 to 8, which leads to the simplification that a front of CO_2 molecules is passing through the concrete. At this front, there is a constant concentration gradient. This concentration gradient is equal to $c_1 – c_2$ with c_1 and c_2, respectively, the concentrations at the concrete surface and in the uncarbonated zone. Equation (10.19) can be written as:

$$J = D\frac{c_1 - c_2}{x} \tag{10.20}$$

The flux J can also be written as the amount of diffusing CO_2 molecules dQ (mol) divided by the time t (s) and the exposed surface S (m^2).

$$J = \frac{dQ}{S\,dt} \tag{10.21}$$

This amount of diffusing CO_2 molecules, dQ, reacts with a certain amount of hydration products in the concrete. This amount of reacting hydration products in a unit volume of concrete is called a (mol/m^3), so:

$$dQ = a\,S\,dx \tag{10.22}$$

Combining these equations leads to:

$$x\,dx = \frac{D}{a}(c_1 - c_2)\,dt \qquad (10.23)$$

If D, a, c_1 and c_2 are assumed to be constant with time, integration leads to:

$$x = A\sqrt{t} \qquad (10.24)$$

where

$$A = \sqrt{2\frac{D}{a}(c_1 - c_2)} \qquad (10.25)$$

and thus, A is a constant with time (m/s$^{0.5}$), and is called the carbonation coefficient. In reality A is not constant in time, due to:

- The time dependent hydration: ongoing hydration changes the pore structure and the amount of hydration products.
- The 'wall effect': a larger amount of mortar is present at the concrete surface, consisting of more hydration products per volume unit and thus binding more diffusing CO_2 molecules. Also the pore volume is larger in this zone, leading to a higher diffusion coefficient.
- The diffusion coefficient is strongly dependent on the pore humidity. An increase in pore humidity causes a strong decrease in the diffusion coefficient.
- The curing of the concrete: due to a lack of curing a coarser pore structure is developed with less hydration products, leading to an increase in the carbonation coefficient.

From Equation (10.25) it can be seen that the material properties governing the carbonation process are the diffusion velocity of CO_2 molecules through the concrete and the amount of carbonatable material present in the concrete. The diffusion velocity is mainly influenced by the porosity of the concrete and the amount of moisture in the pores. The larger the porosity and the lower the amount of water in the pores, the higher the diffusion velocity. The porosity is influenced by the w/c ratio, the curing etc. The amount of carbonatable material depends on the binder system including filler material. The more CSH and CH present in the concrete, the lower the carbonation depth. This is valid both for TVC and SCC.

Experimental results for self-compacting concrete

Many test programmes have been described[14, 15, 21, 23, 44–50]. From the experimental results, SCC sometimes has a larger and sometimes a smaller carbonation depth in comparison with traditional concrete with the same water and cement content, although the differences are small. It should be noted that:

- It is very difficult to compare with traditional concrete since it is not clear which basis of comparison should be used: the same amount of water and cement, the same compressive strength etc.

- If SCC is properly cured the pore structure is denser and less permeable than that of TVC.
- The amount of carbonatable material is very important and is normally higher in SCC.

Conclusions

From the test results described in the literature, the carbonation behaviour of SCC is seen to be similar to that of TVC. However, only a few fundamental experimental programmes have been carried out so far. In this way, the fundamental insight in the differences in carbonation behaviour between SCC and TVC is lacking and further fundamental research is needed. Nevertheless, it seems that the carbonation of SCC is not significantly deviating from the carbonation of TVC.

Furthermore, some general conclusions can be formulated concerning the carbonation of SCC. The carbonation process is influenced by the diffusion velocity of CO_2 molecules through the concrete and by the amount of carbonatable material present in the concrete.

- The diffusion velocity is mainly influenced by the porosity of the concrete and the amount of moisture in the pores. The larger the porosity and the lower the amount of water in the pores, the higher the diffusion velocity.
- The porosity is influenced by factors such as the w/c ratio and the curing. Due to the large amount of fine particles in SCC, the pore structure is denser.
- The amount of carbonatable material depends on the binder system including filler. The more CSH and $Ca(OH)_2$ present in the concrete, the lower the carbonation depth. Depending on the composition of the binder system, including filler, a high amount of $Ca(OH)_2$ and CSH could be present in SCC, reducing the risk of carbonation.

Only one fundamental research project to date has described the modelling of the carbonation behaviour in SCC. This is based on the capillary porosity and the amount of carbonatable material, leading to good predictions of the carbonation coefficient, for both TVC and SCC.

10.3.2 Chloride penetration

Theoretical background

Chloride ions can penetrate into concrete in several ways: by capillary absorption of a chloride-containing solution, by diffusion, by permeability, by the influence of an electrical field etc. Part of the penetrating chlorides will be bound to the cement matrix, and only the 'free' ions will be transported further into the concrete. If the chlorides are penetrating in the concrete by capillary absorption, by permeability or by diffusion, the equations given above can be used. The influence of the presence of chloride ions in the water on the flow can be neglected for typical concentrations of 3.5% NaCl[23]. The equations describing the penetration of chlorides by the influence of an electrical field are described below.

The transport of ions through concrete was discussed in Section 10.2.3. Fick's first law describes this transport caused by the differences in chemical potential. However, the penetration of chlorides is important in order to calculate the time at which the chlorides reach the reinforcement. This is a nonstationary problem, and so Fick's first law should be combined with:

$$\frac{\partial c}{\partial t} = -\frac{\partial J}{\partial x} \tag{10.26}$$

expressing the conservation of mass as a function of time. Both equations can be combined in Fick's second law:

$$\frac{\partial c}{\partial t} = \frac{\partial}{\partial x}\left(D\frac{\partial c}{\partial x}\right) \tag{10.27}$$

This equation can be solved with the following initial and boundary conditions:

$$c = 0 \quad \text{for } x > 0 \text{ and } t = 0 \tag{10.28}$$

$$c = c_0 \quad \text{for } x = 0 \text{ and } t > 0 \tag{10.29}$$

If D, the diffusion coefficient, is considered to be constant, the solution of Equation (10.27) is as follows:

$$c = c_0\left(1 - \text{erf}\left(\frac{x}{2\sqrt{Dt}}\right)\right) \tag{10.30}$$

In some test methods, an electrical field is applied to increase the penetration velocity of the ions into the concrete test specimen. In this case, an electrical potential and a chemical potential are present. If both potentials are working in the same direction, the following equation should be used:

$$J = -D\left(\frac{dc}{dx} - \frac{zFE}{RT}c\right) \tag{10.31}$$

In this equation, R is the universal gas constant, T the temperature (K), F the Faraday constant, z the ion-valence (for chloride ions, $z = -1$) and E the electrical field (V/m). Fick's second law, supposing the diffusion coefficient D is a constant, becomes:

$$\frac{\partial c}{\partial t} = D\left(\frac{\partial^2 c}{\partial x^2} - \frac{zFE}{RT}\frac{\partial c}{\partial x}\right) \tag{10.32}$$

with the same initial and boundary conditions, the solution becomes:

$$c = \frac{c_0}{2}\left(e^{ax}\text{erfc}\left(\frac{x + aDt}{2\sqrt{Dt}}\right) + \text{erfc}\left(\frac{x - aDt}{2\sqrt{Dt}}\right)\right) \tag{10.33}$$

where

$$a = \frac{zFE}{RT} \tag{10.34}$$

and

$$\text{erf}c(x) = 1 - \text{erf}(x) \tag{10.35}$$

These equations are valid if the diffusion coefficient is assumed constant, but in reality it is time-dependent. This time-dependency is caused by:

- Some ions, penetrating into the concrete, are bound to the cement matrix and form certain products that are precipitated in the pores. This decreases the porosity of the concrete and also the diffusion coefficient.
- $CaCO_3$ and $Mg(OH)_2$ are deposited on the concrete surface in the tidal and splash zone, thus decreasing the ease of chloride penetration when exposed to sea water.
- The continuing hydration of the cement leads to a continuous reduction of the porosity and, consequently, of the diffusion coefficient. This effect is more important if fly ashes and blast furnace slags are used.

Experimental results for self-compacting concrete

Early in the development of SCC, some alarming test results were published showing that the chloride penetration was 2–3 times larger than in TVC[51]. This did not immediately lead to fundamental research in order to find the reason for this different behaviour, but for each project where chloride penetration was relevant, the chloride penetration behaviour was verified.

It has been found that the rate of chloride penetration is somewhat faster in SCC[36] or equal[12–15, 25, 34, 37, 38, 52] or slower than in TVC[21, 23, 35, 53–55]. However, the basis of comparison between the self-compacting and the traditional concrete is not always clear.

In some research, there was no comparison with traditional concrete. The influence of viscosity modifying agents on the chloride penetration was investigated in[56]. If a viscosity agent is used in SCC, the amount of bound chlorides does not change and a greater corrosion resistance is exhibited. Suksawang *et al.*[57] and Petersson[58] have studied the influence, respectively, of fly ash and limestone filler. The chloride diffusivity of SCC with limestone filler or fly ash was lower in comparison to SCC without limestone filler or fly ash. The use of steel fibres in SCC has been studied[46], leading to the conclusion that a good penetration resistance is obtained.

Conclusions

Some general conclusions about the chloride penetration in concrete and more specifically SCC could be phrased as follows:

- The process of chloride penetration in concrete structures is mostly the combination of capillary suction and diffusion from a chloride-containing solution.
- The chloride diffusion coefficient is strongly time dependent.
- The presence of blast furnace slag, fly ash or limestone filler reduces the chloride penetration.
- A decrease in the total porosity of the concrete will decrease the penetration depth of the chlorides.

- The conclusion from experimental results comparing TVC and SCC is that the diffusion coefficient of TVC is larger than the diffusion coefficient of SCC at a constant w/c ratio.

10.3.3 Freeze–thaw

Theoretical background

The resistance to freezing–thawing is an important parameter for the durability of a concrete structure under the influence of the weather. When concrete is not sufficiently resistant to frost attack with or without deicing salts, two types of deterioration can occur: internal damage and scaling[59]. Internal damage, inside the concrete, is mostly cracking along the aggregate grains. This can induce a change in the concrete properties[60]. Scaling is cracks just beneath the surface of the concrete, inducing scaling of the cement stone. In low resistant porous aggregates or aggregates sensitive to freezing, the cracks can also run through the aggregate grains. When porous aggregate grains are near the surface, the cement stone above the grain can be pushed off.

The basic deterioration mechanism for internal damage or scaling is the same. Freeze–thaw attack can be considered as a physical phenomenon where freezing of water induces excessive internal pressure in the concrete. This is based on the fact that freezing of water is associated with an increase in volume of 9%. Several theories explaining freeze–thaw attack and freeze–thaw attack in combination with deicing salts have been summarised in[18, 61]. The influencing factors are the microstructure, w/c ratio, cement content, type of cement, additions, freezing temperature, rate of freezing, amount of freeze–thaw cycle, air entrainment or not etc.[61–70].

Experimental results for self-compacting concrete

Some test methods used to investigate the freeze–thaw resistance in combination with deicing salts are: the 'slab-test'[71, 72], the 'cube-test'[73], the 'CDF-test'[74, 75] and the 'beam-test'[76, 77].

Gram and Piiparinen[78] tested the resistance to freeze–thaw in combination with deicing salts according the Swedish Standard SS 13 72 44. The mixes have a w/c ratio between 0.41 and 0.44. Air entrainment is added, except to the mixes with the highest cement content. The cement content (ordinary Portland cement) is between 360 kg/m^3 and 390 kg/m^3 for the self-compacting mixes and between 400 kg/m^3 and 420 kg/m^3 for the traditional mixes. The self-compacting properties were obtained by adding a superplasticiser and 210–230 kg/m^3 limestone filler. The SCC mixes showed excellent frost resistant. The maximum mass loss after 28 days was 0.02 kg/m^3. In the SCC mixes without air entrainment, a higher air content was noticed compared to the traditional mixes without air entrainment, namely 2.5–1.7%. According to the authors, this is probably due to the formation of air voids because of the superplasticiser. This gives an extra protection against freeze–thaw attack. The traditional mixes without air entrainment had a mass loss of 1.10 kg/m^2 after 28 days.

According to Rougeau *et al.*[14], because of air void formation, due to the superplasticiser, the SCC mixes have a better air void system and consequently a better resistance to freezing and thawing in combination with deicing salts. In addition, the vibration energy can disturb the air void system in the case of traditional concrete. Two mixes of SCC and one mix of HPC were tested according the specifications of the French Standard NF P 18-420. The SCC mixes have a w/c ratio of 0.49 and 0.47. The cement content (CEM I 52.5 R) of both mixes is 350 kg/m^3 and the amount of limestone filler is 100 kg/m^3. The mass loss of the SCC mixes does not further increase after 21 freeze–thaw cycles, in contrast with the traditional mixes in which after 56 cycles there is still an increasing mass loss. At 28 days the SCC mixes have a mass loss of 0.85 kg/m^3 and 1.4 kg/m^3 at a w/c ratio of 0.49 and 0.47, respectively.

Jacobs and Hunkeler[25] have also attributed the improved freeze–thaw resistance of SCC to a good air void system. A poor performance is linked with a relatively low air content of air voids included in the range 20–300 μm, and a rather high content of larger air voids.

A higher initial mass loss was noticed by Petersson[79] for the SCC mixes exposed to freeze–thaw in combination with deicing salts. This was due to the flocculation of the fine limestone filler inducing a weaker, more porous zone in the cement matrix. The problem was remedied by a more appropriate mixing procedure and as such more acceptable mass losses were obtained. At 28 days the mass loss reached a maximum of 0.08 kg/m^3. The tests were performed according to the Swedish Standard SS 13 72 44[71] on mixes with a w/c ratio of 0.40, 415 kg/m^3 cement and 180 kg/m^3 limestone filler.

Reinhardt *et al.*[80] have given an overview of some of the literature concerning freeze–thaw resistance of SCC. From the results it follows that the addition of fly ash has a positive effect on the resistance. The connecting thread between all the results that have been mentioned is the need for a sufficiently high air content and an adequate air void system. This is explicitly mentioned by Makishima *et al.*[44].

Persson[81] compared TVC to SCC by means of mixes with a w/c ratio of 0.39 and an air content of 6%. The spalling at freeze–thaw in combination with deicing salts is equal for both concrete types. Concerning the internal damage, SCC seems to be more resistant. No connection was found between the air void structure and the frost resistance. Some of the most important conclusions were as follows: spalling increases with the use of blast furnace slag cement and decreases with the addition of 12% fly ash and 5% silica fume or by the use of a finer filler.

Several authors[18, 82, 83] have concluded that, especially a lower w/c ratio and a lower w/p ratio, the use of fly ash, a cement type with a higher fineness and blast furnace slag cement all increase the resistance against freeze–thaw in combination with deicing salts. There was no explicit influence of the air content which could be due to the fact that no air entrainment was used. A rather high w/c ratio has more harmful effect on SCC than on traditional concrete (Figure 10.6). On the other hand when the standard requirements (maximum w/c ratio, minimum cement content, minimum strength) are met, SCC exhibits a better resistance compared to traditional concrete. This means that it is possible to create an SCC with good performance, without air entrainment, when the w/c ratio is low enough. As soon as higher w/c ratios are used, the use of air entrainment becomes very important to achieve good performance. It was also found that the resistance should not be regarded in terms of compressive strength, but rather as a function of the capillary porosity, as it reflects the possibility to take up water in the microstructure. The same conclusions were also found elsewhere[84, 85].

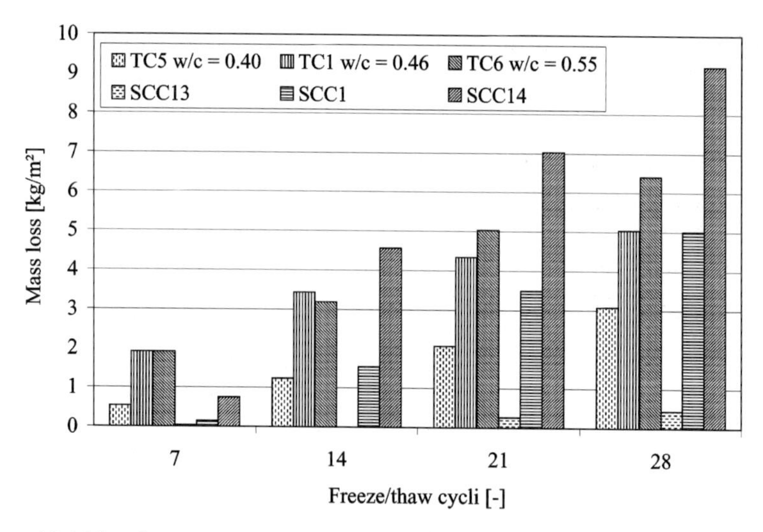

Figure 10.6 *Mass loss, variation of w/c ratio (see also Table 8.6).*

Conclusions

A lower w/c ratio, a lower w/p ratio, the use of fly ash, a cement type with a higher fineness and blast furnace slag cement increase the resistance against freeze–thaw. A higher w/c ratio has more harmful effect on SCC than on traditional concrete. On the other hand, when the requirements (maximum w/c ratio, minimum cement content, minimum strength) are met, SCC exhibits a better resistance compared to traditional concrete. This means that it is possible to create an SCC with good performance, without air entrainment, when the w/c ratio is low enough. As soon as higher w/c ratios are used, the use of air entrainment becomes very important to achieve good performance. The deterioration can be linked with the water transport properties and the capillary porosity.

10.3.4 Alkali–aggregate reactions

Theoretical background

Alkali–aggregate reactions (AAR) are internal chemical reactions occurring in the concrete (internal chemical attack). They occur when alkalis (potassium and sodium oxides from cement) in the pore solution of concrete react chemically with certain potentially reactive components in the aggregates. When the cement hydrates, the alkalis are released into the pore water of the concrete where they may be augmented by alkalis from the environment or alkalis contributed by some aggregates. There are two types of alkali–aggregate reactivity: alkali–silica reactivity (ASR) and alkali–carbonate reactivity (ACR). In the case of ASR, the reaction of alkalis with siliceous material produces an alkali–silica gel, capable of absorbing water and swelling. When the swelling pressure exceeds the tensile strength of the cement matrix in the concrete, cracking occurs. As more gel is produced, it forces its way into the cracks, thus extending them and ultimately disrupting the entire concrete mass [86, 87]. ACR occurs between the alkalis in the pore water of concrete and some dolomitic limestones. The exact mechanism of this reaction is still uncertain, however it is known that it involves dedolomitisation, which causes the aggregate particles to swell. As a result, extensive cracking and very large expansions are observed[87].

Experimental results for self-compacting concrete

At present very few experimental results on alkali–aggregate reactions in SCC can be found. Several mixes of SCC and traditional concrete have been tested on AAR by means of Oberholster tests[88]. All the SCC mixes, made with limestone filler, showed some expansion. The expansion increases for an increasing w/c and an increasing w/p for a constant c/p. This is possibly due to a less dense structure and for this a greater mobility of the alkali ions. When the w/p is kept constant the expansion increases with decreasing w/c ratio. This may be due to the higher cement content, which brings along more alkali ions. When the w/c is kept constant (variation of water and cement), the expansion increases with increasing c/p, meaning a higher cement content. This is due to the presence of more alkali ions and a less dense structure. A significant difference was noticed between the SCC and traditional concrete mixes, as almost no expansion was noticed for traditional concrete. Petrographical analysis could not prove the appearance of AAR in the expanded material. This makes it uncertain which reaction really occurred in the SCC to cause the expansion.

Leemann and Vecsei[89] have tested one SCC and one traditional concrete (vibrated and lower content of superplasticiser compared to SCC). It was found that SCC showed lower expansion. However, the lack of vibration causing a lower permeability and a higher compressive strength of the SCC should also be considered.

Conclusion

No decisive conclusions can be made, based on the limited amount of results available. However, it is clear that the influencing factors for traditional concrete will be the same as those for SCC. The most important influencing factors are: (1) the alkali content of the cement and the cement content of the concrete; (2) the alkali ion contribution from sources other than Portland cement, such as admixtures, salt-contaminated aggregates, and penetration of seawater or deicing salt solution into concrete; (3) the amount, size and reactivity of the alkali-reactive constituent present in the aggregate; (4) the availability of moisture to the concrete structure; and (5) the ambient temperature[87]. There is no indication that the connection between the alkalinity of the pore solution, the presence of reactive aggregates and expansion is fundamentally different for SCC compared to traditional concrete. Therefore, the same measures that have been proven effective for traditional concrete should be used for SCC[90].

10.3.5 External chemical attack

Ingress of dissolved deleterious substances from external sources may induce various forms of deterioration by chemically reacting with the cement paste or aggregate constituents[87]. The main type of aggressive fluids to which concrete generally will be exposed include mineral and organic acids, solutions of sulfates, chlorides, sugars, nitrates, phenols and ammonium compounds[91]. Aggressive fluids can penetrate into the concrete by diffusion, by hydraulic pressure, by a capillary mechanism or by a combination of the mentioned forces[92]. Essentially there are five principal groupings of the effects of aggressive solutions: (1) simple leaching of free calcium hydroxide; (2) reaction between the aggressive solutions and cement compounds producing secondary compounds which are either leached from the concrete or remain in a nonbinding form; (3) reaction similar to the previous one, but crystallisation of secondary compounds giving rise to expansive forces which disrupt the concrete; (4) crystallisation of salts directly from the attacking solution causing disruption of the concrete; (5) corrosion of the embedded steel reinforcement arising from the breakdown of the passivation zone[87]. From all the possible forms of external chemical attack only acid, ammonium and sulfate attack will be discussed below.

Acid attack

Theoretical background

In a well-hydrated Portland cement paste, the solid phase exists in a state of stable equilibrium with a high-pH pore fluid (12.5–13.5). It is clear that concrete would be in a state of chemical disequilibrium when it comes in contact with an acidic environment[43]. Generally, during acid attack CH and CSH are attacked vigorously and all Portland cement compounds are susceptible to degradation. One of the possible reactions in concrete is the conversion of calcium to soluble products. This reaction results in the decalcification of CSH in concrete, leading to a progressive opening of the structure and reduction in its pH[93]. This can lead to strength loss, expansion, spalling of surface layers and ultimately disintegration[94]. Acid solutions include both mineral acids such as sulphuric, hydrochloric, nitric and phosphoric acids, and organic acids such as lactic, acetic, formic, tannic and propionic acids and other acids produced in decomposing silage[87].

Lactic acid ($C_3H_6O_3$) and acetic acid (CH_3COOH) (HX) react easily with the free lime ($Ca(OH)_2$) and create very soluble calcium salts (CaX_2) that leach out of the concrete very easily. At that moment the hydration products start to decompose and the concrete starts to disintegrate (Equation (10.36)). As a consequence the strength of the concrete decreases because of the higher porosity of the concrete and the decomposing of the hardened cement paste.

$$2HX + Ca(OH)_2 \rightarrow CaX_2 + 2H_2O \qquad (10.36)$$

Besides CH, other hydration products can also be attacked, as is shown in the following reactions concerning attack by acetic acid:

$$2CH_3COOH + Ca(OH)_2 \rightarrow (CH_3COO)_2Ca + 2H_2O \qquad (10.37)$$

$$6CH_3COOH + 3CaO.2SiO_2.3H_2O \rightarrow 3(CH_3COO)_2Ca + 2SiO_2 + nH_2O \qquad (10.38)$$

$$6CH_3COOH + 3CaO.Al_2O_3.6H_2O \rightarrow 3(CH_3COO)_2Ca + Al_2O_3 + nH_2O \qquad (10.39)$$

Attack due to sulphuric acid is a combination of sulfate attack and acid attack and will be discussed in the subsection below on sulfate attack.

Experimental results for self-compacting concrete

Al Tamimi and Sonebi[95] investigated the resistance of SCC and conventional concrete (CC) to a 1% hydrochloric solution (HCl). The SCC mixes contained 47% of carboniferous limestone powder. The mixes, w/c of 0.36 and 0.46, were immersed in the solution for a period of 18 weeks at 20°C. In the hydrochloric acid solution SCC showed less deterioration than the conventional concrete. After 18 weeks of immersion, a mass loss of 9% was observed for SCC compared to 21% for conventional concrete. In the hydrochloric solution, the amount of ettringite formed in SCC and conventional concrete was substantially lower than in the case of a 1% sulphuric solution. Neither thaumasite nor gypsum were found in either SCC or conventional concrete.

An experimental project was set up by Boel[18] and Boel and De Schutter[96] to investigate the effect of immersing concrete samples (SCC and TVC: see Table 8.6 in Chapter 8) in an acid solution of acetic acid combined with lactic acid (pH = 2.5). The damage was evaluated by measurements of mass and compressive strength. It was found that the deterioration due to acetic and lactic acid strongly depends on the concrete composition and the penetrability of the cement matrix. Some measures which could be taken to decrease the deterioration are: the use of fly ash, the use of a filler enhancing a dense structure, a high c/p, a low powder content, a low w/c and avoidance of calcareous rubble. The water transport properties of SCC have been studied experimentally and theoretically by Audenaert[23]. Several parameters derived from tests like capillary suction, water permeability and water immersion were presented. The link of these parameters with the mass loss due to submersion in the solution of acetic and lactic acid has been studied by Boel[18]. It has been found that the best correlation occurred when the parameters derived from capillary suction were used. As those parameters are a function of the calculated capillary porosity, according to Powers' model, the mass loss has also been plotted against the capillary porosity (Figure 10.7). From this figure it can be noted that the capillary porosity influences the deterioration of traditional concrete and SCC mixes due to acid attack as a combination of acetic and lactic acid. Figure 10.7 also illustrates that the acid attack of SCC is similar to that of traditional concrete, as the capillary porosity is similar.

Conclusion

Only a few results on the acid attack of SCC can so far be found in the literature. Based on the limited amount of results for exposure to acetic and lactic acid, the use of fly ash or limestone powder with a finer grading leads to a reduction of mass loss and to a better general strength. Acid attack of SCC seems to be similar to that of traditional concrete if the capillary porosity is similar.

Ammonium attack

Theoretical background

Ammonium sulfate and ammonium nitrate are usually considered to be the most aggressive of the ammonium salts. Ammonium nitrate provokes a solubilisation and leaching of lime in the cement paste. The reaction forms calcium nitrate and calcium alumina-nitrate. This is accompanied by a reduction of the pH of the concrete. The

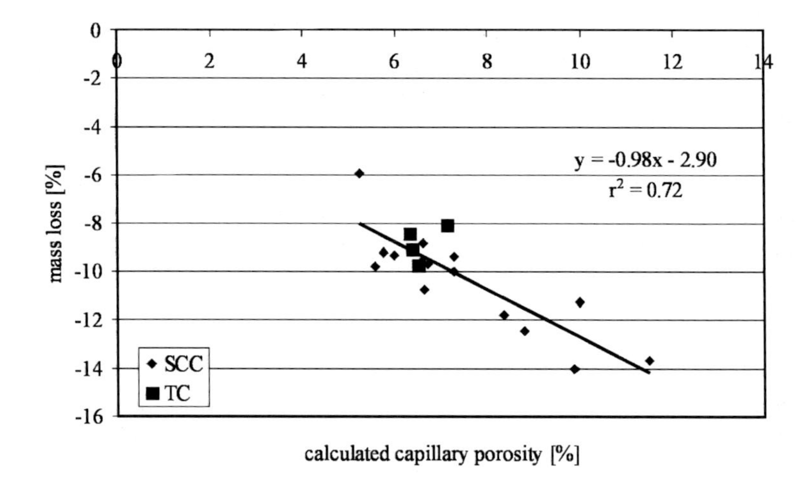

Figure 10.7 *Mass loss as a function of the calculated capillary porosity.*

leaching of lime also leads to a weakening of the concrete matrix, inducing some cracking. Ammonium sulphate is the most aggressive of the sulphate salts. Considerable swelling and cracking is induced, causing loss of strength of the concrete. An ammoniac formation leads to reduction of the pH and avoids the equilibrium state of the reaction[97].

$$Ca(OH)_2 + 2NH_4Y \rightarrow CaY_2 + 2H_2O + 2NH_3 \qquad (10.40)$$

Y is the anion associated with NH_4^+

Experimental results for self-compacting concrete

Assié *et al.*[98] have performed an accelerated leaching test on low strength SCC and TVC of an equivalent strength grade. The solution applied was a 500 g/l ammonium nitrate. When the nitrate leaching depth is plotted against the square root of time, the results show that the deterioration and the reaction kinetics of SCC and TVC are similar.

Conclusions

One study has shown that ammonium nitrate leaching of SCC is similar to that of TVC.

Sulfate attack

Theoretical background

Portland cement concrete can deteriorate in solutions containing sulfate, such as some natural or polluted groundwaters. Sodium, potassium, magnesium and calcium sulfate in soil are the main source of sulfate ions in groundwater. Sulfate attack can lead to strength loss, expansion, spalling of surface layers and ultimately disintegration. The deterioration products mostly include: gypsum, ettringite and thaumasite, with ettringite found mainly in the deteriorated concrete. However, as explained further, in limestone-powder based SCC thaumasite may also be noticed.

The sulfate attack reactions depend on the availability of the sulfate ion, the cations present in the sulfate solution (sodium, potassium, ammonium, magnesium etc.), the C_3A content of cement, the density and the permeability of the concrete[87]. Each of these is important in determining not only the rate and degree of sulfate attack, but also the nature and mechanism of the attack.

Five types of sulfate attack have been described by Brown[99]: the classic form of sulfate attack associated with ettringite and/or gypsum formation, the thaumasite form of sulfate attack, physical sulfate attack associated with crystallisation at or near an evaporation surface, delayed ettringite formation and sulfate attack associated with the Afm-phases.

Experimental results for self-compacting concrete

Important differences between TVC and SCC are the addition of high amounts of mineral fillers (e.g. limestone, fly ash) and the increased impermeability of concrete. As discussed in Chapter 6, limestone powder (calcium carbonate) is often used in SCC, at amounts often approaching those of cement. This makes the concrete potentially vulnerable to the thaumasite form of sulfate attack. Several test methods lead to different test results. Depending on the duration of the test, the sulfate concentration and the kind of sulfate, different conclusions can be drawn.

Classic form and thaumasite form of sulfate attack

The classic form of sulfate attack is determined by the chemical interaction of sulfate-rich soil or water with calcium aluminate hydration products (CAH) formed during the hydration of cement. Gypsum ($CaSO_4.2H_2O$) and ettringite ($C_3A.3CaSO_4.32H_2O$) are mainly formed.

Thaumasite may be formed through the reaction of water with calcium carbonate, any sulfate salt, and hydrated calcium silicates. The formation is said to be enhanced at low temperatures (< 5°C), although it has also been noticed at higher temperatures. There is a strong correlation between the C_3A and aluminium content of the cement and the amount and rate by which thaumasite is formed[100, 101]. The increased danger of thaumasite formation has been reported in the case of SCC when limestone filler is used. This finding has been both endorsed and denied internationally.

Chemical tests have been applied by Boel[18] and Boel *et al.*[102] to evaluate the resistance of SCC subjected for six weeks to a 1.5 g/l sulphuric acid (H_2SO_4) with pH = 1. Some contradictory results were found. A lower w/c or the use of fly ash causes a less permeable concrete. In spite of this, more deterioration is found. These (apparently) contradictory results are also found in the presence of magnesium and sodium sulfate[103–105]. This is probably due to the limited possibility for expansion in a dense structure. Generally, with immersion in a sulphuric acid solution, there is a first mass gain and a subsequent mass loss. The mass increase can be explained by the formation of expansive products. The residual strength is sometimes higher than 100%. This could be explained by the filling of space by the expansive products which strengthen the structure. As more deterioration occurs and more products are formed, more microcracks will be formed and the strength will decrease. This was also mentioned in[106]. Boel[18] found that the mixes with less space for expansion, were more deteriorated after six weeks of immersion. The mass loss after six weeks decreases with the use of high sulfate resistant (HSR) cement (a lower C_3A-content induces less aluminate phases to be converted into ettringite), a lower c/p ratio and a higher w/c ratio. It was unexpected that a higher w/c ratio leads to a lower mass loss. It was also mentioned in[107] that researchers did not reach a consensus concerning the w/c ratio. A lower w/c ratio induces a less permeable concrete, but also the amount of available expansion space is reduced.

Al Tamimi and Sonebi[95] and Al Tamimi *et al.*[108] have investigated the resistance of SCC and traditional concrete, at w/c ratios of 0.36 and 0.46 by a submersion in a 1% H_2SO_4-solution. The SCC mixes contained limestone powder. The immersion was continued for 18 weeks at 20°C, after which the mass loss of the SCC was smaller than that of the traditional concrete. In both mixes gypsum was found after the test. The presence of thaumasite together with ettringite was only found in SCC. Obviously thaumasite was not only formed at low temperatures, but also at higher temperatures, as the laboratory tests were carried out at 20°C.

SCC specimens containing different amounts of limestone powder (w/c of 0.40) were exposed to a weak to moderately strong magnesium sulfate solution by Kalinowski and Trägårdh[109]. It was observed that lower deterioration correlates with a denser microstructure and a finer capillary pore size. The reaction products formed were primarly thaumasite and gypsum.

Trägårdh and Kalinowski[110] investigated SCC mixes containing two different Portland limestone cements with different C_3A contents. The mixes were exposed to a magnesium sulfate solution for 1500 days. A faster rate of deterioration due to thaumasite formation was found in the case of the highest C_3A content in the cement. Tests have also been performed on mixes with and without extra limestone powder (50 kg/m^3). Those mixes had the same w/c, w/p, f/c and cement content. The limestone powder was added at the expense of the fine aggregate content. It was found that the degradation was worse in the case where extra limestone powder was added. However, severe thaumasite formation was also observed when there was no extra limestone powder added, but when the cement contained 10–12% calcite.

The influence of different fillers on the sulfate resistance due to a magnesium sulfate solution was studied in[101]. The influence of the addition of fly ash, silica fume and slag at limestone cement pastes was investigated. It was found that addition of silica fume significantly improved the sulfate resistance. Fly ash and slag did not significantly change the sulfate resistance of the limestone cement pastes.

Delayed ettringite formation

The presence of limestone powder in SCC could induce a somewhat higher cumulative heat production[111]. Because of this, temperatures higher than 65°C could be obtained in the hardening concrete, which could result in the formation of delayed ettringite. Poppe and De Schutter[111] have noted that when quartzite powder was added, no additional hydration heat was observed. Another factor which can possibly also influence the heat of hydration to a greater extent in SCC than in traditional concrete is the type and amount of chemical admixtures[112]. Trägårdh and Bellmann[112] also showed, based on a performance-based comparison, that the increased risk concerning delayed ettringite formation is small. An extra heat production from limestone filler or superplasticisers in SCC is overshadowed by the dominant factors such as cement content, type of cement and volume of concrete. However, when the cement content is equal in SCC and traditional concrete and the SCC contains high amounts of limestone filler, an increased risk of delayed ettringite formation in SCC is possible. This conclusion was made considering that the maximum heat production increases with decreasing c/p[111].

Conclusions

Depending on the duration of the test, the sulfate concentration and the kind of sulfate, different conclusions can be drawn.

The positive factors influencing the sulfate resistance in traditional concrete remain the same for SCC. The resistance of concrete to sulfate attack can be increased by lowering the permeability (low w/c, high cement content, good grading curve of the aggregate) and by the use of sulfate-resisting cements. Cements with a high chemical resistance to sulfate attack are: Portland cements with a low C_3A content and binders containing pozzolanic and latently hydraulic admixtures. The low content of C_3A in the cement limits the amount of ettringite that can be formed during sulfate attack. The pozzolanic reaction lowers the amount of portlandite in the microstructure and makes the hardened cement paste less vulnerable to the formation of gypsum. However, the materials can contain reactive components such as silica and aluminium which could provide a source of reactants for thaumasite formation. Contradictory results have been reported in the literature regarding thaumasite formation and pozzolanic materials.

The influence of the addition of limestone powder depends on its influence on the pore structure and microstructure. Several authors have found that a more dense structure, and as such a lower permeability, is achieved when large amounts of limestone powder are used (see Chapter 8). On the other hand it was also found that concrete with limestone powder is more prone to detrimental thaumasite formation.

The presence of limestone powder and superplasticiser in SCC could induce a somewhat higher cumulated heat production and thus an increased risk of delayed ettringite formation. This extra heat production can, however, be overshadowed by dominant factors such as the cement content, type of cement and volume of concrete.

10.3.6 Fire resistance

Theoretical background

Damage to concrete due to fire is mostly not considered as a degradation mechanism, but rather as an accidental action. However, the driving forces causing explosive spalling when the concrete element is exposed to fire conditions are related to the microstructure and the transport mechanisms of the cementitious material, as it is the case for the degradation mechanisms. The main difference is that fire damage happens on a much smaller time scale and within one hour, or even less, the concrete can be severely damaged.

When concrete is exposed to very high temperatures, some conversions may occur. As an example, for SCC based on limestone powder, the limestone powder ($CaCO_3$) will be decomposed at around 700°C. However, the problem of explosive spalling, which happens at temperatures as low as 300°C and within a short time (about 20 min) after the initiation of the fire, is more important.

The mechanisms responsible for the explosive fire spalling are not fully understood. However, it is clear that the moisture content of the concrete and the microstructure of the material are the most important factors. Above 100°C, the water in the concrete is transformed into water vapour. As the temperature rises, the vapour pressure in the concrete increases. When the microstructure of the concrete is rather open (high w/c ratio), the vapour can escape quite rapidly, relieving the vapour pressures. However, when the concrete has a dense microstructure, the vapour pressure can reach significant values (values above 3 MPa have been measured[113]). Due to the high internal pressure, a small layer of concrete can be suddenly pushed off, this is known as explosive spalling.

Explosive spalling has been reported to be a problem in HPC. HPC has a very dense microstructure, and vapour diffusion is very limited. Traditional concrete with moderate strength is not prone to explosive spalling, because of the somewhat more open pore structure. As the pore structure of SCC is different from the pore structure of traditional concrete, and is more related to the pore structure of HPC, SCC might be sensitive to explosive fire spalling. The thermal properties of SCC are similar to those of HPC[114]. Hence, the thermal properties of HSC can be applied to estimate the thermal performance of SCC structures.

Experimental results for self-compacting concrete

The experimental research concerning fire resistance and explosive fire spalling of SCC is still rather limited, but a few studies can be found[113, 115–122]. When comparing the results obtained within the different experimental studies, significant differences can be noticed concerning the degree of explosive spalling of SCC when exposed to fire conditions. Several aspects might be responsible for these differences. The composition of SCC might be quite different in the different studies, which might lead to a different microstructure depending on the type and amount of filler material. The age of the specimens exposed to the fire conditions and the moisture conditions are also different, leading to more severe explosive spalling at earlier age, on specimens with higher humidity. Last but not least, the heating regimes applied in the different experimental programmes are not always identical. Sometimes electrical furnaces are considered and different regulations permit different heating curves to be followed for the fire conditions.

In some experimental programmes[117] the addition of small amount of polypropylene fibres has shown a positive influence in reducing the explosive fire spalling of SCC, as is also the case in HPC. The mechanisms behind the positive effect of polypropylene fibres are not fully understood. According to some researchers, an increased porosity would be obtained when the polypropylene fibres melt at around 170°C. However, recent results obtained by Liu indicate that the increase in pore volume is not very significant. A more significant effect is the increased connectivity of the pore system after melting of the fibres, yielding a significantly higher diffusivity of the material[117].

Conclusions

In spite of the diversity in the experimental programmes reported, some general conclusions can be formulated concerning the explosive fire spalling of SCC. In comparison with traditional concrete, SCC has a higher probability of explosive spalling when exposed to fire conditions. As the microstructure of SCC more resembles the microstructure of HPC, the spalling behaviour of SCC is comparable to the spalling behaviour of HPC. The moisture content, however, is a very important factor. Since SCC shows lower transport properties, the drying time after casting is longer than for TVC. As a consequence, the risk of explosive spalling of SCC is higher for a longer time. In cases where no fire spalling is accepted, it is recommended that protective measures be taken by placing thermal barriers between the fire and the concrete, or by adding polypropylene fibres to the SCC. The required amount of polypropylene fibres depends on several parameters, like the geometry of the fibres, the type of concrete, loading conditions, moisture content etc. Based on the available experimental results in the literature, it seems that, for polypropylene fibres with diameters below 30 μm, 1–2 kg fibres per m^3 concrete is enough for many applications.

10.4 Conclusions

It was shown in Chapter 8 that the microstructure and porosity of SCC differ from those of TVC and so we can expect differences in the durability behaviour. This chapter has analysed results and information from a number of programmes which have studied the transport of various materials through SCC and its behavior in most of the important exposure conditions. Detailed conclusions have been given at the end of each section, but the most important can be summarised as follows.

- The denser microstructure of SCC leads to lower gas permeability and diffusion (although not all evidence is conclusive in the latter case) and reduced the capillary absorption compared to the equivalent TVC. The water permeability seems to be similar to that of TVC of the same strength class.
- Data on carbonation rates are conflicting, but the differences between SCC and TVC are small.
- Chloride diffusion is lower in SCC than in TVC of the same w/p ratio.
- The low w/p ratios of SCC can lead to good freeze–thaw resistance without the use of air-entraining agents.
- Results on the ASR in SCC are limited, but similar preventive measures to those for TVC are suggested.
- Acid attack on SCC is similar to that on TVC with similar porosity.
- As in TVC, the sulfate resistance of SCC is improved by the low w/p ratios and the use of additions and fillers, but if significant quantities of limestone filler are used then the risk of thaumasite formation may increase.
- The denser microstructure of SCC will increase the risk of explosive spalling in fires, so the use of polypropylene fibres is important when fire is a significant hazard.

Studies on many aspects of durability of SCC are ongoing, and hence more results and conclusions can be expected to be available in due course.

References

[1] Taylor H. *Cement Chemistry*, 2nd edn., Thomas Telford, London, UK, 1997.

[2] Basheer L., Kropp L. and Cleland D. Assessment of the durability of concrete from its permeation properties: a review. *Construction and Building Materials* 2001, 15, 93–103.

[3] Figg J. Methods of measuring the air and water permeability of concrete. *Magazine of Concrete Research* 1973, 25(85), 213–219.

[4] Lydon F.D. Effect of coarse aggregate and water cement ratio on intrinsic permeability of concrete subject to drying. *Cement and Concrete Research* 1995, 25(8), 1737–1746.

[5] Dhir R.K., Hewlett P.C. and Chan Y.N. Near surface characteristics of concrete, Intrinsic permeability. *Magazine of Concrete Research* 1989, 41, 87–97.

[6] Kropp J. and Hilsdorf H.K. Performance criteria for concrete durability. RILEM Report 12, 1995.

[7] Martys N.S. Survey of concrete transport properties and their measurement, NISTIR 5592, NIST Gaithersburg, MD, USA, 1995.

[8] RILEM TC 116-PCD Permeability of concrete as a criterion of its durability, recommendations. *Materials and Structures* 1999, 32, 174–179.

[9] Carcassès M., Abbas A., Ollivier J.-P. and Verdier J. An optimised preconditioning procedure for gas permeability measurement. *Materials and Structures* 2002, 35, 22–27.

[10] Willem X. Study of the effect of auto-consolidation on the durability indicators of concrete (in French). Ph.D. thesis, University of Liège, Belgium, 2005.

[11] Kollek J.J. The determination of the permeability of concrete to oxygen by the Cembureau method – a recommendation. *Materials and Structures* 1989, 22(129) 225–230.

[12] Zhu W., Quinn J. and Bartos P.J.M. Transport properties and durability of self-compacting concrete. In: Ozawa K. and Ouchi M. (Eds.) *Proceedings of the Second International symposium on Self-compacting Concrete*, Tokyo, Japan, 2001, pp 451–458.

[13] Zhu W. and Bartos P.J.M. Permeation properties of self-compacting concrete. *Cement and Concrete Research* 2003, 33(6) 921–926.

[14] Rougeau P., Maillard J.L. and Mary-Dippe C. Comparative study on properties of self-compacting and high performance concrete used in precast construction. In: *Proceedings of the First International RILEM Symposium on Self-compacting Concrete*, Stockholm, Sweden, 1999, pp 251–261.

[15] Assié S. Durability of self-compacting concrete (in French). Ph.D. thesis, INSA, Toulouse, France, 2004.

[16] Sonebi M. and Ibrahim M.S.R. Assessment of the durability of medium strength SCC from its permeation properties. In: *Proceedings of the Fifth International RILEM Symposium on Self-compacting Concrete*, Ghent, Belgium, 2007.

[17] Assié S., Escadeillas G. and Marchese G. Durability of self-compacting concrete. In: *Proceedings of Third International RILEM Symposium on Self-compacting Concrete*, Reykjavik, Iceland, 2003, pp 655–662.

[18] Boel V. Microstructure of self-compacting concrete in relation with gas permeability and durability aspects (in Dutch). Ph.D. thesis, Ghent University, Belgium, 2006.

[19] Powers T. and Brownyard T. Studies of the physical properties of hardened cement paste (nine parts). *Journal of the American Concrete Institute*, 1946–1947, 43.

[20] Verdier J. and Carcassès M. Equivalent gas permeability of concrete samples subjected to drying. *Magazine of Concrete Research* 2004, 56(4) 223–230.

[21] Trägårdh J. Microstructural features and related properties of self-compacting concrete. In: Skarendahl Å. and Petersson Ö. (Eds.) *Proceedings of the First International RILEM Symposium on Self-compacting Concrete*, Sweden, 1999, pp 175–186.

[22] Reinhardt H. and Jooss M. Permeability and diffusivity of self-compacting concrete as function of temperature. In: Wallevik O. and Nielsson I. (Eds.) *Proceedings of the Third International RILEM Symposium on Self-compacting Concrete*, Reykjavik, Iceland, 2003, pp 808–817.

[23] Audenaert K. Transport mechanisms in self-compacting concrete in relation to carbonation and chloride penetration (in Dutch). Ph.D. thesis, Ghent University, Belgium, 2006.

[24] CEB 238 New approach to durability design – an example method for carbonation induced corrosion, Comite Euro-International du Beton, Lausanne, Switzerland, 1997.

[25] Jacobs F. and Hunkeler F. Design of self-compacting concrete for durable concrete structures. In: Skarendahl Å. and Petersson Ö. (Eds.) *Proceedings of the First International RILEM Symposium on Self-compacting Concrete*, Sweden, 1999, pp 397–407.

[26] Jianxiong C., Xincheng P. and Yubin H. A study of self-compacting HPC with superfine sand and pozzolanic additives. In: Skarendahl Å. and Petersson Ö. (Eds.) *Proceedings of the First International RILEM Symposium on Self-compacting Concrete*, Sweden, 1999, pp 549–560.

[27] Brouwers H. and Radix H. Self-compacting concrete: theoretical and experimental study. *Cement and Concrete Research* 2005, 35, 2116–2136.

[28] DIN 1048 part 5: Test methods for concrete–hardened concrete: separately made test specimens (in German).

[29] GBJ 82-85 Test methods of aging performance and durability for ordinary concrete. Standardization Administration of the People's Republic of China, 1985.

[30] DIN EN 12390-8 Testing of hardened concrete – Part 8: Water penetration depth under pressure (in German).

[31] Hall C. and Hoff W. *Water Transport in Brick, Stone and Concrete*, E&FN Spon, London, UK, 2002.

[32] Dhir R., Hewlett P. and Chan Y. Near-surface characteristics of concrete: assessment and development of in situ test methods. *Magazine of Concrete Research* 1987, 39(141) 183–195.

[33] BS 1881, part 208 Testing concrete. Recommendations for the determination of the initial surface absorption of concrete, 1996.

[34] Nunes S., Coutinho J., Sampaio J. and Figueiras J. Laboratory tests on SCC with Portuguese materials. In: Ozawa K. and Ouchi M. (Eds.) *Proceedings of the Second International RILEM Symposium on Self-compacting Concrete*, Tokyo, Japan, pp 393–402.

[35] Jacobs F. and Hunkeler F. Ecological performance of self-compacting concrete. In: Ozawa K. and Ouchi M. (Eds.) *Proceedings of the Second International RILEM Symposium on Self-compacting Concrete*, Tokyo, Japan, 2001, 715–722.

[36] Fornasier G., Fava C. and Zitzer L. Self-compacting concrete in Argentina: the first experience. In: Ozawa K. and Ouchi M. (Eds.) *Proceedings of the Second International RILEM Symposium on Self-compacting Concrete*, Tokyo, Japan, 2001, pp 309–318.

[37] Mortsell E. and Rodum E. Mechanical and durability aspects of SCC for road structures. In: Ozawa K. and Ouchi M. (Eds.) *Proceedings of the Second International RILEM Symposium on Self-compacting Concrete*, Tokyo, Japan, 2001, pp 459–468.

[38] Jacobs F. and Hunkeler F. SCC for rehabilitation of a tunnel in Zurich/Switzerland. In: Ozawa K. and Ouchi M. (Eds.) *Proceedings of the Second International RILEM Symposium on Self-compacting Concrete*, Tokyo, Japan, 2001, pp 707–714.

[39] Ushiyama H. and Goto S. Diffusion of various ions in hardened Portland cement paste. *Sixth International Congress on the Chemistry of Cement*, Moscow, USSR, 1974.

[40] Roy S., Poh K. and Northwood D. Durability of concrete – accelerated carbonation and weathering studies. *Building and Environment* 1999, 34, 597–606.
[41] Tang L. and Zhu W. Chloride penetration. In: De Schutter G. and Audenaert K. (Eds.) *Durability of Self-compacting Concrete*, RILEM State-of-the-art Report, 2007, pp 76–88.
[42] Tang L. Chloride transport in concrete – measurement and prediction. Ph.D. thesis, Chalmers University of Technology, Sweden, 1996.
[43] Mehta P.M. and Monteiro P.J.M. *Concrete Microstructure, Properties and Materials*, 2nd edn., University of California, CA, USA, 2001.
[44] Makishima O., Tanaka H., Itoh Y., Komada K. and Satoh F. Evaluation of mechanical properties and durability of super quality concrete. In: *Proceedings of the Second International Symposium on Self-compacting Concrete*, Tokyo, Japan, 2001, pp 475–482.
[45] Sideris K., Kiritsas S. and Haniotakis E. Mechanical characteristics and durability of self-compacting concretes produced with Greek materials (in Greek). *Proceedings of the 14th Greek Concrete Conference*, Kos, Greece, 2003, Vol. II, pp 187–193.
[46] Corinaldesi V. and Moriconi G. Durable fiber reinforced self-compacting concrete. *Cement and Concrete Research* 2004, 34, 249–254.
[47] Brunner M. Durability of SCC with high water content. In: *Proceedings of the Second North American Conference on the Design and Use of Self-consolidating Concrete and the Fourth International RILEM Symposium on Self-compacting Concrete*, Chicago, IL, 2005, on CD.
[48] Cioffi R., Colangelo F. and Marroccoli M. *Durability of self-compacting concrete. Proceedings of the Eighth National Congress AIMAT*, Palermo, Italy, 2006.
[49] Sideris K. Durability of self-compacting concretes of different strength categories (in Greek), Internal Report, Laboratory of Building Materials, Democritus University of Thrace, Greece, 2006.
[50] Zhu W. Private communication, 2006.
[51] Persson B. Chloride diffusion coefficient and salt frost scaling of self-compacting concrete and of normal concrete. *Nordic Mini-seminar on Prediction Models for Chloride Ingress and Corrosion Initiation in Concrete Structures*, Göteborg, Sweden, 2001.
[52] Hwang C. and Chen Y. The property of self-consolidating concrete designed by densified mixture design algorithm. In: *Proceedings of the First North American Conference on the Design and Use of Self-consolidating Concrete*, Evanston, IL, USA, 2002, pp 121–126.
[53] Raghavan K., Sarma B. and Chattopadhyay D. Creep, shrinkage and chloride permeability properties of self-consolidating concrete. In: *Proceedings of the First North American Conference on the Design and Use of Self-consolidating Concrete*, Evanston, IL, USA, 2002, pp 341–347.
[54] Trägårdh J., Skoglund P. and Westerholm M. Frost resistance, chloride transport and related microstructure of field self-compacting concrete. In: *Proceedings of the Third International Conference on SCC*, Reykjavik, Iceland, 2003, pp 881–891.
[55] Westerholm M., Skoglund P. and Trägårdh J. Chloride transport and related microstructure of self-consolidating concrete. In: *Proceedings of the First North American Conference on the Design and Use of Self-consolidating Concrete*, Evanston, IL, 2002, pp 355–361.
[56] Petrov N., Khayat K. and Tagnit-Hamou A. Effect of stability of self-consolidating concrete on the distribution of steel corrosion characteristics along experimental wall elements. In: Ozawa K. and Ouchi M. (Eds.) *Proceedings of the Second International RILEM Symposium on Self-compacting Concrete*, Japan, 2001, pp 441–450.

[57] Suksawang N., Nassif H. and Najm H. Durability of self-compacting concrete with pozzolanic materials. In: *Proceedings of the Second North American Conference on the Design and Use of Self-consolidating Concrete and the Fourth International RILEM Symposium on Self-compacting Concrete*, Chicago, IL, USA, 2005, on CD.
[58] Petersson Ö. Limestone powder as filler in self-compacting concrete – frost resistance, compressive strength and chloride diffusivity. In: *Proceedings of the First North American Conference on the Design and Use of Self-consolidating Concrete*, Evanston, IL, USA, 2002, pp 391–396.
[59] Siebel E. and Breit W. Results of a European Round Robin analysis frost and frost-scaling resistance (in German). Betonwerk + Fertigteil-Technik, 1999, pp 85–92.
[60] Lohaus L. and Petersen L. Festigkeit, dynamisches E-Modul und Diffusionswiderstand, Einfluss der Frostschädigung auf Betoneigenschaften, Beton, Vol. 12/2002.
[61] Copuroğlu O. The characterisation, improvement and modelling aspects of frost salt scaling of cement-based materials with a high slag content. Doctoral thesis, Technical University of Delft, Delft, the Netherlands, 2006.
[62] Rasmussen T.H. Long-term durability of concrete. *Nordic Concrete Research* 1987, 6, 159–178.
[63] Bakker R.F. Regradation of concrete by frost and frost-scaling (in Dutch). *Cement* 1986, No. 12, pp 25–27.
[64] Schäfer A. (1964), Frostwiderstand und Porengefüge des Betons – Beziehungen und Prüfverfahren. Deutscher Ausschuss für Stahlbeton, Vol. 167, Beuth, Berlin, Germany, 1964.
[65] Palecki S. and Setzer M.J. Durability of high performance concrete under frost attack. *6th International Sympsium on High Strength/High Performance Concrete*, Leipzig, Germany, 2002.
[66] Visser J.H.M. Damage in laboratory versus damage in practice – frost resistance of concrete (in Dutch). *Cement* 2002, nr. 2, pp 99–105.
[67] Bremner T.W., Bilodeau A. and Malhotra V.M. Performance of concrete in marine environment – long term study results. *Sixth CANMET/ACI International Conference on Recent Advances in Concrete Technology*, Bucharest, Romania, 2003.
[68] Neville A. *Properties of Concrete*, 4th edn., Longman, Harlow, UK, 1995.
[69] Siebel E. Extrapolation of frost laboratory testing to behaviour in practice (in German). Deutscher Ausschuss für Stahlbeton, Vol. 560, Beuth, Berlin, 2005.
[70] Kivekäs L. Durability of concrete in arctic offshore structures. *Nordic Concrete Research* 1987, 6, 129–178.
[71] Swedish Standard SS 13 72 44 Method for determining the frost resistance of concrete, Borås-method, Swedish National Testing and Research Institute, 1995.
[72] Tang L. and Petersson P.-E. Slab test – Freeze/thaw resistance of concrete – internal deterioration. *Materials and Structures* 2001, 34, 526–531.
[73] Bunke N. Testing of concrete – Guidelines and references in addition to DIN 1048 (in German). Deutscher Ausschuss für Stahlbeton, Vol. 422, Beuth, Berlin, 1991.
[74] Setzer M.J., Fagerlund G. and Janssen D.J. CDF-test – Test method for the freeze-thaw resistance of concrete-tests with sodium chloride solution. *Materials and Structures* 1996, 29, 523–528.
[75] RILEM TC 117-FDC CDF-test – Test method for the freeze-thaw resistance of concrete-tests with sodium chloride solution (CDF). *Materials and Structures* 1996, 29, 523–528.
[76] ÖNORM B 3303 Testing of concrete (in German), Betonprüfung, Austrian Normalisation Institute, Vienna, Austria, 1983.

[77] ÖNORM B 3306 Testing of frost-scaling resistance of precast concrete elements (in German), Austrian Normalisation Institute, Vienna, Austria, 1982.

[78] Gram H.-E. and Piiparinen P. Properties of SCC – especially early age and long term shrinkage and salt frost resistance. In: *Proceedings of the First International RILEM Symposium on Self-compacting Concrete*, Stockholm, Sweden, September 1999, pp 211–226.

[79] Petersson Ö. Limestone powder as filler in self-compacting concrete – frost resistance and compressive strength. In: *Proceedings of the Second International RILEM Symposium on Self-compacting Concrete*, Tokyo, Japan, October 2001, pp 277–284.

[80] Reinhardt H.-W. *et al.* State-of-the-art Report on Self-compacting Concrete (in German). Deutscher Ausschuss für Stahlbeton, Vol. 516, Beuth, Berlin, Germany, 2001.

[81] Persson B. Internal frost resistance and salt frost scaling of self-compacting concrete. *Cement and Concrete Research* 2003, 33, 373–379.

[82] Boel V., Audenaert K. and De Schutter G. Behaviour of self-compacting concrete concerning frost action with deicing salts. In: *Third International Conference on SCC,* Reykjavik, Iceland, RILEM Publications, Cachan, France, pp 837–843.

[83] Boel V., Audenaert K., De Schutter G., Heirman G., Vandewalle L., Desmet B., Vantomme J., d'Hemricourt J. and Ndambi J.M. Experimental durability evaluation of self-compacting concrete with limestone filler. *Presented at: Second North American Conference on the Design and Use of Self-consolidating Concrete and the Fourth International RILEM Symposium on Self-compacting Concrete*, Chicago, IL, USA, October–November 2005.

[84] Persson B. On the internal frost resistance of SCC, with and without polypropylene fibres. *Materials and Structures* 2006, 39(7) 705–714.

[85] Friebert M. Durability of HPC containing fly ash and silica fume. In: Bager D. (Ed.) *Durability of Exposed Concrete Containing Secondary Cementitious Materials*, Hirtshals, 2001, pp 113–142.

[86] Tang M., Yinnon L. and Sufen H. Kinetics of alkali-carbonate reaction. In: Okada K., Nishibayashi S. and Kawamura M. (Eds.) *Proceedings of the Eighth International Conference on Alkali-aggregate Reactions*, Kyoto, Japan, 1989, pp 147–152.

[87] Mailvaganam N.P., Grattan-Bellow P.E. and Pernica G. Deterioration of concrete: symptoms, causes and investigation. The Institute for Research in Construction, Ottawa, Ontario, Canada, 2000.

[88] Boel V. and De Schutter G. Pore structure of SCC in comparison with traditional concrete. In: *Proceedings of the Sixth CANMET/ACI Conference on Recent Advances in Concrete Technology*, Bucharest, Romania, 2003, CANMET/ACI, pp 159–173.

[89] Leemann A. and Vecsei A. Influence of compaction and temperature changes on AAR of concrete (in preparation).

[90] Leemann A. Alkali–silica reaction. In: De Schutter G. and Audenaert K. (Eds.) *Durability of Self-compacting Concrete*, RILEM State-of-the-art report, RILEM Publications, Cachan, France, 2007, pp 137–141.

[91] Plum D.R. Concrete attack in an industrial environment. *Concrete Repairs*, Vol. 1, Concrete Publications, 1986, pp 11–14.

[92] Biczok I. *Concrete Corrosion, Concrete Protection*. Hungarian Academy of Science, 1964, pp 117–126.

[93] Monteny J., Vincke E., Beeldens A., De Belie N., Taerwe L., Van Gemert D. and Verstraete W. Chemical, microbiological, and in situ test methods for biogenic sulfuric acid corrosion of concrete. *Cement and Concrete Research* 2000, 30(4) 623–634.

[94] Collepardi M. Ettringite formation and sulphate attack on concrete. In: Malhotra V.M. (Ed.) *Proceedings of the Fifth CANMET/ACI International Conference on Recent Advances in Concrete Technology*, ACI SP-200, pp 21–37.
[95] Al-Tamimi A. and Sonebi M. Assessment of self-compacting concrete immersed in acidic solution. *ASCE Journal of Materials in Civil Engineering* 2003, 15(4) 354–357.
[96] Boel V. and De Schutter G. Resistance of SCC to acetic and lactic acid. In: De Schutter, G. and Boel V. (Eds.) *Proceedings of the Fifth International RILEM Symposium on SCC*, Ghent, Belgium, September 2007.
[97] Sonebi M. Chemical resistance. In: De Schutter G. and Audenaert K. (Eds.) *Durability of Self-compacting Concrete*, RILEM State-of-the-art Report, RILEM Publications, Cachan, France, 2007, 152–160.
[98] Assié S., Escadeillas G., Marchese G. and Waller V. Durability properties of low-resistance self-compacting concrete. *Magazine of Concrete Research* 2006, 58(1) 1–7.
[99] Brown P.W. Thaumasite formation and other forms of sulfate attack. *Cement and Concrete Composites* 2002, 24, 301–303.
[100] Khöler S., Heinz D. and Urbonas L. Effect of ettringite on thaumasite formation. *Cement and Concrete Research* 2006, 36, 697–706.
[101] Vuk T., Gabrovsek R. and Kaucic V. Influence of mineral admixtures on sulfate resistance of limestone cement pastes aged in cold $MgSO_4$ solution. *Cement and Concrete Research* 2002, 32, 943–948.
[102] Boel V., Audenaert K. and De Schutter G. Acid attack of self compacting concrete. *ICCRRR 2005, the International Conference on Concrete Repair, Rehabilitation and Retrofitting*, Cape Town, South Africa, November 2005.
[103] Al-Amoudi O.S.B. Attack on plain and blended cements exposed to aggressive sulfate environments. *Cement and Concrete Composites* 2002, 24, 305–316.
[104] Kalousek G.R., Porter L.C. and Benton E.J. Concrete for long time service in sulphate environment. *Cement and Concrete Research* 1972, 2(1) 79–89.
[105] Hughes D.C. Sulfate resistance of OPC, OPC/fly ash and SRPC pastes: pore structure and permeability. *Cement and Concrete Research* 1985, 15(6) 1003–1012.
[106] González M.A. and Irassar E.F. Effect of limestone filler on the sulphate resistance of low C3A Portland cement. *Cement and Concrete Research* 1998, 28(11) 1655–1667.
[107] Monteny J. Invloed van polymeermodificatie en cementtype op de resistentie van beton tegen chemische en biogene zwavelzuuraantasting. Ph.D. thesis, University of Ghent, Belgium, 2002.
[108] Al-Tamimi A., Sonebi M., Tagnit-Hamou A. and Saric Coric M. Durability investigation of self-compacting concrete using scanning electron microscope. *Second International IMS Conference on Applications of Traditional and High-performance Materials in Harsh Environment*, Dubai, 2006.
[109] Kalinowski M. and Trägårdh J. Thaumasite and gypsum formation in SCC with sulfate resistant cement exposed to a moderate sulphate concentration. In: *Proceedings of the 2nd North American Conference on the Design and Use of Self-consolidating Concrete and the 4th International RILEM Symposium on Self-compacting Concrete*, Chicago, IL, USA, 2005, pp 319–327.
[110] Trägårdh J. and Kalinowski M. Investigation of the conditions for a thaumasite form of sulfate attack in SCC with limestone filler. In: *Proceedings of the Third International Conference on Self-compacting Concrete*, Reykjavik, Iceland, 2003.
[111] Poppe A.-M. and De Schutter G. Analytical hydration model for filler rich self-compacting concrete. *Journal of Advanced Concrete Technology* 2006, 4(2) 259–266.
[112] Trägårdh J. and Bellmann F. Sulphate attack. In: De Schutter G. and Audenaert K. (Eds.) *Durability of Self-compacting Concrete*, RILEM State-of-the-art Report, 2007, 89–119.

[113] Ye G., De Schutter G. and Taerwe L. Spalling behaviour of small self-compacting concrete slabs under standard fire conditions. In: De Schutter G. and Boel V. (Eds.) *Proceedings of the Fifth International RILEM Symposium on SCC*, Ghent, Belgium, September 2007.

[114] Boström L. and Jansson R. Fire resistance. In: De Schutter G. and Audenaert K. (Eds.) *Durability of Self-compacting Concrete*, RILEM State-of-the-art Report, 2007, 142–151.

[115] Persson B. Self-compacting concrete at fire temperatures. Report TVBM-3110, Lund Institute of Technology, Sweden, 2003.

[116] Noumowé A., Carré H., Daoud A. and Toutanji H. High-strength self-compacting concrete exposed to fire test. *Journal of Materials in Civil Engineering* 2006, 18(6) 754–758.

[117] Liu X. Microstructural investigation of self-compacting concrete and high-performance concrete during hydration and after exposure to high temperatures. Ph.D. thesis, University of Ghent, Belgium, 2006.

[118] Boström L. The performance of some self-compacting concretes when exposed to fire. SP report 2002: 23, Sweden, 2002.

[119] Boström L. Innovative self-compacting concrete – development of test methodology for determination of fire spalling. SP report 2004:6, Borås, Sweden, 2004.

[120] Boström L. and Jansson R. (Eds.) *Spalling of self-compacting concrete, Proceedings of the Fourth International Workshop on Structures in Fire*, Aveiro, Portugal, 2006.

[121] Persson B. Fire resistance of self-compacting concrete, SCC. *Materials and Structures* 2004, 37, 575–584.

[122] Horvath J., Hertel C., Dehn F. and Schneider U. Influence of pre-ageing on the temperature behaviour of self-compacting concrete (in German). *Beton- und Stahl betonbau* 2004, 99(10).

Chapter 11
Standards and specifications

11.1 Introduction

Most of the material in this chapter deals with the European experience, but some attempt has been made to describe what is happening in the rest of the world. Chapter 2 has described the history of SCC and its use. It is clear that whether described as a material or a method of placing, it is new. For this reason, if the worldwide position is examined, in general, it is found that standards and specifications have not yet been fully developed. The standards have not kept pace with the advances of technology. Part of the reason for this is discussed in Chapters 2 and 4. The question of how to conveniently and reliably define SCC in practice, on site, has not yet been fully answered, though progress is undoubtedly being made.

There is a 'circular', self-referential difficulty – SCC is not yet (in practical terms) defined; it is not defined because standardised test methods have not been available to use in any definition; and there are (generally) no standard test methods because it is not completely clear what the methods have to test/identify. However, this obstacle will eventually be surmounted and sensible relationships established.

Italy has proper standards, both for test methods and for SCC itself[1, 2], although they appear to simply reflect current practice and familiarity, rather than being underpinned by a rigorous prenormative evaluation. Otherwise, at best, there may be a limited addendum to an existing national standard. In the Danish provisions for the application of European standard EN 206-1, 'Concrete', for example, there is an informative annex for SCC, which specifies a test method for 'flowability'. In several countries, however, 'guidelines', data sheets, or similar advisory documents have been written e.g. Australia, Austria, Belgium, China, France, Germany, Italy, Japan, Norway and the UK. These may refer to SCC differently, even in English (e.g. in Australia as 'super-workable concrete', or in the USA as 'self-consolidating concrete'), but all deal with SCC. Other countries, e.g. the Netherlands, USA, the Czech Republic are currently engaged in drawing up similar advice. These documents have been written by bodies like national associations of civil engineers, concrete societies, trade organisations (e.g. ready-mixed concrete, precast concrete, admixtures) and semi-national research organisations, for use by engineers, producers and specifiers. They have different levels of authority and acceptance.

11.2 Specification of self-compacting concrete

It is important to make it clear at the outset that just as SCC does not differ greatly from TVC, other than in its very different fresh properties, so its specification should be broadly similar. There is no wholly new approach, and most of the ways of specifying normal concrete, following normal standard procedures, are necessary. Understood here, is the need to specify strength, exposure conditions etc. and the need to meet chloride and alkali limits if these are standardised. The question for consideration is how to specify the self-compactability of the concrete.

The concrete has been satisfactorily designed, produced and placed, but its basic property, self-compactability, is not yet defined: the minimum 'degree' of compaction required, the 'adequate' compaction, remains an elusive parameter. It is accepted that a genuinely complete, 'full', 100% compaction, is probably not practical or economically viable in concrete construction, but questions of what degree of 'undercompaction' is tolerable, and how it is to be measured remain unresolved.

The normal procedure for specifying concrete is often modified in one or both of two ways. First, the ready-mixed supplier may have ready, off the shelf, one or more standard mix designs of 'proprietary' concrete, which may be accepted by the purchaser. These are usually commercially confidential. In fact, proprietary mixes are defined in the European guidelines for self-compacting concrete[3] as 'concrete for which the producer assures the performance subject to good practice in placing, compacting and curing, and for which the producer is not required to declare the composition'. However, because of the use of the word 'compacting', the definition is clearly not restricted to SCC. Naturally such a mix would have to comply with any specified parameters (e.g. maximum water/cement ratio, minimum cement content), but otherwise the purchaser accepts the mix as offered by the supplier. It is possible that proprietary mixes, either generally, or specifically for SCC, will be introduced in the 2011 revision of European Standard EN 206-1[4].

The other way of specifying SCC is for the requirements for the concrete to be developed in discussion between supplier, purchaser, and often (given the importance of superplasticisers and VMAs to SCC) the admixture supplier, and for the mix to be designed and agreed dependent on the particular requirements of the job (e.g. reinforcement details, shape of section). Normally the specification is a part of the tender documents for a contract, decided well before the start of a job, before the award of the contracts to contractor and supplier. With SCC on the other hand, the development and refinement of the specification are more likely to go hand-in-hand with the development and refinement of the mix design, in a way that is not usual for TVC.

This approach is often adopted when the reason for using SCC is the need to solve a particular problem: of access, or surface finish, perhaps. In these circumstances input to the specification may also be needed from the designer (engineer or architect). In these circumstances, there will often be a need for site experimentation: the use of one or more proposed SCC designs either in trial panels, or in noncritical areas that are incorporated in the final works. This is the situation that obtains in the USA, UK, and many other countries.

It is important for the contractor to be involved in the development of a specification (and mix design) for a number of reasons:

- The degree of detail necessary in the SCC specification will depend on the application and placing process. 'Over-specification' should always be avoided.

- The design of a suitably robust mix is heavily dependent on placing methods, height of free fall, required length of flow, facilities and skill/experience of personnel, etc.
- The need for the supplier to know the required consistence retention time.
- The construction programme is important: the season (i.e. essentially temperature) at which the concrete will be placed affects mix design and production. The dosage of the admixture, for example, may depend on the temperature.

A further approach, used in the USA[5] with public projects, is for government agencies to work closely with state universities to help them develop specifications for public work in their state.

For precast products the situation is different. The basic specification refers to the purchaser's performance requirement for the product. It is then up to the producer to determine the properties of the fresh concrete, taking into account the requirements of the product and the characteristics of the production process.

11.3 International advisory documents for self-compacting concrete

Because the standardisation process, including, particularly, standardisation of methods of specification is lagging, construction with SCC that follows any recognised procedure is on the basis of these national, or other guidelines. In Europe, attempts have been made to coordinate the writing of these documents, often explicitly with future standardisation in mind, and these have been written in the accepted format of CEN standards. The first of these was produced in 2002 by EFNARC, the European Federation of Specialist Construction Chemicals and Concrete Systems[6]. EFNARC itself, and four other European organisations, BIBM, CEMBUREAU, EFCA and ERMCO produced a more comprehensive and up-to-date document in 2005[3].

11.3.1 European guidelines

The organisations involved in writing this document represent, respectively, the European precast concrete, cement, concrete admixtures and ready-mixed concrete industries. The guidelines have been translated into at least three other languages (German, Swedish and Spanish) and are used in Europe and as far afield as New Zealand[7]. They are likely to form the basis of that part of the revision of EN 206-1 which deals with SCC. They contain important lists of definitions and standards, and chapters on the engineering properties of SCC, materials, mix design, production, site requirements, placing and finishing, SCC in precast, and on appearance and surface finish. However, the chapters relating to standardisation and particularly specification will be considered here. First, because SCC is characterised by its fresh properties, and because these must be measured and tested in some way in order to specify the concrete, the European guidelines currently recommend the following tests:

- for filling ability – slump-flow
- for flow-rate (viscosity) – T_{500} or V-funnel

- for passing ability – L-box
- for segregation resistance – sieve segregation test

Detailed procedures have been given, and these test methods have been discussed more fully in Chapter 5 of this book.

Otherwise, the European guidelines have intentionally followed the structure of the existing EN 206-1 in many ways. Part 6, 'Specifying SCC for ready-mixed and site-mixed concrete', and Annex A, 'Requirements for SCC', together detail what needs to be specified. The definitions given in EN 206 apply, together with others relevant to SCC, e.g. for proprietary concrete. The guidelines differentiate between basic and additional requirements for SCC. Basic requirements are:

- compressive strength class (EN 206-1)
- exposure class (national provisions)
- limiting values e.g. w/c ratio, minimum cement content (national provisions)
- aggregate size
- chloride class (EN 206-1)
- slump-flow class or, exceptionally, a target value

Apart from slump-flow (which replaces the usual consistence test for TVC), these requirements are exactly the same as for ordinary concrete, emphasising the fact that SCC may be considered a new way of placing concrete, rather than a new type of concrete.

Additional requirements are:

- T_{500} value, or a V-funnel class
- *L*-box class, or, exceptionally, a target value
- segregation resistance class or, exceptionally, a target value
- requirements for temperature of concrete, if different from those of EN 206-1 other technical requirements (e.g. washout resistance)

It should be noted that references to the specific clauses in EN 206-1 are given in the guidelines[3].

Following the methodology in EN 206-1, the guidelines establish a number of classes (Tables 11.1–11.4) for consistence, i.e. for filling ability (referred to as 'flow'), flow rate (referred to as viscosity), passing ability and segregation resistance, which are proposed for specification. These are also discussed in more detail in Chapter 5.

Table 11.1 *Slump-flow classes.*

Class	*Slump-flow spread (mm)*
SF 1	550–650
SF 2	660–750
SF 3	760–850

Table 11.2 *Flow rate (viscosity) classes.*

Class	T_{500} *(s)*	*V-funnel (s)*
VS 1/VF 1	≤ 2	≤ 8
VS 2/VF 2	> 2	9–25

Table 11.3 *Passing ability classes.*

Class	*Passing ratio*
PA 1	≥ 0.8 (2 bars)
PA 2	≥ 0.8 (3 bars)

Table 11.4 *Segregation resistance classes.*

Class	*Segregation Index (referred to as resistance) %*
SR 1	≤ 20
SR 2	≤ 15

It is also important to note, that of the fresh properties of SCC, only filling ability is considered always and everywhere as essential in these guidelines. Of the tests, only slump-flow is a 'basic' requirement for SCC in the fresh state. This requirement reflects the greatest amount of existing practical experience with this test, not necessarily that it is the best test for the purpose. The document makes it clear that the additional requirements are not normally specified, and should only be so where needed; it goes on to provide advice on where they should, indeed, be specified. This depends on:

- confinement conditions – related to size and shape of element, reinforcement density and detail, etc.
- placing method: pump, skip etc.
- number and position of concrete placing points
- finishing method

The construction situations, for which the different classifications are suitable, and necessary, are described in some detail, and a useful diagrammatic summary representation of the required properties of SCC, based on the application, is shown in Figure 11.1.

In relation to these consistence classes the question of conformity and identity testing arises, as it affects both hardened and fresh properties. Generally speaking, SCC and TVC of similar strength have comparable hardened properties, and if there are differences, these are usually covered by the safe assumptions on which the design

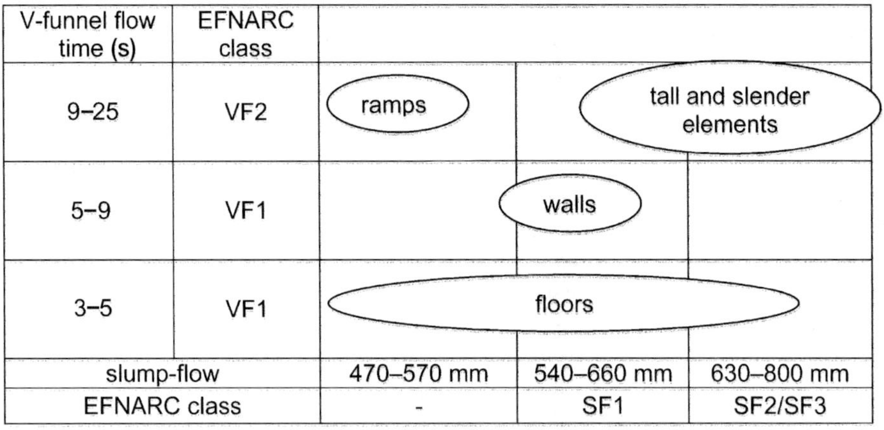

V-funnel flow time (s)	EFNARC class			
9–25	VF2	ramps		tall and slender elements
5–9	VF1		walls	
3–5	VF1		floors	
slump-flow		470–570 mm	540–660 mm	630–800 mm
EFNARC class		-	SF1	SF2/SF3

Figure 11.1 *Classes of fresh SCC and their practical applications (adapted from Walraven).*

codes are based. It is therefore reasonable to apply to SCC the normal detailed EN 206-1 conformity rules and procedures for traditional vibrated hardened concrete. It should be noted that if suppliers of proprietary mixes operate a third party-certified quality assurance scheme, they are probably justified in arguing that identity testing for strength is unnecessary, since they guarantee the performance of the concrete, and are independently accredited.

Assessment of the conformity of the fresh properties is less well developed, even for TVC, and practices differ considerably in different European countries and worldwide. It is, however, in some ways simpler, because in EN 206-1, conformity can be assessed on single test results, rather than on statistical analysis of on-going production results, as for compressive strength. Analysis of long-term conformity can be made, but in most countries conformity testing is the same as identity testing, i.e. based on a single batch. What is clear is that the existing conformity limits (for the consistence of TVC) are wider than the classes themselves (EN 206-1, Tables 11 and 18), and if this remains the case, it is reasonable that the same principle should apply to SCC. The European guidelines make this assumption, and to make the implications of this clear, two examples might be quoted from the document:

- For the middle of the three slump-flow classes, SF2, the class limits are 660–750 mm. The conformity criteria proposed are ≥ 640, ≤ 800 mm, i.e. the specified limits – 20 / + 50 mm.
- For sieve segregation class SR1, the segregation index limit is ≤ 20%. The conformity criterion proposed is ≤ 23%, i.e. the specified limit + 3%.

The document makes no statement about conformity where a target value is specified. If the European guidelines were used contractually as the basis of specification for a contract, these conformity limits would apply. Otherwise it is important that some agreement is reached between supplier and purchaser, as part of the 'additional requirements' of the specification. It should be noted that it may be that, in the revision of EN 206-1 to include SCC, the conformity limits will be simplified, so that the class limits and the conformity limits are the same.

11.3.2 Other national guidelines

Thus far, the European guidelines are the most systematic attempt to write a comprehensive practical specification for SCC. Other guides or similar publications have been written. They all contain much information other than questions of specification, and in fact, other than the European guidelines, specification does not necessarily appear to form a particularly important part of their contents. This is partly due to their being, in many cases, the countries' first attempt to comprehensively document SCC, and partly because of the continued difficulty of actually defining what SCC is, in terms of test methods and their results. This difficulty is clear from the multiplicity of test methods recommended.

Generally, the documents seem to lean heavily on, and quote from each other, which emphasises the fact that the amount of practical experience with SCC is still relatively low. SCC remains in the developmental stage, though advances are rapid. To take just one example: the permitted length of flow for fresh SCC: this is not an essential element of specification, but it may be necessary in some circumstances. Most of the guideline documents surveyed mention figures around 10 m, but the question may be asked in each case whether real experience or research underlies the statement, or whether the number is simply quoted from previous publications. The question may even be asked whether the 10 m means 10 m in each direction, or 10 m overall, i.e. 5 m in each direction; this is not clear from a general overview.

A summary of how these documents deal with matters of specification is provided in Table 11.5.

Comments on matters arising from some of these documents follow. The comments are not intended to be exhaustive, but rather to highlight some aspects of specification and identification of SCC that have wider applicability than the country for which these national documents were written. It is worth pointing out that some of the documents are written with the construction process principally in mind, others more directly with the concrete itself.

Since 'modern' SCC was largely 'invented' in Japan, it is perhaps not surprising that the earliest publication was Japanese[8]. For what appear to be specification purposes, it establishes three ranks of construction circumstances, of increasing severity, based on element geometry and reinforcement density. However, there is no attempt to define or specify the type of SCC that is required for these ranks.

An attempt was made to do so six years later in the Australian recommended practice for 'super-workable' concrete (i.e. SCC)[9] which follows the Japanese ranking approach. In this document, except for the most difficult structures (rank 1), where suitable values have to be determined job by job, ranges of values are suggested (for ranks 2 and 3, more routine structures) for the slump-flow, J-ring, L-box and Orimet tests.

In the United Kingdom, the Concrete Society report[10] points out that a performance specification for SCC (that is performance in the fresh state) is not yet

Table 11. 5 *Summary of model specifications for SCC.*

Country	*Europe*	*Japan*	*UK*	*France**	*Belgium*	*Australia*
Date	2005	1999	2005	2000	2005	2005
Test methods proposed	Slump-flow L-b., Siev. segr., V-f	Slump-flow funnels L-b, U-box	Slump-flow L-b, V-f. Siev segr.	Slump-flow L-b, Siev. segr., bleeding, in situ anal.	Slump-flow L-b, U-box	Slump-flow, J-r, L-b, Orimet, sieve segr.
Test methods described	yes	yes	yes	yes	yes	yes
Type of concrete	proprietary	no	proprietary	no	no	no
Principle of specification	Classes or target value detail	3 ranks based on element	Classes or target values	no	no detail	3 ranks based on element
Flow rate (viscosity) mentioned	yes	yes	yes	no	no	no
Based on	EN 206	JSCE standards	EN 206, nat.building codes, Euro guide	RILEM	EN 206 JSCE	EFNARC
Conformity rules	yes	no	Ref euro	no	no	no
Conformity for slump-flow	–80 / +100	no	no	no	≥ 650	no
Other method for conformity	no	U-box	no	Sieve segr.	U-b, sieve segr.	no
Testing frequency	As per EN 206	50 m^3	no	no	no	50 m^3, SF every load
Acceptance test	no	Full-scale	no	slump-flow	no	no
Other requirements	Advice given	no	Detailed advice given	no	no	no
Value for applications proposed	yes	3 ranks based on element detail**	no	no	no	no
Production control	normal	May be tighter	normal	no	no	no

Table 11. 5 *(continued)*

Country	*Europe*	*Japan*	*UK*	*France**	*Belgium*	*Australia*
Date	2005	1999	2005	2000	2005	2005
Verification/ trial panels	no	for surface quality	for surface quality	no	no	no
Batch size	no	80% mixer capacity	no	no	no	no
Formwork pressure	hydrostatic	hydrostatic	hydrostatic	no	hydrostatic	hydrostatic
Length of flow	≤ 10 m	≤ 8–15 m	≤ 10 m	no	≤ 7 m	≤ 10 m
Maximum free fall	no	≤ 5 m	As for vibr. concrete	≤ 5 m	no	≤ 5 m

* The French document does not deal directly with specification
** Specification for SCC for ranks is not defined

possible because the test methods for the verification of performance have not yet been standardised. For this reason, for the moment, most specification will be as proprietary concrete. The introduction of standards will permit the specification of designed concrete, as defined in EN 206-1.

Much SCC that is produced is 'tailor made' for specific circumstances, and the requirements are developed through discussion and trial. In these circumstances, the 'additional requirements' mentioned in the European guidelines are very important. The UK report has an extremely useful checklist of what these additional requirements might be (remembering that they are not necessary in every case), and suggests that they be agreed contractually in writing. Items in this written agreement supplement those laid down in EN 206-1. The suggested items include:

- method of placing
- required consistence retention time
- test method(s)
- procedure for trial panels
- method statement for placing
- maximum length of flow and height of drop
- adjustment of consistence on site
- definition of surface finish required

In Germany a guideline[11] has been published, which has the status of a standard and is implemented by the national building regulations. There is also a guide published by a cross-industry cement 'platform', the result of a national research project to specify and describe SCC[12]. This latter document includes a useful attempt to categorise SCC by slump-flow and flow time only. The diagram is reproduced as Figure 11.2.

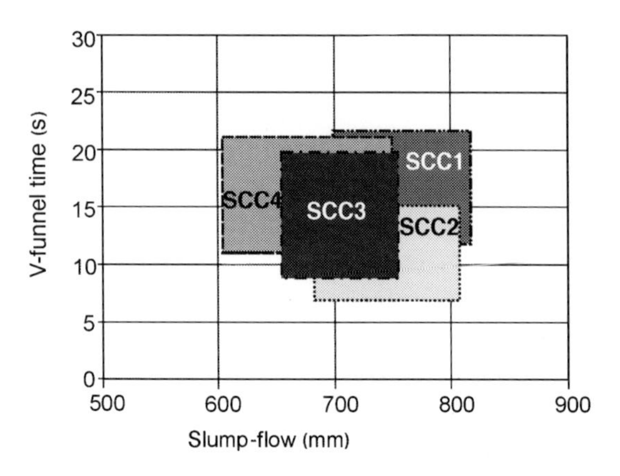

Figure 11.2 *German method for defining acceptable SCC. Flow-rate against spread, measured by slump-flow test.*

This shows the form of typical SCC 'windows', which are defined by the producer, on the producer's responsibility, as acceptable SCC. They could be used for specification, or to identify different SCCs for different applications, though the guideline document does not attempt to do so. V-funnel time could be used, but a special test has been developed whereby funnel time and flow are combined into a single procedure, though two measurements, of flow and time, are made. This is typical of efforts made to simplify the specification procedure by reducing the number of fresh properties that have to be specified.

The Norwegian guide[13] takes a similar approach, making a positive effort to make specifying easier for engineers who may have no experience at all with SCC. It is felt that this combination of lack of experience (of using SCC, and of the test methods) and perceived difficulty of specification is one of the reasons for the surprisingly slow take-up of SCC outside the precast industry. The Norwegian document is at pains to make clear, at several points, that 'SCC (is) an alternative form of execution, not an alternative material'. It also clearly states the principle that for the moment the specification of SCC remains to be decided on a job-by-job basis: 'the contractor and the producer must jointly determine intended properties and possible limit values for the properties of the fresh concrete, based on the area of use for the concrete and on the choice of local materials'.

The use of a combination of J-ring and slump-flow is recommended, and limiting values are given for walls/columns and for slabs, different values in each case. Interestingly, the document suggests that the acceptance values will be narrower than the overall limits. This document was published in 2002. A final point that might be mentioned, identified in the Norwegian document, is that among the 'additional requirements' that may need to be specified are the producer's quality assurance procedures. This may be more important for the production of SCC because of their responsibility for the wider aspects of concrete performance, site aspects that are usually the contractor's responsibility.

11.4 Advisory documents available from concrete and admixture suppliers

Concrete and admixture suppliers have made efforts to promote and market SCC. Usually, they have produced short documents, or information sheets, describing the benefits, uses, applications and properties of their product, whether admixtures or proprietary concrete. These documents do not normally give detailed information (often commercially confidential) but are intended as an introduction to the product and to its specification. The suppliers have encouraged specification as a dialogue between them and the purchaser, as described above.

11.5 Standards for test methods

What essentially distinguishes SCC from TVC are its fresh properties, and the principal difficulty faced by engineers at present is that these properties have to be specified without recourse to standard test methods. (Tests on hardened SCC are the same as for TVC.) In Europe the absence of standardised tests for fresh concrete is now being addressed. The EU funded collaborative research into test methods[14] to simplify the standardisation process (described in Chapter 5), and taking into account that work and the European guidelines, CEN (the European standardisation body) subcommittee responsible for concrete standards has decided on the standardisation of five test methods for SCC: slump-flow, L-box, V-funnel, sieve segregation and J-ring. These are in the process of being written, and will be published as additions to the EN 12350 series of tests for fresh concrete.

In the USA, the standardisation process for SCC test methods has also begun. ASTM standards for a segregation test (column technique) and methods for flow and passing ability (slump-flow and J-ring) have been published[15–17].

11.5.1 European EN standards

In Europe, the general standard for concrete, EN206-1[4] was published in 2000, after many years of work. Because it is not a 'harmonised' standard, each country has a set of 'National Application Documents' (NAD) in addition to the standard itself. The current EN 206-1 itself, and most of the national provisions, do not include SCC in any way. The CEN subcommittee for concrete has established a task group to make recommendations on 'Provisions for SCC' so that they can be included in the revision of EN 206-1, which is due in 2011. These requirements for SCC will include testing, product requirements, and the relationship with the execution standard.

Part of the preparatory work of this subcommittee was a survey of the regulatory situation for SCC throughout Europe. It was clear that very different 'rules' are applied, either through the NADs, or the terms of the national guidelines, which may exist, and/or through the building regulations which may be in force. Differences in the applicable rules include:

- restrictions on mix design
- use of admixtures
- questions of durability
- classes of SCC

- test methods in use
- conformity and identity testing

In revising EN 206-1, some new aspects of the standard, which may have to be considered include, for SCC:

- establishment of classes for consistence
- tolerances on target values for consistence
- reference to viscosity modifying admixtures, which are not covered by the European standard for admixtures
- any changes to the rules for factory production control (part of the existing requirements of EN 206-1, and other European standards) and conformity rules
- use of lightweight and heavyweight aggregates
- any necessity for different production control requirements, e.g. for control of consistence

This list is based on the existing provisions of the current EN 206-1, and assumes that they will largely continue to apply after the revision, which does not solely concern SCC.

Another CEN technical committee is concerned with the precast concrete standard, EN 13369[18], for precast concrete products. The committee is considering what modifications are necessary to the standard for the inclusion of SCC. Much of this is likely to be similar to the topics listed above for site-produced or ready-mixed concrete. Since the standard is about actual concrete elements, produced in a factory, not simply concrete itself, attention might be drawn to factory production control. This may to have to include any particular requirements for placing and finishing of SCC.

11.5.2 EN 13670, the execution standard

The series of European standards now includes EN 13670 Execution of Concrete Structures[19], expected to be accepted very shortly, which codifies on-site procedures for concreting work. In general, it underpins the need for specifiers of SCC to sometimes define 'additional requirements' related to placing and other site matters, because it states (for all concrete, not just SCC) that the concrete specification shall include the requirements of the standard, together with those related to the actual method of execution. There is a short section on SCC in the draft, which explicitly states that 'working procedures for the actual cast shall be established and additional requirements to those in EN 206-1 (for) fresh concrete and its conformity criteria shall be agreed with the producer'. This is critical for SCC because, perhaps more so than in the case of TVC, the product requirements depend upon the application. This confirms the de facto situation on specification of SCC as described in this chapter: that it often, or usually, remains necessary for supplier and purchaser to jointly agree the specification.

11.5.3 International standards

The International Standards Organisation (ISO) is currently writing a standard (DIS 22965) on concrete, largely based on EN 206-1. The current draft does not include any reference to SCC.

References

[1] Norma Italiana UNI 11140, Self-compacting concrete. Specification, characteristics and checking, 2003.

[2] Norma Italiana UNI 11141–11145, Test methods for SCC, 2003.

[3] BIBM, CEMBUREAU, EFCA, EFNARC and ERMCO, The European Guidelines for Self-compacting Concrete, 2005.

[4] EN 206-1: 2000, Concrete – Part 1: Specification, performance, production and conformity.

[5] Daczco, J. Private communication, BASF Admixtures Inc, USA.

[6] EFNARC, Specification and Guidelines for Self-compacting Concrete, 2002.

[7] Khrapko M. Private communication, CBE Consultancy, New Zealand.

[8] Japanese Society of Civil Engineers, Recommendation for self-compacting concrete, 1999.

[9] Concrete Institute of Australia, *Recommended practice, Super-Workable Concrete,* Concrete Institute of Australia, Rhodes, NSW, 2005.

[10] Concrete Society, Self-compacting concrete – a review, Technical Report 62, Concrete Society, Camberley, UK, 2005.

[11] Deutscher Ausschuss Fur Stahlbeton, Guidelines for self-compacting concrete (in German), 2003.

[12] Verein Deutscher Zementwerke, Self-compacting concrete properties and hosting (in German), Zement – Mekblatt B29, 2006.

[13] Norwegian Concrete Association, Guidelines for production and use of self-compacting concrete, Publication no. 29, 2002.

[14] Gibbs J.C. Self-compacting concrete: getting it right. *Concrete*, 2004, 38(6) 10–14.

[15] ASTM C1611/C1611M-06: Standard test method for static segregation of self-consolidating concrete using column technique.

[16] ASTM, C1611/C1611M-05 Standard test method for slump flow of self-consolidating concrete, 2007.

[17] ASTM, C1621/C1621M-06 Standard test method for passing ability of self-consolidating concrete by J-Ring, 2007.

[18] EN 13369 Common rules for precast concrete products, 2004.

[19] prEN 13670 Execution of concrete structures (in preparation).

Chapter 12
Benefits of using self-compacting concrete

A change from TVC to SCC brings a number of tangible benefits in different aspects of concrete construction. Many of the individual benefits alone are significant, but a greater benefit can be obtained when the use of SCC is already envisaged in the planning and design stages. Maximum total benefit arises when the management of the construction process itself is already adapted to exploit the new technology. Such an approach creates genuine synergy and the overall benefit obtained is much greater than the sum of the individual ones.

12.1 Working conditions

Use of SCC leads to elimination of vibrators, which significantly improves health and safety on the concrete construction site. It leads to a major reduction of exposure of the workers to noise and vibration[1, 2]. In addition, placing SCC is much less strenuous than placing TVC[1].

The placing and compaction of fresh concrete is generally recognised as the physically most demanding and unpleasant activity in the concrete construction process. In some countries it has become difficult to recruit workers prepared to undertake this work. Those workers, who place and vibrate fresh concrete, are unlikely to do so for long. The task is therefore very often in the hands of a transient workforce, often inadequately trained, but responsible for the compaction process. Inevitably, this has a negative impact on the overall quality of the final product and its costs.

The very substantial improvement of working conditions of personnel during concrete construction, an elimination of the worst activity by replacement of TVC with SCC has been recognised by the European Commission. The contribution of the European research on SCC[1], namely the improvement of the working environment was rated so highly that it became the first engineering-related project to progress into the final of the 2002 EU Descartes Prize. (An annual prize of Euro 1M is awarded to the best overall EU supported research project of any type in any subject area.)

12.1.1 Noise

Exposure to excessive levels of noise and vibration has been considered a serious health problem for the construction workforce for a considerable time. Measures aimed at reducing this health hazard were introduced[3, 4], and these are now becoming more widespread and stricter. This brings into focus the selection of construction methods and materials, where environmental aspects are considered increasingly side by side with economy.

Compaction of concrete is a significant source of noise during concrete construction. It is therefore not surprising to find the introduction of SCC sometimes called a 'quiet revolution'. The exposure of workers to noise and vibration is already limited by legislation in many countries. In the European Union the original directives regarding exposure to noise[3, 4] have been updated and tightened[5, 6], particularly with regard to the effects of noise on hearing. The level of exposure of the workers at which protection becomes mandatory has been reduced by 5 dB[5, 6]. The employers have increased responsibilities, not only for avoiding excessive exposures and providing adequate protection to workers but also for carrying out formal risk assessments prior to any activity in which noise was a significant factor and to monitor health (hearing) of any workers who are regularly exposed to noise levels above 85 dB[6].

A programme of tests and trials, both in a laboratory and during practical placing of SCC and TVC on construction sites was carried out as part of the European project on SCC[1, 2]. The aim of the task was to obtain a quantitative assessment of the reductions of exposure to noise when the traditional vibration was eliminated and SCC used instead. Identical full-scale structural elements or identical parts of a multistorey reinforced-concrete in situ framework were cast using both SCC and TVC mixes. Three types of vibrators, which are typical noise sources associated with compaction of TVC, were selected and the noise levels tested in accordance with[7–10]:

- External/clamp-on vibrators: these are attached to formwork. Such types and arrangements of vibrators are normally used in precast concrete production and in cases where concrete elements of complex shapes are being cast. Surface vibrators were attached to a vertical plywood formwork assembly for casting of reinforced concrete columns of 300 × 450 mm in cross-section, 3000 mm tall. All the vibrators were fixed onto horizontal timber elements of the formwork at heights of 570 mm and 1570 mm above the base. The vibrators were in pairs, one vibrator above the other. The trial was held in a large hall similar in size to a typical precast concrete production plant. Three measuring points at each of the measuring distances, i.e. 1 m, 2 m and 4 m, were used during the measurement of noise from the surface vibrators. The sound field in the workplace remained undisturbed during the course of the measurements.
- Internal/poker vibrators: These are the most common tools for compaction of the traditional fresh concrete when placed in situ. Two internal/immersion poker vibrators of 35 mm diameter, powered by compressed air, were used. The noise from the two vibrators handled simultaneously by two operators was measured during compaction of concrete poured into a horizontal plywood formwork for a single reinforced concrete beam (approx. 200 × 300 mm in cross-section, 3800 mm long). The casting was carried out outdoors, simulating a typical in situ concrete placement.
- Vibrating table: This type is used regularly for compaction of standard test specimens for assessment of properties of hardened traditional concrete and in precast concrete works. A medium-size vibrating table with a top plate of 1265 × 620 mm was used, placed inside the testing hall.

The noise was measured as a time-varying noise. The fluctuations have been expressed using the standard deviations of L_{pA} from the mean value measured over the total time period of 15 s. The results of the measurements are given in Table 12.1, inclusive of the results of measurement of background noise.

The elimination of the primary sources of noise and vibration, such as poker and other vibrators, produces benefits in addition to those primarily related to potential damage to hearing. Vibrators are mostly driven by compressed air, which requires additional, sometimes very noisy compressor plant for its production, with hoses for distribution to the vibrators. The absence of vibrators therefore produces secondary, but significant benefits: it reduces the background noise and provides a safer site, uncluttered with hoses and pipes, which could cause falls and injuries. Experience from full-scale practical trials, where SCC was used as an alternative to TVC showed that the working conditions improved so much that the workers did not wish to return to placing TVC instead of SCC[1, 2].

The sound intensities at a construction site where TVC is placed are normally at levels, which make it necessary to wear hearing protection to avoid damaging human hearing. The sound levels at an ordinary concrete construction site are in many places so high that it becomes necessary to wear hearing protection. Such protection is needed to avoid damaging human hearing when the maximum safe daily exposures of a worker obtained in different activities and operations are not evaluated or added up. Provided the level of background noise is low, the use of SCC reduces the exposure of the workers to sound intensities that are as low as one-tenth of those produced when placing TVC[2,11].

A much quieter construction site or a precast concrete workshop produces a better environment for effective and accurate communication. It is not uncommon for problems to arise from incomplete and poorly heard communications.

12.1.2 Vibration

Added to the hazard from exposure to noise is the effect of exposure to vibration generated by hand-tools such a poker vibrators. Vibration from concrete poker

Table 12.1 *Equivalent A-weighted sound-pressure levels, $L_{pA\ eq}$ and the mean instantaneous A-weighted sound-pressure levels, $L_{pA\ m}$ inclusive of the standard deviations over the same total period, at distances r from sound sources under test* [2].

Sound sources	*Indices*	*Values measured at distances r from sound source*			
		Operating			*Switched off*
		r = 1 m	*r = 2 m*	*r = 4 m*	*r = 1–4 m*
2 surface vibrators	$L_{pA\ eq}$	101.9 dB	98.4 dB	94.8 dB	68.6 dB
	$L_{pA\ m}$	101.8 ± 0.7 dB	98.3 ± 0.6 dB	94.8 ± 0.6 dB	68.5 ± 0.5 dB
2 internal vibrators *	$L_{pA\ eq}$	82.9 dB	82.7 dB	76.9 dB	62.2 dB
	$L_{pA\ m}$	82.0 ± 3.9 dB	81.9 ± 3.1 dB	76.8 ± 1.7 dB	62.2 ± 0.5 dB
1 vibration table	$L_{pA\ eq}$	102.1 dB	98.0 dB	94.1 dB	81.0 dB
	$L_{pA\ m}$	102.1 ± 0.4 dB	98.0 ± 0.3 dB	94.1 ± 0.4 dB	77.1 ± 6.2dB

* Values measured at the position of one operator: $L_{pA\ eq}$ = 87.7 dB $L_{pA\ m}$ = 84.9 ± 7.6 dB.

vibrators is transferred to the operator's hands. It affects nerve systems, impairs blood circulation and causes serious musculoskeletal problems. Over-exposure leads to a condition known as 'hand-arm vibration syndrome' (HAVS), which is cumulative and irreversible. Until recently, the scale of the condition, often referred to as the 'vibration white finger' has not been fully recognised, but recent statistics in the UK indicate that it as the most common prescribed disease under the industrial injuries scheme[12]. A strong campaign to reduce or eliminate the vibration-linked health hazard from construction has commenced[13].

Replacement of the TVC by SCC eliminates completely this important source of serious health hazard in construction.

12.1.3 Physical strain

Concreting work is strongly linked with musculoskeletal problems, most often regarding back/spine, joints and muscle in the lumbo-sacral region. A survey[14] indicated that 26% of concrete workers in Sweden suffered either often or very often from back problems. Use of SCC eliminates the need for vibration and alters the way the fresh mix is handled during placing. Ergonomic assessments and trials in which physical strain/exertion was measured were carried out when both TVC and SCC were placed. The trials involved typical reinforced concrete construction, namely casting of columns and floor slabs in a multistorey building. Results of the survey confirmed that placing of SCC was much easier. It required only a minimum of assistance by raking and moving of concrete by shovels. Any additional screeding (floor slabs) in which additional movement of the fresh mix was needed became much less strenuous.

Physical strain was investigated in typical conditions and on routine tasks during construction of reinforced concrete walls and floors on multistorey building sites. Exertion of workers carrying out placing and compaction of both TVC and SCC was assessed on a Borg scale (0–20, from very light to very heavy/limit). Figure 12.1 shows the great difference when TVC (dotted lines) is replaced by SCC and vibration is eliminated (two activities show zero exertion levels). It is possible to estimate the

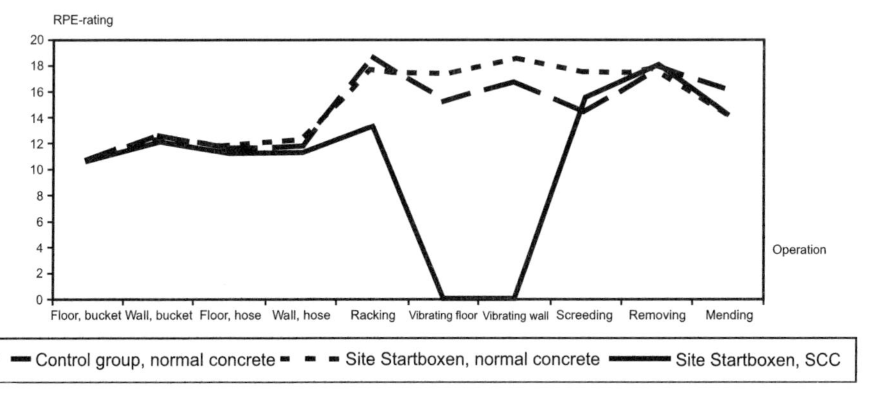

Figure 12.1 *Physical strain of an operator while placing and compaction of concrete during construction of a typical reinforced concrete building (casting of concrete walls and slabs)* [14].

total physical exertion/load by the size of the area 'under the exertion curve'. SCC dramatically reduces the physical load.

All evidence available supports a clear conclusion that working with SCC is invariably and significantly less arduous overall than that with TVC.

12.2 Environmental benefits

12.2.1 Noise

Use of SCC and the substantial reduction of noise during concreting also benefit the environment around the site. The reduction of noise emanating from a concrete construction site is sufficient to permit extended hours of construction activity in urban, city-centre, sites, where severe limits on noise often apply. This, in turn, shortens the construction times and lowers the secondary costs.

12.2.2 Recycled and waste materials

Typical mix designs for SCC differ from the TVC in its higher content of additions (fine fillers). Such very fine-grained materials were often deemed unsuitable for normal concrete. Their content was usually limited to less than 2% of volume, preferably to be avoided altogether. Successful trials have been carried out on unitisation of quarry dusts and crushed aggregate, which would have required washing to remove excessive amounts of dust. More information has been provided in Chapter 3.

12.3 Quality and economy

12.3.1 Quality

It is well established that the quality of finished concrete products depends not only on the properties of the concrete and the production process used, it is strongly dependent on the skills and training of all workers involved, and on their supervision and overall site control. Compaction of traditional concrete is the most unpleasant and physically demanding task in concrete construction, which meant that the least trained, often a casual construction workforce, has usually carried it out. As the compaction process is inherently not amenable to supervision, the quality of hardened TVC in practical applications was often below that which could potentially have been achieved.

The use of SCC leads to well-compacted structural concrete, virtually free of voids and honeycombing. In turn, this leads to a significant reduction, often a complete elimination of 'making good' and other remedial measures. The cost of 'making good' is never included in cost estimates for concrete construction as it is ubiquitous. It represents a factor permanently reducing profit margins and reducing the competitivity of concrete construction. The strength and durability of concrete and its surface finish may become considerably better than those of good hardened TVC. The generally denser microstructure of 'ordinary' and 'high-performance', hardened SCC improves its quality overall.

12.3.2 Direct costs

The use of SCC influences several components of overall construction costs. It leads to both increases and decreases of the direct cost. The decreases are:

- The cost of labour is reduced by the elimination of the need for labourers to compact freshly placed concrete, and for their supervision. In some practical applications of SCC a further reduction in the cost of labour may come from a reduction in the manpower needed to place the mix, for raking and shovelling. This can be sometimes eliminated completely. Comparisons of labour requirements when a traditional in situ concrete is placed with the required compaction by vibration, and placing of SCC, assisted/supervised by just one person are clearly illustrated by Figures 12.2 and 12.3.
- The cost of vibrators of all types together with the cost of running them, such as the supply of energy via electricity or compressed air, are entirely eliminated.
- The formwork, namely for precast concrete production can be lighter and more re-uses are possible. The cost of formwork can be reduced.
- Placing can be very much faster than when TVC, which needs compaction, is used. Greater and longer nonstop pours become technically feasible. Shorter casting time substantially increases productivity of precast concrete yards/plants of all types.
- The cost of labour and materials required for repairs and making good are greatly reduced, or entirely eliminated.

The decreases or increases are:

- Cost of constituent materials and of the fresh SCC as supplied tends to be higher than that of a TVC of similar strength grade. This may reflect one or more of factors such as a greater amount of more expensive admixtures, greater cost of specialist high-performance additives (fine fillers). However, potentially very inexpensive additives can be used (such as quarry dusts) and inexpensive admixtures may be used or their cost reduced when a continued supply is required for a larger project or for permanent supply to a ready-mixed or precast concrete producer.

Maximum economy is obtained when the use of SCC is pre-planned and the construction process adjusted beforehand. An example of such an approach is illustrated by Figure 12.3. A change from TVC to high-performance (strength) SCC has allowed small–medium span concrete bridges to be cast as integral structures, much faster and with less manpower than would have been achieved with TVC[15]. Planned use of SCC for construction of anchorage blocks for the record holding Akashi-Kaikyo bridge in Japan shortened the completion of the structures by six to eight months.

Recent studies in Denmark[16, 17], in which the whole of the 'value-chain' in concrete construction was investigated, have confirmed that the greatest reductions in overall costs can be achieved when casting horizontal elements such as floors and slabs, and in most of precast concrete (reductions of up to 25%). Productivity in both these areas has improved significantly. This is not difficult to foresee when Figures 12.2 and 12.3 are compared. Compaction of floors/slabs, with thicknesses varying in the range 100–250 mm by poker vibrators is particularly inefficient as very many,

Figure 12.2 *A team of eight workers placing and finishing a slab made of TVC. A change to SCC would reduce the workforce to approximately one nozzleman and one or two finishers. (Reproduced with permission of The Concrete Journal.)*

Figure 12.3 *Typical placing arrangement for SCC*[18]*. Note the reduction of workforce down to one 'nozzleman' who directs the flow. A supervisor may occasionally call to inspect the process.*

closely spaced, vertical insertions must be carried out to ensure that all of the concrete is subjected to sufficient vibration to allow it to compact.

Heavy precasting is often associated with large structures, such as prestressed segmental bridges. A changeover from TVC into SCC shortens casting times of the segments, which can be sufficient to permit a faster turnover of moulds, and a considerable shortening of the overall construction time.

12.3.3 Indirect costs

A change to SCC also has both positive and negative effect on indirect costs:

The decreases are associated with:

- Better working environment on and around construction site. There are fewer direct and indirect health problems, less time lost through injury, and reduced chances of injury-related compensation being required.
- Better communications on-site reduce misunderstandings and consequent remedial actions.

The increases arising out of the adoption of SCC include:

- The costs of additional education and training of existing staff at all levels, from the engineer to the labourer. It is expected that future new staff, at all levels, would have already received education and training which included technology of SCC, and this type of indirect cost will diminish with time.
- New test equipment may have to be purchased by contractors and concrete suppliers.
- Producers of concrete may have to invest into additional silos to hold the different additives and improve their means of monitoring and adjusting of moisture content of aggregate.

A changeover from TVC to SCC brings particularly strong and immediate benefits to the producers of precast concrete.

The direct benefits include:

- Increased productivity. Casting time is shorter and unformed surface finishing is decreased. This leads to moulds being turned faster between castings.
- Lower labour costs per product due to reduced need for placing and finishing and making good/repairs.
- Longer production runs are possible due to a much improved working environment (large reduction in levels of noise).
- Lower costs of replacement of mould and plant (less wear and tear), elimination of vibrators, lower energy consumption.

Indirect benefits include:

- Health and safety risks from exposure to noise and vibration are significantly reduced.

- There is a cleaner working environment with less wastage.
- Lower cost of admixtures through long-term contracts or bulk purchases. Specialist, potentially lower-cost, admixtures are available, as the self-compaction needs to be maintained for only a short period of time.

The benefits listed above soon attracted the interest of the precast concrete industry, which has started a rapid conversion to SCC, as seen from Figure 12.4.

12.4 Design and management

The introduction of SCC as a replacement for TVC offers a number of specific benefits. However, the maximum benefit, which will be greater than a simple sum of the individual ones, is obtained when the use of SCC is already adopted in early design stages and the construction/management process itself is adapted to this new technology[19].

Elimination of compaction from the concrete construction process opens up the possibility to develop robotisation/automation, leading to still higher productivity and a better and more consistent quality of the concrete product. This can be envisaged first in precast concrete production; however, some in situ concreting operations also have a potential for automation/robotisation.

The selection (specification) of SCC for a given project/structure can be done at different stages of a project, depending on which advantages or aspects of SCC are to be exploited. Specification of SCC will also differ in detail, reflecting the stage at which it is produced. The initiative may come from:

- The architect: The selection may reflect considerations of factors such as an overall environmental impact and better potential quality; and specific ones

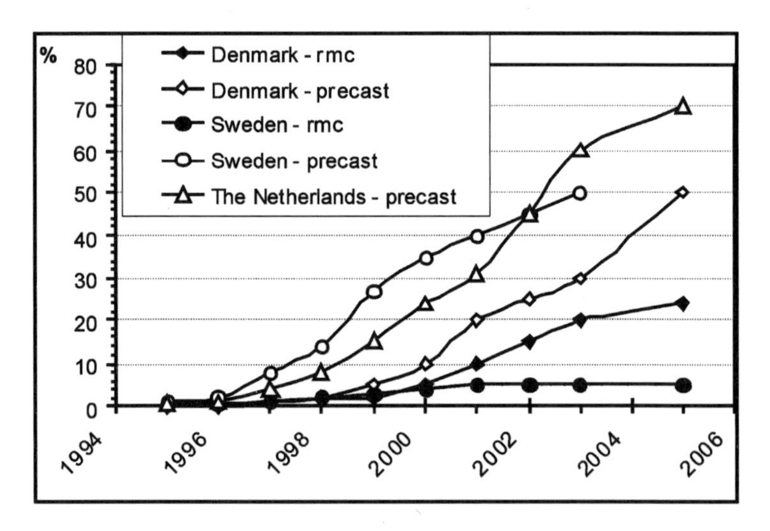

Figure 12.4 *Share of SCC out of total concrete production in Sweden, Denmark and the Netherlands. Note the much faster uptake by the precast concrete sector (after Glavind et al.[16]).*

such as a greater freedom of shape and form of structures and demand for exceptional surface finishes or use of novel construction elements.

- The structural engineer/designer: Design, which allows the use of structural elements with congested reinforcement, use of high/ultra-high strength concrete, other high-performance and special concretes, improving durability, extending life-cycle, leading to more slender structures saving materials etc. The range of structural steel–concrete composite elements widens considerably when SCC is used.
- The builder/contractor: The contractor goes further, and requires better knowledge of what SCC can and cannot do. Specification of SCC automatically changes the construction process and influences costs. The degree of detail of the SCC specification will depend on the specific application. It is emphasised that there is no 'over-specification' and there is good two-way communication with the supplier.

SCC can be produced by the contractor 'in-house', and this is almost always the case with precast-concrete production. However, in all other situations, it is usually ordered by the contractor from a supplier, who takes on the responsibility for the mix design, which must lead to a mix of adequate performance both in fresh and hardened states. In order to produce an optimum and as robust as practicable mix of required performance, the supplier of SCC must be adequately informed of the proposed application and of the construction process selected. The latter largely determines the required characteristics of the SCC when fresh. Relevant information includes data such as the distance the mix is expected to flow and any obstructions to flow, density of reinforcement etc. Methods of placing, including expected heights of a free-fall, if any, must be also indicated. Environmental factors, such as the expected weather during construction, namely temperature, also have to be advised. Verification/ acceptance of SCC and conformance testing are required, these have been reviewed in Chapters 11 and 5.

Initially, most SCC appears to have been specified for special applications, where compaction was either known to be impossible or when the originally intended use TVC was discovered to be unsuitable during construction. In such early cases, the use of SCC was immediately very beneficial, usually regardless of the extra cost of the change from TVC to SCC. It enabled a project to be constructed, or it enabled the completion of a project, which was found to be impossible to complete satisfactorily with TVC (see Figure 12.5). In the specific applications described above, the benefit of using SCC was mainly in the mitigation of large potential losses (penalties for late completion, massive remedial works, removal and replacement of poor quality concrete placed etc.), rather than a positive net gain for the contractor, the consultant and the client.

The elimination of compaction from concrete construction process suggests that a re-adjustment of traditional responsibilities and division of management at that stage is potentially beneficial. Responsibility of the supplier of fresh concrete can be extended, to include not only transport but also the placing itself (see Figure 12.3)[18].

In such cases, the contractor only ensures that the formwork and reinforcement are ready, adequate access is provided, and it is then up to the supplier of concrete to

Figure 12.5 *Heavily reinforced bridge deck: It was impractical to concrete it using TVC, SCC had to be used. (Reproduced with permission of J.L. Vitek, Metrostav a.s.)*

deliver it, place it and hand over completed concrete elements. The supplier is well able to ensure that the mix has properties adequate for a reliable and efficient placing of concrete by given means and in given conditions, and that specified performance parameters are reached when hardened. Problems encountered during placing can be more easily resolved in such cases, as they are internal to the supplier alone.

Introduction of SCC brings potential benefits to architects and structural engineers, and ultimately to the users of buildings. New types of structural elements, which were not possible with TVC, can be produced using SCC. Such elements include different types of steel–concrete structural elements.

Typical examples are long steel tubes filled with concrete and acting in a composite manner. There are many examples of practical applications, which tend to be increasing in popularity. The size of the steel tubes or other elements with hollow cross-sections can vary, from small to very large diameter[20], and from short to several storeys high. An early example is the 'concrete-filled-tube' system used in Japan was a tubular framework of a skyscraper where up to 60 m tall tubular steel columns were filled with SCC by pumping from the bottom[21].

Another type of a steel–concrete composite element is a 'sandwich', sometimes called 'bi-steel'. SCC in this case fills a cavity between two parallel steel plates, with or without an additional reinforcement within the concrete core.

A completely novel application of SCC has been used in development of a 'ConRam' system. The system exploits the liquid-like behaviour of SCC when fresh, with its capability to fill structural steel tubes[22]. The fresh SCC mix acts at first as a 'hydraulic fluid': it is pumped into telescoping steel tubes, which function as hydraulic cylinders. The 'pressurised' mix is then allowed to harden within the steel tubes, which then function as heavy duty load bearing structural elements. The system can be used to support sides of an excavation, ensuring there was no movement, and then form part of the permanent structure when the concrete hardens. The system relies on the fresh mix being not only self-compacting but also as much as possible incompressible (preferably with zero air content). This has been difficult to achieve with ordinary SCC mixes, which did compress and caused substantial loss in the 'hydraulic' load transmission. Interestingly, samples of SCC, which hardened under pressure within the steel pipes showed much denser microstructure and very much higher compressive strength compared with those obtained from the same concrete using cube specimens cured in normal conditions.

In general, use of SCC enables the architect or structural engineer to use more complex concrete shapes, thinner and more highly reinforced cross-section; it significantly extends the range of 'what is buildable' in concrete.

References

[1] Grauers M *et al.* Rational production and improved working environment through using self-compacting concrete. EC Brite-EuRam Contract No. BRPR-CT96-0366, 1997–2000.

[2] Bartos P.J.M. and Cechura J. Improvement of working environment in concrete construction by the use of self-compacting concrete. *Structural Concrete*, 2001, 2(3) 127–131.

[3] 86/188/ EEC, Council Directive of 12 May 1986 on Protection of workers from the risk related to exposure to noise at work. *Official Journal of the European Communities* 24.5.1986, No L 137, 28–34.

[4] Health and Safety Executive, *Reducing Noise at Work, Guidance on Noise at Work Regulations*, 1989, L108, HSE Books, UK, 2000.

[5] European Union Directive 2003/10/EC of 6th February 2003 on The Minimum health and safety requirements regarding the exposure of workers to risks arising from physical agents (noise). *Official Journal of the European Union* (Seventeenth individual Directive arising from Art.16(1) of Directive 89/391/EEC).

[6] UK Health and Safety Executive: The control of noise at work regulations 2005. UK H.M. Stationery Office.

[7] ISO 1996-1, Acoustics: Description and measurement of environmental noise. Part 1: Basic quantities and procedures.

[8] ISO 1999, Acoustics: Determination of occupational noise exposure and estimation of noise induced hearing impairment.

[9] ISO 2204, Acoustics: Guide to International Standards on the measurement of air-borne acoustical noise and evaluation of its effects on human beings.

[10] ISO 11202, Acoustics: Noise emitted by machinery and equipment – Measurement of emission sound pressure levels at a workstation and at other specified positions, Survey methods in-situ.

[11] Soderlind L. Comparison of sound levels when casting with normal concrete and self-compacting concrete. Final Report Task 8.2. In: Grauers M. *et al.* Rational production and improved working environment through using self-compacting concrete. EC Brite-EuRam Project BE96-3801, Contract No. BRPR-CT96-0366, 2000.

[12] Thompson N. Getting a grip on hand-arm vibration. *Proceedings of the Institution of the Civil Engineers – Civil Engineering*, 2007, 160, 15.

[13] UK Health and Safety Executive: *Control of vibration at work regulations*, H.M. Stationery Office, London, 2005.

[14] Bartos P.J.M. and Soderlind L. Ergonomic studies, Final Report Task 8.5. In: Grauers M *et al.* Rational production and improved working environment through using self-compacting concrete. EC Brite-EuRam Project BE96-3801, Contract No. BRPR-CT96-0366, 2000.

[15] Petersson Ö., Billberg P., *et al.* Viberingsfri betong for brogjutningar enligt Bro 94, CBI Report, Stockholm, Sweden, 1999 (in Swedish).

[16] Glavind M., Nielsen C.V., Gredsted L. and Hansen C.N. SCC – A technical breakthrough and a success for the Danish concrete industry. In: *Proceedings of the Fifth International Symposium on SCC*, Ghent, Belgium, September 2007, RILEM Publications, Cachan, France, pp 993–999.

[17] Poulsen A. Cost savings by use of SCC in floors. Presented at Nordic Workshop SCC – Vision and Reality, 19 June 2006, Copenhagen, Denmark. www.NordicSCC.net

[18] Petersson Ö. and Skarendahl Å. (Eds.) Self-compacting Concrete, RILEM Publications, Cachan, France, 2000.

[19] Stubbs A. Self-compacting concrete: if not, why not? *Concrete*, V41, No 4., May 2007.

[20] Bartos P.J.M. Self-compacting concrete in Bridge construction – guide for design and construction. Concrete Bridge Development Group, Technical Guide 7, Camberley, UK, 2005.

[21] Hayakawa M. Development and application of super workable concrete. In: Bartos P.J.M. and Cleland D.J. (Eds.) *Special Concretes – Workability and Mixing*, E&FN Spon, London, UK, 1993, pp 183–190.

[22] Zhu W.Z. Private communication, ACM Centre, University of Paisley, Scotland, 2006.

Chapter 13
Practical applications

13.1 Industrial structures

13.1.1 Gas tanks in Texas, USA

A very wide range of applications of SCC have been reported since its first uses in the early 1990s. The examples given in this chapter are all more recent and have been chosen to illustrate the various types of applications in which SCC has been particularly successful. One of the first applications of SCC worldwide was a gas tank in Osaka Japan, which has been widely reported in the literature[1]. Two LNG tanks have also been constructed (2005–2007) near Freeport, Southern Texas, USA[2]. In a classical approach, this kind of construction is normally produced with conventional concrete and involves a substantial number of workers. In this case however, SCC has been used, reducing the direct labour costs as well as the construction period. Besides these economical advantages, it is clear that self-compacting concrete provides some substantial technical advantages concerning concrete placement in high wall lifts containing dense reinforcement as well as post-tensioning ducts.

The two tanks built in Freeport are of the full containment type. The liquid natural gas is stored at –62°C in an insulated steel container tank which is surrounded by an outer containment tank made of concrete, 0.8 m thick, about 90 m in diameter and 40 m in height. About 20000 m^3 of pumped SCC was needed for the tank. Figure 13.1 shows one of the LNG tanks during construction, while Figure 13.2 gives a more general view of the two gas tanks under construction.

The concrete tank structurally supports the inner steel tank and the dome-shaped roof. The concrete also acts as a barrier to LNG leakage. For this barrier the durability of the SCC was a key point in addition, of course, to the target strength (characteristic strength of 40 MPa). The durability aspects included permeability to chlorides and resistance against cryogenic conditions.

Class F fly ash was incorporated in the mix composition at 25% of the total binder content and a sulfate resisting cement (ASTM C150 type II) was used. The water/powder ratio was kept at 0.37, using a polycarboxylate type superplasticiser. The total powder content was limited to 460 kg/m^3 of concrete in order to limit the heat of hydration. Part of the mixing water was added as flaked ice to further reduce the risk of early age cracking.

The slump-flow spread of the concrete was 750 mm with no visible segregation at the edges. The L-box passing ratio was 0.83. The SCC-mix was designed to have a 40 min consistence retention time at 30°C. Quality control of the concrete was mainly based on slump-flow measurements.

Figure 13.1 *Construction of LNG tank in Freeport, Southern Texas, USA*[1].

Due to the limited amount of fine particles, a VMA was added to the concrete to avoid segregation. Large-scale tests were carried out to determine the segregation characteristics of the concrete. The casting rate was about 0.5 m vertical rise per hour.

Formwork pressure was also measured during the large-scale tests. It was concluded that this did not exceed 50% of the hydrostatic pressure. Furthermore, the tests revealed that a smooth surface finish could be obtained with no significant blowholes. Finally, the tests also showed the importance of studying the casting

Figure 13.2 *The two SCC Freeport LNG tank walls under construction*[1].

sequence in detail in order to avoid dynamic segregation if the mix has to flow for more than 10 m horizontally.

The application of SCC produced some advantages during the construction of the two tanks:

- A crew of only 9 men was needed to fill the formwork, instead of 25 men in the case of TVC.
- A SCC lift of 4.5 m high could be achieved in 10 h, compared to a lift of 3.2 m in 12–15 hours in the case of TVC.
- 30% time reduction was obtained for preparing the formwork, due to the application of higher forms.
- 60% time reduction was obtained for casting activities.
- A significant noise reduction was obtained during casting.

In conclusion, SCC was shown to be a very attractive material for the construction of these two gas tanks.

13.1.2 Industrial hall in Veurne, Belgium

SCC with a slump-flow spread of about 800 mm was used for the construction of an industrial hall in Veurne. The total powder content was about 550 kg/m^3, consisting of 350 kg of blast furnace slag cement (CEM III/A 42.5) and 200 kg fly ash. A polycarboxylate type of superplasticiser was used to provide a long consistence retention time. The water/cement ratio of the concrete was 0.5. The V-funnel time was about 6–8 s, and the air content of the fresh concrete was 1.7%. The 28-days compressive strength determined on 150 mm cubes was about 60 MPa.

Walls with a height up to 8 m were cast in one operation. Steel formwork was used, consisting of large steel panels 8 m long (Figure 13.3). The concrete was placed by pumping from the top.

In order to avoid defects in the finished concrete surface, the formwork was carefully prepared with demoulding oil (Figure 13.4). The thickness of the layer of demoulding oil had to be controlled with great care as an excess could lead to the formation of air bubbles on the finished surface.

Despite the great care, which was taken to prepare the formwork, another problem occurred. Black spots appeared on the surface, as can be seen in Figure 13.5. A regular pattern of rectangular spots was noticed, as well as some irregular stripes around the centre pen holes. Detailed investigation into the occurrence of these remarkable black spots showed that they had been caused by magnetisation of the steel formwork. The large panels were handled in the formwork factory by means of strong magnets, causing some residual magnetisation of the steel surface. Also the surface finishing around the centre pen holes also caused some residual magnetisation. Fly ash particles, present in the fresh concrete, were attracted by the magnetised areas and produced areas of black colouring on the finished concrete surface. Fortunately, it was not a serious problem to solve: the black spots were easily removed by rubbing them with emery paper. The phenomenon showed that the finished surface of SCC can be so smooth and perfect that every defect is easily noticed.

Figure 13.3 *8 m high formwork for an industrial hall.*

Figure 13.4 *Careful preparation of formwork in order to avoid blowholes.*

Figure 13.5 *Finished surface, showing black traces due to magnetisation of the formwork linked to the use of fly ash in the SCC.*

Prefabrication

It is apparent worldwide that the application of SCC in the precast industry is developing faster than the application of SCC for in situ casting. It is clear that a factory is providing an easier environment to deal with a material that requires a greater attention to quality control. Furthermore, factories are also benefiting from the significant advantages regarding health and safety, and working environment (see Chapter 12). In some concrete precasting plants, the radio has been recently rediscovered!

Beams and columns, heavily reinforced or often prestressed, are typical prefabricated products (Figure 13.6) made with SCC[3]. A compressive strength in the range 60–80 MPa is quite common for precast SCC. Technically, it is quite easy to reach these strength levels with powder-type SCC.

The absence of the vibration equipment, to which the moulds were traditionally attached, leads to a significant increase in the service life of the moulds, and permits new, more efficient, casting techniques to be developed. The effects on costs are discussed in Chapter 12.

With SCC, high quality smooth surfaces can be obtained, as shown in Figure 13.7. SCC has proved to be an excellent choice for the prefabrication of architectural concrete elements, with the most beautiful colours obtained by adding pigments into the mix.

13.2 Public buildings and housing

13.2.1 Residential buildings in Brazil

In Brazil, SCC has been evaluated as a replacement for ordinary concrete in the everyday construction of reinforced concrete structures, like beams and slabs of a residential building. The application of SCC and a 100 mm slump ordinary concrete

Figure 13.6 *Precast column made of SCC*[3].

Figure 13.7 *Surface of a prestressed slab after casting and finishing*[3].

Figure 13.8 *Application of SCC for slabs and beams of a building floor. Note that only two men are actually engaged in placing the SCC*[4].

were monitored during the construction of two floors of a building floors[4]. The total floor area was about 500 m^2.

Both SCC and TVC had a characteristic compressive strength of 25 MPa at 18 days. The SCC mix proportions (per m^3) were as follows: 406 kg cement, 820 kg sand, 714 kg coarse aggregate, 4 l polycarboxylate superplasticiser (33% solid content), and 0.4 l viscosity enhancing admixture.

On site, each batch of SCC was tested by means of slump-flow, V-funnel, and L-box. If needed, small mix adjustments were made. Both SCC and TVC were placed by pump. Figure 13.8 shows the casting of the SCC.

In order to prevent SCC flowing away to the lower level, low dividing 'walls' were constructed using traditional concrete, as shown in Figure 13.9. In the areas in which SCC could flow into the ceramic bricks, the openings had to be closed with mortar or plastic sheets.

In the case of the TVC, the floor-slab, containing about 64 m^3 of concrete, was cast in about two and a half hours. Eleven workers were needed during the casting, leading to a productivity of 2.35 m^3/h/worker. In the case of the SCC, 57 m^3 was cast in a similar time period with, on the average, only 2.5 workers. This has led to a significant increase in productivity, to 9 m^3/h/worker.

13.2.2 Trump Tower, Chicago, IL, USA

SCC is also now finding application in high-rise buildings. In Chicago, the city of skyscrapers, a new tower (Trump International Hotel and Tower) is under construction[5]. When completed in 2009 (Figure 13.10), this tower will reach a

Figure 13.9 *Low dividing 'wall' of TVC in order to prevent SCC flowing into the neighbouring section of the floor slab*[4].

Figure 13.10 *Computer rendering of Trump International Hotel and Tower, Chicago, IL, USA*[5]. *(Reproduced with permission of Skidmore, Owings & Merrill LLP.)*

Figure 13.11 *Casting of the caisson mat slab*[5]*. Note the presence of only two men, each controlling one discharge point. (Reproduced with permission of Skidmore, Owings & Merrill LLP.)*

height of 345 m (415 m including the spire). It will be the tallest concrete building in the United States, and the tallest building built in North America since the completion of the Sears Tower in 1974.

SCC with a compressive strength of 110 MPa at 90 days was required for the tower, due to very severe stress conditions and locally very congested reinforcement. In order to avoid thermal cracking in some massive outrigger elements, the heat of hydration of the concrete had to be limited. Most probably this is the first application of 110 MPa SCC pumped and placed to an elevation up to 200 m above ground. The mix included a blended binder system consisting of Portland cement, slag cement, fly ash and silica fume.

For the foundation of the tower, a 3 m deep caisson mat slab at the base of the core walls was required. This slab was also cast with SCC, due to its ease of placing and finishing in a confined area. The SCC had a minimum cylinder compressive strength of 69 MPa at 56 days, and the mix included crushed dolomite limestone aggregate (up to size 20 mm), natural fine sand, a mix of Portland cement and slag cement, class C fly ash, chemical retarder, polycarboxylate superplasticiser, and VMA. The target slump-flow spread was 660–710 mm and the water/powder ratio was 0.26 to 0.28. The mat was cast in one single operation lasting 22 h with 3800 m^3 of concrete placed. The concrete was placed by conveyors, as shown in Figure 13.11. The horizontal flow of SCC within the slab was up to about 15 m. After casting, the top surface was covered with insulation in order to limit the heat loss and reduce the internal temperature gradient.

As of February 2007, the Trump Tower structure has risen to the second setback zone (Level 25), as illustrated in Figure 13.12.

13.2.3 World Financial Centre, Shanghai, China

Another high-rise tower under construction is the World Financial Centre, Shanghai, China. This tower will consist of 101 storeys and reach a total height of about 492 m, which would make it one of the very highest towers in the world. According to the plan the tower should be finished in 2008. SCC was used for the central core, the perimeter walls located at the lower levels from floor 1 to floor 5, and the megacolumns at the corners of the building from floor 6 upwards. The perimeter walls, corner columns and central core can easily be seen in Figure 13.13. SCC has also been used in the foundation slab. As can be seen in Figure 13.14, the reinforcement was locally very congested, making traditional casting and vibrating very difficult.

For casting of the tower above ground level, the SCC was pumped onto the level where it was needed by means of very powerful pumping equipment. Pressures up to 400 bar were needed during these pumping operations.

Figure 13.12 *Progress in the construction of the Trump tower as of February 2007*[5]*. (Reproduced with permission of Skidmore, Owings & Merrill LLP.)*

Figure 13.13 *Construction of the World Financial Centre, Shanghai, China.*

Figure 13.14 *Foundation slab with dense reinforcement ready for casting.*

Table 13.1 *SCC mix composition (in kg/m^3) used for World Financial Centre, Shanghai.*

Level	*Cement*	*Water*	*Sand*	*Stone*	*Fly ash*	*Slag*	*Superplasticiser*
1–6	350	170	780	940	80	80	5.4
7–60	350	170	720	1020	70	70	4.2
>61	300	170	800	980	70	80	3.6

The mix proportion of the SCC is given in Table 13.1. Ordinary Portland cement CEM I 42.5 was applied. Fly ash and slag were added to increase the powder content. Coarse aggregate size up to 25 mm in size was used. A polycarboxylic ether superplasticiser was added to the mix.

13.2.4 Villa Gistel, Belgium

A house was built in Gistel, Belgium (see Figure 13.15) which relied almost entirely on SCC[6]. Four parallel walls made from SCC and two glazed walls provided the architectural and structural concept of the building. The inner walls, with a thickness of 40 cm, ensured the transverse stability of the structure, due to the rigid connection with the basement walls. The outer walls of the villa were built in two phases, consisting of two walls with thickness 17 cm, with 6 cm insulation in between.

As it was decided to cast the 20 m long and 8 m high walls in one operation, classical casting from the top would yield several risks, related to the large drop height of the

Figure 13.15 *Villa in Gistel, Belgium. (Reproduced with permission of Engineering Office Mouton, Ghent, Belgium.)*

concrete. Consequently, it was decided to pump the SCC into the formwork from beneath, under pressure. Due to the high expected formwork pressures, a metal formwork system was applied. Figure 13.16 shows the pumping operation, with one inlet point at either side of the formwork. The 8 m high walls were cast in less than an hour.

The SCC, having a strength grade C30/37, was made with blast furnace slag cement. The maximum particle size of the aggregates was 14 mm. Limestone powder was added in order to obtain an adequate powder content. The slump-flow spread of the concrete was in the range 700–800 mm and the V-funnel time was in the range 4–7 s. A filling degree of at least 95% was obtained with the Kajima box (filling) test. The water/cement ratio of the SCC was kept below 0.50.

Great care was taken to produce a pristine finished concrete surface. The formwork was cleaned carefully and prepared with a layer of demoulding oil. On site trials checked the effect of the demoulding oil and optimised the required dosage. When mounting the formwork, ladders were covered with clean cloths. Dirty footprints on the formwork of the slabs were also avoided. Despite all these measures, some traces of dirt and corrosion were visible after demoulding. Happily, these were all easily removed with emery paper.

13.3 Bridges

Shortly after the development of SCC, it was used in the construction of the Akashi-Kaikyo Bridge near Kobe, Japan [1]. This is the world's longest suspension bridge, with a main span of 1991 m. SCC was used for the anchor blocks on both abutments

Figure 13.16 *Filling the formwork from beneath, with two inlet points. (Reproduced with permission of Engineering Office Mouton, Ghent, Belgium.)*

of the bridge and the caissons supporting the pylons. Figure 13.17 shows the southern abutment. The use of SCC reduced the construction time of the abutments by 20%.

In Europe, the first bridges in SCC were produced in Sweden[1]. Figure 13.18 shows one of these during construction. SCC with a slump-flow spread in the range 680–720 mm was used and the quality control of the concrete on site was mainly based on slump-flow measurements.

There have been numerous subsequent applications of SCC in bridges[7].

13.4 Other examples

13.4.1 Multifunction sports complex for SK Slavia Praha football club a.s. in the Czech Republic

Construction of a new 21000-seat stadium for the Slavia football club is underway in Prague, the Czech Republic[8]. Following a previous, very successful, use of precast SCC with lightweight aggregate (LWSCC) for a new medium-sized stadium for Volkswagen in Wolfsburg, Germany (Figure 13.19), the project team selected the same approach for this stadium in Prague, with an even greater exploitation of precast LWSCC. All of the terracing and seating, and all staircases are made of LWSCC.

Figure 13.17 *Southern abutment (height approximately 50 m) of the Akashi-Kaikyo Bridge, Japan.*

Figure 13.18 *Casing of one of Europe's first SCC bridges, Nyneshamn, Sweden (1999).*

Figure 13.19 *Erection of a stand at Wolfsburg using lightweight aggregate SCC. (Reproduced with permission of R Hela and M. Hubertova.)*

Figure 13.20 *Casting of in situ LWSCC during renovation of an existing building. Note the 'cohesion' of the mix. (Reproduced with permission of R Hela and M. Hubertova.)*

Lightweight expanded clay aggregate 'Liapor' was produced by Lias Vintirov and used to produce a range of precast SCC elements by Prefa-Praha a.s. The main contractor was Hochtief CZ a.s.

The choice of the Liapor lightweight aggregate reduced the density of hardened LWSCC to either 1730 kg/m^3 or 1480 kg/m^3. The corresponding compressive strengths (moduli of elasticity) were 44 MPa (25 GPa) and 38 MPa (21 GPa), respectively. The self-weight of the precast superstructure was reduced by about 20%, which saved on the dimensions of the supporting structures. The compressive strength of the LWSCC was adequate for structural elements of substantial size to be designed, while their lower weight made their handling and installation easier and less expensive. The panels were up to 10 m long, 4 m high, with a minimum thickness of 120 mm.

The LWSCC mix was specified according to EN206-1 as LC 25/28 D 1:6 XF4 and LC 35/38 D 1:8 XF4. The water/cement ratio was 0.45 for both mixes, with the powder content varying in the range 580–640 kg/m^3. The air content was 6%. The reduced weight of LWSCC makes it very suitable for reconstruction (see Figure 13.20).

The use of lightweight aggregate did not prevent the manufacturer of the precast element from achieving precise shapes and a uniformly excellent surface finish, as if normal density aggregate has been used, as illustrated in Figure 13.21.

13.4.2 Citytunnel, Malmö, Sweden

Another interesting application of SCC is the Citytunnel project[9], consisting of 17 km of railway connecting Malmö Central Station and the Öresund Bridge. The underground section of the Citytunnel runs over a distance of 6 km, containing a 4.5

Figure 13.21 *Erection of stands from precast LWSCC during construction of a football stadium in Prague, Czech Republic. (Reproduced with permission of O. Zlamal and M. Maly, www.stadioneden.cz.)*

km long section of two bored parallel tunnel tubes. The Triangeln station in Malmö city centre will be about 25 m below ground level. 29 columns and a head beam nearly 200 m long had to be cast in the pillar tunnel of this station. Due to the very complicated boundary conditions for the heavily reinforced head beam, with difficult access, SCC was chosen for both the columns and the beam.

Three different concrete types were defined in the specification depending on the impact of water, salt and frost. The strength classes ranged from C30/37 to C40/50, with a water/cement ratio in the range 0.50–0.40. The cement content was in the range 280–340 kg/m^3.

For the SCC due attention had to be given to the possible occurrence of the thaumasite form of sulfate attack (see also Chapter 10). When using limestone powder in the SCC, the sulfate content was limited to 100 mg/l in the surrounding water and less than 1000 mg/l in the surrounding soil. Furthermore, in order to obtain the required fire resistance, polypropylene fibres were added to the concrete. The actual fire resistance of the SCC was experimentally verified.

The mix design was based on the 'Okamura-principle'. One m^3 of SCC consisted of 850 kg coarse aggregate with maximum size 16 mm, 680 kg sand, 425 kg cement, 160 kg limestone filler, 25.5 kg microsilica, 1.2 kg polypropylene fibres, 150 kg water, 3.8 kg superplasticiser, and 2 kg air entraining agent. A workability retention time of at least 1.5 h was obtained with this composition.

References

[1] Skarendahl Å. and Petterson Ö. (Eds.) Self-compacting Concrete – state-of-the-art report of RILEM Technical Committee 174-SCC. RILEM Publications, Cachan, France, 2000.

[2] Bernabeu O. and Redon C. Self-compacting concrete for LNG tanks construction in Texas. In: De Schutter G. and Boel V. (Eds.) *Proceedings of the Fifth International RILEM Symposium on SCC*, Ghent, Belgium, September 2007.

[3] Juvas K.J. Experiences of working with self-compacting concrete in the precast industry. In: De Schutter G. and Boel V. (Eds.) *Proceedings of the Fifth International RILEM Symposium on SCC*, Ghent, September 2007.

[4] Repette W.L. Self-compacting concrete – a labour cost evaluation when used to replace traditional concrete in building construction. In: De Schutter G. and Boel V. (Eds.) *Proceedings of the Fifth International RILEM Symposium on SCC*, Ghent, September 2007.

[5] Baker W.F., Houson M.R., Korista D.S., Rankin D.S. and Sinn R.C. High performance self-consolidating concrete for North America's tallest reinforced concrete building: Trump International Hotel and Tower, Chicago. In: De Schutter G. and Boel V. (Eds.) *Proceedings of the Fifth International RILEM Symposium on SCC*, Ghent, September 2007.

[6] Annerel E. Self-compacting concrete, a practical view (in Dutch), Arch-index, 57, September 2006, pp 23–32.

[7] Bartos P.J.M. Self-compacting concrete in bridge construction – guide for design and construction. Concrete Bridge Development Group, Technical Guide 7, Camberley, UK, 2005.

[8] Hela R. and Hubertova M. Private communication, Technical University of Brno, Czech Republic, 2007.

[9] Abel F. and Willmes F. Use of SCC for a tunnel of the Citytunnel Malmö project with 120 years of life cycle. In: De Schutter G. and Boel V. *Proceedings of the Fifth International RILEM Symposium on SCC*, Ghent, Belgium, September 2007.

Index

Author Index